OXFORD STUDIES IN PROBABILITY

OXFORD STUDIES IN PROBABILITY · 2

Poisson Approximation

A. D. BARBOUR

Department of Applied Mathematics
University of Zurich

LARS HOLST

Department of Mathematics
Royal Institute of Technology
Stockholm

SVANTE JANSON

Department of Mathematics
Uppsala University

CLARENDON PRESS · OXFORD
1992

Oxford University Press, Walton Street, Oxford OX2 6DP
Oxford New York Toronto
Delhi Bombay Calcutta Madras Karachi
Petaling Jaya Singapore Hong Kong Tokyo
Nairobi Dar es Salaam Cape Town
Melbourne Auckland
and associated companies in
Berlin Ibadan

Oxford is a trade mark of Oxford University Press

Published in the United States
by Oxford University Press, New York

A catalogue record for this book is available from the British Library

Library of Congress Cataloging in Publication Data
(Data available)

ISBN 0–19–852235–5

Text typeset by the authors using Lam S-TEX
Printed and bound in Great Britain
by Biddles Ltd, Guildford and King's Lynn

This book is dedicated
to Charles Stein

Preface

Poisson approximation is a subject with a long history and many ramifications. It can be used as a tool for estimating probabilities connected with rare or exceptional occurrences almost as widely as normal approximation can be used for the common or typical. By comparison, the scope of this book is very much narrower. An appropriate subtitle would be 'using the Stein–Chen method and coupling', because the enterprise grew out of our common enthusiasm for the beauty and power of the Stein–Chen magic formula when combined with coupling methods, and, since even this limited aim has taken more time and pages to achieve than we had ever imagined, it is as well that we did not attempt more.

The Introduction is intended to give a description of the main ideas behind the Stein–Chen method, enough to motivate the more detailed results involving the use of couplings presented in Chapters 2 and 3, as well as to encourage the reader to go on to explore the extensions to Poisson process approximation in Chapter 10. The intervening chapters are primarily concerned with applications of the method in a variety of contexts, many of them classical, illustrating that good estimates of the accuracy of Poisson approximation can frequently be obtained quite simply. The bulk of the material can be understood with no more than a knowledge of elementary discrete probability theory.

In accordance with the contractual provisions, we wish to emphasize that it may in practice be hazardous to apply any of the theorems stated in this work without first checking that their aforementioned conditions are fulfilled.

Acknowledgements

We wish in particular to thank David Aldous, Tim Brown, Monika Bütler, Louis Chen, Ourania Chryssaphinou, Geoff Eagleson, Geoffrey Grimmett, Gesine Reinert, and Małgorzata Roos for their assistance, advice and encouragement.

The work was supported in part by the Schweizerischer Nationalfonds Grant No. 21–25579.88.

Contents

1
Introduction

One of the oldest limit theorems of probability is the Poisson 'law of small numbers'. In its simplest form, it states that the binomial distribution $\mathrm{Bi}\,(n,p)$ converges to the Poisson distribution $\mathrm{Po}\,(\lambda)$ as $n \to \infty$, if $p = \lambda/n$ for some $\lambda > 0$, as was proved by Poisson (1837), pp. 205–207. The first person to have appreciated its use in the statistical analysis of rare events seems to have been von Bortkewitsch (1898), who emphasized that statistical methods were not only applicable to data consisting of large numbers, even though normal approximation might no longer be appropriate. His analysis of the numbers of Prussian soldiers killed each year as a result of being kicked by horses has proved particularly memorable, although it was only one of several sets of data that he considered. What is more important is that he showed that the Poisson distribution fitted the data very closely, by tabulating the expected and observed frequencies and by comparing the sample variance with the sample mean.

The most natural way of proving binomial convergence to the Poisson is to write down the point probabilities: if W has distribution $\mathrm{Bi}\,(n,p)$, an elementary but somewhat involved calculation leads to the estimate

$$\mathbb{P}[W=k] = \binom{n}{k} p^k (1-p)^{n-k} = \frac{(np)^k}{k!} e^{-np}\{1 + O(np^2, k^2 n^{-1})\} \qquad (1.1)$$

for each $k \in \mathbb{Z}^+$. This formula of course shows more, in that the relative error in the approximation of the point probabilities by the corresponding point probabilities from the $\mathrm{Po}\,(np)$ distribution can also be explicitly bounded, for any given pair n, p. In particular, the approximation remains good, provided that np^2 and k^2/n are both small. Thus, for example, if A is any subset of $\mathbb{Z}^+$, it follows that for some c not depending on A, n or p,

$$\begin{aligned} |\mathbb{P}[W \in A] - \mathrm{Po}\,(np)\{A\}| &\le c \sum_{k \in A} \frac{(np)^k}{k!} e^{-np}\{np^2 + k^2 n^{-1}\} \\ &\le c(2np^2 + p). \end{aligned} \qquad (1.2)$$

The total variation distance d_{TV} on probability measures over $\mathbb{Z}^+$ is defined by

$$d_{TV}(\lambda, \mu) = \sup\{|\lambda(A) - \mu(A)| : A \subset \mathbb{Z}^+\} = \frac{1}{2} \sum_{j \ge 0} |\lambda\{j\} - \mu\{j\}| \qquad (1.3)$$

(see Section A.1 of the appendix). Thus, from (1.2), the total variation distance between the Bi (n,p) and the Po (np) distributions is at most of order $\max(np^2, p)$, and faithful approximation of the binomial distribution by the Poisson is possible provided that $p = o(n^{-1/2})$ as $n \to \infty$.

It is not immediately obvious, in view of the simplicity of the above argument, that the order np^2 estimate it yields when np is large can actually be improved. However, for large np, the bulk of the probability in the Bi (n,p) and Po (np) distributions lies within $O(\sqrt{np})$ of np, so that the interesting values of k are around np. Now, considering in more detail the terms neglected in (1.1), the factor multiplying the Poisson Po (np) probability is, to the next order, $\exp\{-\frac{k^2}{2n} + kp - \frac{1}{2}np^2\}$, which is precisely 1 when $k = np$, as a result of cancellation. This suggests that the order np^2 error estimate may be pessimistic, as turns out to be the case: Prohorov (1953) showed that

$$d_{TV}(\mathrm{Bi}\,(n,p), \mathrm{Po}\,(np)) \le p[(2\pi e)^{-1/2} + O(\min(1, \{np\}^{-1/2} + p))]$$

(our formulation corrects a minor error in the original). Thus the law of small numbers continues to hold good, so long as p is small.

The practical significance of the binomial distribution Bi (n,p) arises from a model of a sequence of n repeated experiments, conducted under identical conditions and independently of one another. If A denotes a particular outcome, and if $\mathbb{P}[A] = p$, the number of times that A occurs in the n experiments has the Bi (n,p) distribution. However, obtaining identical conditions and independence between experiments is rather difficult, and it is natural to ask how much the law of small numbers really depends on these assumptions.

The simplest assumption to relax is that of identical conditions. Let $I_1, \ldots, I_n$ denote a sequence of independent 0–1 random variables such that

$$\mathbb{P}[I_i = 1] = p_i; \quad \mathbb{P}[I_i = 0] = q_i; \quad p_i + q_i = 1$$

for each i, and let $W = \sum_{i=1}^n I_i$. If all the p_i's were equal to p, the distribution of W would be Bi (n,p), as above. We are now interested in the more general question of how well a Poisson distribution approximates the distribution of W, even when the p_i's are allowed to be different. Experience with the binomial case suggests that a good place to start might be with the point probabilities, although it is clear that there are no longer succinct formulae available. In fact, we have

$$\mathbb{P}[W = k] = \prod_i q_i \sum_{i_1,\ldots,i_k}{}^{1} \left(\prod_{j=1}^{k} (p_{i_j}/q_{i_j})\right), \tag{1.4}$$

where $\sum^1$ sums over all distinct k-subsets $\{i_1, \ldots, i_k\}$ of $\{1, 2, \ldots, n\}$. Now, letting $\lambda = \sum_{i=1}^n p_i$, we have

$$\lambda^k = (\sum_{i=1}^n p_i)^k = \sum_{i_1, \ldots, i_k}^2 (\prod_{j=1}^k p_{i_j}),$$

where $\sum^2$ sums over independent indices, each ranging from 1 to n. Hence, comparing the two types of summation, it follows that

$$\begin{aligned} 0 &\le \lambda^k - k! \sum_{i_1, \ldots, i_k}^1 (\prod_{j=1}^k p_{i_j}) \\ &\le \binom{k}{2} \sum_{i=1}^n p_i^2 \sum_{i_1, \ldots, i_{k-2}}^2 (\prod_{j=1}^{k-2} p_{i_j}) \le (\max_i p_i) \binom{k}{2} \lambda^{k-1}, \end{aligned}$$

and thence that

$$\sum_{i_1, \ldots, i_k}^1 (\prod_{j=1}^k p_{i_j}) = \frac{\lambda^k}{k!} (1 + O\{k^2 \lambda^{-1} \max_i p_i\}).$$

Thus, returning to (1.4) and making the easier estimations, it follows that

$$\mathbb{P}[W = k] = \text{Po}(\lambda)\{k\} \exp\{O(\lambda \max_i p_i, k^2 \lambda^{-1} \max_i p_i)\}, \qquad (1.5)$$

uniformly in $\max_i p_i \le c$, for any $0 < c < 1$. Note that the order terms reduce to the earlier quantities np^2, $k^2 n^{-1}$ in the case of equal probabilities.

As in the case of equal p's, the estimate (1.5) of the point probabilities can in principle be translated into an estimate of the total variation distance between the distribution of W and the Poisson $\text{Po}(\lambda)$ distribution, which is small if $\max(\lambda, 1) \max_i p_i$ is small. In particular, Poisson approximation can be expected to be good, provided that $\max_i p_i = o(n^{-1/2})$, a result analogous to that obtained by the simple argument in the case of equal p's. However, better results are also obtainable here, showing that it is enough that $\max_i p_i$ should be small to ensure good Poisson approximation. The earliest result of this kind is that of Hodges and Le Cam (1960), who proved that

$$\max_{j \ge 0} |\mathbb{P}[W \le j] - \text{Po}(\lambda)\{[0, j]\}| \le 3(\max_i p_i)^{1/3}.$$

Almost immediately, Le Cam (1960) established two results in the spirit of Prohorov (1953):

$$d_{TV}(\mathcal{L}(W), \text{Po}(\lambda)) \le 4.5 \max_i p_i, \qquad (1.6)$$

and

$$d_{TV}(\mathcal{L}(W), \mathrm{Po}(\lambda)) \leq 8\lambda^{-1} \sum_{i=1}^{n} p_i^2, \tag{1.7}$$

both of which reduce to estimates of order p when the p_i's are equal: for the last of these, it is assumed that $\max_i p_i \leq 1/4$. Kerstan (1964) sharpened (1.7) by reducing the factor 8 to 1.05, under the same restriction on the p_i's.

It is interesting to observe that Le Cam's result (1.7) is already far from obvious. Suppose that one considers the accuracy of approximation in the case $n = 1$. Then the total variation distance between the distribution of W — that is, $\mathrm{Bi}\,(1, p)$ — and $\mathrm{Po}\,(p)$ is exactly $p(1 - e^{-p}) \approx p^2$. Now, for general n, W is composed of a sum of independent Bernoulli $\mathrm{Bi}\,(1, p_i)$ random variables, and a sum of independent $\mathrm{Po}\,(p_i)$ random variables has the Poisson $\mathrm{Po}\,(\lambda)$ distribution, suggesting that one way to demonstrate the Poisson approximation is merely to match the summands as well as possible, and let their sums take care of themselves. Formally, a simple coupling inequality allows one to deduce from this procedure that

$$d_{TV}(\mathcal{L}(W), \mathrm{Po}(\lambda)) \leq \sum_{i=1}^{n} p_i^2, \tag{1.8}$$

an inequality also obtained by Le Cam (1960): the argument given comes from Serfling (1975). The extra factor λ^{-1} in (1.7), which improves the approximation dramatically when λ is large, reflects the fact that one can do better than just throwing away the errors as they arise: the truth is that they tend to cancel each other to some degree, and estimates such as (1.7) are exploiting this phenomenon.

The methods used by Kerstan and Le Cam, although quite different, both rely on the multiplicative properties of convolution. Kerstan compares the probability generating function of the sum of indicators with the Poisson probability generating function in an ingenious way, and then uses Cauchy's formula as a basic tool in obtaining the estimate of total variation distance from the difference of the probability generating functions. Le Cam also uses complex variables, in this case Fourier methods, for the detailed estimates, but his basic approach is to represent distributions over the non-negative integers $\mathbb{Z}$ as operators on the space of bounded measurable functions over $\mathbb{Z}$, with multiplication of operators corresponding to convolution. This method has since been developed by Deheuvels and Pfeifer, who have used it to obtain the sharpest results yet for the Poisson approximation of sums of independent indicators: in particular, in Deheuvels and Pfeifer (1988), the complex variable techniques of Uspensky (1931) and Shorgin (1977) are combined with the operator method to yield asymptotic expansions of arbitrary order for set probabilities, together with explicit bounds on the error involved. Indeed, by choosing the space of operators appropriately, the method gives good estimates of the distance between the distribution of W and $\mathrm{Po}(\lambda)$ in at least three different metrics. For a more detailed discussion, see Witte (1990).

The preceding methods all rely heavily on the assumption of independence of the summands: without independence, the multiplicative structure associated with convolution is no longer available. Thus, in order to be able to tackle problems where the summands are dependent, new techniques are required. In the classical central limit problem, the most effective way of generalizing results to dependent settings has been the use of martingales. Indeed, in many stochastic models, the flow of time provides an obvious filtration, and conditions describing the distribution of the future evolution of a process in terms of its past behaviour are extremely natural. It is then but a short step to using martingales as a basic tool. In the case of sums of indicators, this kind of approach was taken by Freedman (1974) and Serfling (1975). The basic element is a sequence of (indicator) random variables $I_1, \ldots, I_n$, and the error estimates for Poisson approximation of their sum W are expressed in terms of the distribution of the random variables

$$p_i = \mathbb{P}[I_i = 1 | I_1, \ldots, I_{i-1}].$$

For instance, Theorem 1 of Serfling (1975) includes the estimate

$$d_{TV}(\mathcal{L}(W), \mathrm{Po}(\lambda)) \leq \sum_{i=1}^{n} \{\mathbb{E}^2(p_i) + \mathbb{E}|p_i - \mathbb{E}(p_i)|\},$$

where $\lambda = \sum_{i=1}^{n} \mathbb{E}(p_i)$. Although the proof is based on an appropriate coupling, the connection with compensator methods for point processes, and hence with martingale ideas, is very strong. Further developments in this direction have been made by Brown (1983), where, in addition, total variation comparison of the whole process with a Poisson process is accomplished. Note, however, that in the case of independent summands Serfling's estimate reduces to the simple estimate (1.8), so that the magic factor $1/\lambda$ has been lost.

In 1970, Stein introduced an entirely different method of approach to approximating the distribution of random variables, in the context of the central limit theorem for dependent summands. His method is, however, in no way restricted to normal approximation, and the form that it takes for Poisson approximation was worked out by Chen (1975a). To start with, note that, if $g : \mathbb{Z}^+ \to \mathbb{R}$ is any bounded function and $Z \sim \mathrm{Po}(\lambda)$,

$$\mathbb{E}\{\lambda g(Z+1) - Zg(Z)\} = 0, \tag{1.9}$$

because the sum defining the expectation telescopes. What is less evident, but none the less true, is that any function $f : \mathbb{Z}^+ \to \mathbb{R}$ for which $\mathbb{E}f(Z) = 0$ can be expressed in the form

$$f(j) = \lambda g(j+1) - jg(j)$$

for a bounded function g. Thus, conversely, if (1.9) holds for all bounded g, Z has the Poisson distribution $\mathrm{Po}(\lambda)$. Similarly, if $\mathbb{E}\{\lambda g(W+1)-Wg(W)\}$ is nearly zero for all g, then W is nearly distributed as $\mathrm{Po}(\lambda)$. The hope is that it may be easier to establish this latter fact than to directly estimate the difference between $\mathbb{E}f(W)$ and $\mathbb{E}f(Z)$: see Stein (1986) Lecture VIII.

In detail, the argument runs as follows. For any $A \subset \mathbb{Z}^+$, a function $g = g_{\lambda,A} : \mathbb{Z}^+ \to \mathbb{R}$ is constructed to solve the equation

$$\lambda g(j+1) - jg(j) = I[j \in A] - \mathrm{Po}(\lambda)\{A\}, \quad j \geq 0; \tag{1.10}$$

the value $g(0)$ is in fact irrelevant, and is conventionally taken to be zero. The solution of (1.10) is easily accomplished recursively, starting with $j = 0$ and working upwards. Thus, if W is any random element of $\mathbb{Z}^+$, it follows immediately that

$$\mathbb{P}[W \in A] - \mathrm{Po}(\lambda)\{A\} = \mathbb{E}\{\lambda g(W+1) - Wg(W)\}, \tag{1.11}$$

so that estimates of the total variation distance between the distribution of W and $\mathrm{Po}(\lambda)$ can be found, provided that the right hand side of (1.11) can be uniformly estimated for all the $g_{\lambda,A}$. To see how this works, consider the case of independent indicator summands $I_1, \ldots, I_n$. Then

$$\mathbb{E}(I_i g(W)) = \mathbb{E}(I_i g(W_i + 1)) = p_i \mathbb{E}g(W_i + 1), \tag{1.12}$$

where $W_i = \sum_{j \neq i} I_j$, because of the independence of I_i and W_i. Thus

$$\mathbb{E}\{\lambda g(W+1) - Wg(W)\} = \sum_{i=1}^{n} p_i \mathbb{E}\{g(W+1) - g(W_i+1)\}, \tag{1.13}$$

and now, since W and W_i are equal unless $I_i = 1$, an event of probability p_i, it becomes clear that the right hand side of (1.13) can be usefully estimated in terms of simple properties of the functions g. In particular, it follows that

$$|\mathbb{P}[W \in A] - \mathrm{Po}(\lambda)\{A\}| \leq 2 \sup_j |g_{\lambda,A}(j)| \sum_{i=1}^{n} p_i^2 \tag{1.14}$$

and that

$$|\mathbb{P}[W \in A] - \mathrm{Po}(\lambda)\{A\}| \leq \sup_j |g_{\lambda,A}(j+1) - g_{\lambda,A}(j)| \sum_{i=1}^{n} p_i^2, \tag{1.15}$$

where the suprema range over $j \geq 1$.

All that remains to be done is to compute bounds for the quantities $\sup_j |g_{\lambda,A}(j)|$ and $\sup_j |g_{\lambda,A}(j+1) - g_{\lambda,A}(j)|$ in $j \geq 1$. Fortunately, it is possible to obtain rather simple estimates which are valid uniformly for all A. Let

$$\|g\| = \|g_{\lambda,A}\| = \sup_j |g_{\lambda,A}(j)|; \quad \Delta g = \Delta g_{\lambda,A} = \sup_j |g_{\lambda,A}(j+1) - g_{\lambda,A}(j)|. \tag{1.16}$$

Then the following lemma can be proved.

Lemma 1.1.1 *If g solves (1.10), the following estimates are valid:*

$$\|g\| \le \min(1, \lambda^{-1/2}), \qquad \Delta g \le \lambda^{-1}(1 - e^{-\lambda}) \le \min(1, \lambda^{-1}). \tag{1.17}$$

Remark 1.1.2 The argument given below is actually only enough to show that $\|g\| \le 2\min(1, \lambda^{-1/2})$, but one can do better. In Barbour and Eagleson (1983), a variant of the argument yields $\|g\| \le \min(1, 1.4\lambda^{-1/2})$, and in Remark 10.2.4 below an entirely different argument, using probabilistic methods, shows that $\|g\| \le \min\{1, \sqrt{2/e\lambda}\}$.

Remark 1.1.3 Both estimates depend on λ in the correct way.

Proof Let U_m denote the set $\{0, 1, \dots, m\}$. Then it is easy to verify that the solution $g = g_{\lambda,A}$ to (1.10) is given by

$$\begin{aligned} g(j+1) &= \lambda^{-j-1} j! e^{\lambda} (\mathrm{Po}(\lambda)\{A \cap U_j\} - \mathrm{Po}(\lambda)\{A\}\mathrm{Po}(\lambda)\{U_j\}) \\ &= \lambda^{-j-1} j! e^{\lambda} (\mathrm{Po}(\lambda)\{A \cap U_j\}\mathrm{Po}(\lambda)\{U_j^c\} \\ &\qquad - \mathrm{Po}(\lambda)\{A \cap U_j^c\}\mathrm{Po}(\lambda)\{U_j\}), \quad j \ge 0, \end{aligned} \tag{1.18}$$

where U_j^c denotes the complement of U_j. Hence, for any A,

$$|g(j+1)| \le j! e^{\lambda} \lambda^{-j-1} \mathrm{Po}(\lambda)\{U_j\}\mathrm{Po}(\lambda)\{U_j^c\}, \tag{1.19}$$

with equality for $A = U_j$. Taking $j \approx \lambda$ and using Stirling's formula shows that a rate of $\lambda^{-1/2}$ for large λ is the best one can hope for when estimating $\|g\|$.

Equation (1.19) gives two easy ways of estimating $|g(j+1)|$. First,

$$\begin{aligned} |g(j+1)| &\le j! e^{\lambda} \lambda^{-j-1} \mathrm{Po}(\lambda)\{U_j\} = \lambda^{-1} \sum_{r=0}^{j} \lambda^{-r} j!/(j-r)! \\ &\le \lambda^{-1} \sum_{r=0}^{j} (j/\lambda)^r \le (\lambda - j)^{-1}, \end{aligned} \tag{1.20}$$

useful in $j < \lambda$, and then, by a similar argument,

$$|g(j+1)| \le j! e^{\lambda} \lambda^{-j-1} \mathrm{Po}(\lambda)\{U_j^c\} \le \frac{j+2}{(j+1)(j+2-\lambda)}, \tag{1.21}$$

useful for $j > \lambda - 2$. These two estimates are sufficient to prove that $|g(j+1)| \le 5/4$ for all λ and for all $j \ge 1$: for $j = 0$, it follows directly that

$$|g(1)| \le \lambda^{-1}(1 - e^{-\lambda}) \le 1,$$

and hence that $\|g\| \le 5/4$.

In order to obtain the upper estimate of $\|g\|$ involving the factor $\lambda^{-1/2}$, use the estimates (1.20) and (1.21) except when $|j-\lambda| \le \lambda^{1/2}$. Here, one starts as for (1.20) and (1.21), but only uses a geometric series to bound the sums from the point where the ratio of successive terms becomes less than $1-\lambda^{-1/2}$, using the upper bound 1 for each of the earlier terms. Thus, for instance, for $\lambda - \lambda^{1/2} \le j < \lambda$,

$$\lambda^{-1}\sum_{r=0}^{j}\lambda^{-r}\frac{j!}{(j-r)!} \le \lambda^{-1}\left[1+\lambda^{1/2}+\frac{(1-\lambda^{-1/2})}{(1-\{1-\lambda^{-1/2}\})}\right] \le 2\lambda^{-1/2}.$$

In order to obtain the estimate of Δg, note that the solution (1.18) shows that $g_{\lambda,A} = \sum_{j\in A} g_{\lambda,\{j\}}$. Taking $A=\{j\}$, it follows from (1.18) that $g(i+1)$ is negative and decreasing in $i<j$ and positive and decreasing in $i \ge j$, so that the only positive value taken by $g(i+1)-g(i)$ in $i \ge 1$ is, for $j>0$,

$$\begin{aligned} g(j+1)-g(j) &= e^{-\lambda}\lambda^{-1}\Big[\sum_{r=j+1}^{\infty}(\lambda^r/r!) + \sum_{r=1}^{j}(\lambda^r/r!)\frac{r}{j}\Big] \\ &\le \min\{j^{-1}, \lambda^{-1}(1-e^{-\lambda})\}; \end{aligned} \tag{1.22}$$

for $j=0$ all such differences are negative. Hence, for general A, (1.22) also gives an upper bound for $g(j+1)-g(j)$. Finally, since $g_{\lambda,A} = -g_{\lambda,A^c}$, it follows that $\Delta g \le \lambda^{-1}(1-e^{-\lambda})$ also, as required. Note that the bound is attained by the positive jump of $g_{\lambda,\{1\}}$. □

As a result of Lemma 1.1.1, (1.15) yields an explicit estimate of the total variation distance between the distribution of W and $\mathrm{Po}(\lambda)$, in the form

$$d_{TV}(\mathcal{L}(W), \mathrm{Po}(\lambda)) \le \lambda^{-1}(1-e^{-\lambda})\sum_{i=1}^{n} p_i^2 \le \min(1,\lambda^{-1})\sum_{i=1}^{n} p_i^2. \tag{1.23}$$

This estimate, thanks to the bound on Δg established in Lemma 1.1.1, has the magic factor $1/\lambda$. What is more, the constant is a little better than that in Kerstan's improvement of Le Cam's result (1.7), and there is no restriction on the value of $\max_i p_i$. Given that the argument is also relatively simple — even if far from obvious — the method provides an attractive way of proving Poisson approximation for sums of independent indicators.

However, the main advantage of the method, which far outweighs these considerations, is that very little needs to be changed if the hypothesis of independence is dropped. Indeed, independence is only used once, in Equation (1.12), and all that has to be done is to modify this step of the argument appropriately.

The first way of modifying (1.12) is that used by Chen (1975a). Here, the equation $\mathbb{E}(I_i g(W_i+1)) = p_i \mathbb{E}g(W_i+1)$ is replaced by

$$\mathbb{E}(I_i g(W_i+1)) = \mathbb{E}(I_i g(Y_i+1)) + \mathbb{E}\left[I_i\left(g(Y_i+Z_i+1) - g(Y_i+1)\right)\right], \tag{1.24}$$

where $Z_i = \sum_{j \in \Gamma_i^s} I_j$ and $Y_i = W - I_i - Z_i = \sum_{j \in \Gamma_i^w} I_j$. Here, the set of indices Γ_i^s is chosen to contain all those j, other than i, for which I_j is 'strongly' dependent on I_i, and the set Γ_i^w, containing all other indices apart from i, is thought of as designating those I_j which are 'weakly' dependent on I_i. In the case of independent summands, Γ_i^s would be taken to be empty for each i. The second term on the right hand side can now immediately be estimated by

$$\left|\mathbb{E}\left[I_i\left(g(Y_i+Z_i+1) - g(Y_i+1)\right)\right]\right| \le \mathbb{E}(I_i Z_i)\Delta g. \tag{1.25}$$

On the other hand,

$$\left|p_i \mathbb{E}g(Y_i+1) - p_i \mathbb{E}g(W+1)\right| \le p_i(p_i + \mathbb{E}Z_i)\Delta g, \tag{1.26}$$

and all that is left is the term

$$\mathbb{E}(I_i g(Y_i+1)) - p_i \mathbb{E}g(Y_i+1). \tag{1.27}$$

This last can often be shown to be small, provided that Γ_i^s has been so chosen that Y_i is almost independent of I_i: for instance, Y_i may be precisely independent of I_i, as in the case of independent summands, or their dependence may be controlled by a mixing condition of some kind. Combining (1.25) – (1.27), one thus obtains an estimate of the following form.

Theorem 1.A *With the above definitions, for any choice of the index sets* Γ_i^s,

$$d_{TV}(\mathcal{L}(W), \mathrm{Po}(\lambda)) \le \sum_{i=1}^{n} \left[(p_i^2 + p_i \mathbb{E}Z_i + \mathbb{E}(I_i Z_i))\right] \min(1, \lambda^{-1}) + \sum_{i=1}^{n} \eta_i \min(1, \lambda^{-1/2}), \tag{1.28}$$

where η_i *is any quantity satisfying*

$$\left|\mathbb{E}(I_i g(Y_i+1)) - p_i \mathbb{E}g(Y_i+1)\right| \le \eta_i \|g\| :$$

for instance,

$$\eta_i = \mathbb{E}\left|\mathbb{E}\{I_i \mid (I_j, j \in \Gamma_i^w)\} - p_i\right|. \tag{1.29}$$

□

Note that the contribution to estimate (1.28) arising from (1.27) is, for large λ, multiplied by a factor of order only $\lambda^{-1/2}$ rather than λ^{-1}, and the factor cannot be improved to λ^{-1}, as is shown in Example 3.4.3 below. Hence, in many applications, it is desirable to arrange for the contribution from (1.27) to be zero, by making Y_i independent of I_i. Note, in this connection, that it is not in principle necessary that the random variables Y_i and Z_i which add to give W_i should themselves be expressible as sums of I_j's: all that matters is that Z_i should be small and Y_i almost independent of I_i, in the sense that the right hand side of (1.28) is small.

The two parts of the estimate (1.28) can be rather naturally interpreted in the context of a stationary mixing sequence of indicator random variables. Here, one would typically define Γ_i^s to consist of all those indices $j \neq i$ for which $|j - i| \leq m$, say, for some $m > 0$. The first part of the estimate is then no larger than

$$p + \mathbb{E}Z + p^{-1}\mathbb{E}(IZ) \leq (2m+1)p + \mathbb{E}(Z|I=1), \qquad (1.30)$$

where p is the common value of the p_i's and I and Z have the joint distribution of I_{m+1} and $\sum_{1\leq|j-m-1|\leq m} I_j$. Clearly, m is chosen so that mp is small, and $\mathbb{E}\{Z|I = 1\}$ is then only larger in so far as there is a tendency for there to be clusters of 1's. Thus the first part of the estimate (1.28) quantifies the effect of local dependence on the accuracy of the Poisson approximation. The second part of the estimate can also be simply interpreted, this time in terms of the degree of dependence at a distance: indeed, the quantity η_i can be re-expressed as

$$2p \sup_{B\in\mathcal{F}_i} \left|\mathbb{P}[B|I_i = 1] - \mathbb{P}[B]\right|,$$

where $\mathcal{F}_i$ denotes the σ-field of events generated by the random variables $(I_j, j \in \Gamma_i^w)$, which is just p times a (two-sided) mixing coefficient. Thus, in the terminology of Leadbetter, Lindgren and Rootzén (1983), the first part of (1.28) is controlled by a condition like (D′), and the second part by a condition like (D). These ideas are made precise in Smith (1988).

There is another approach to modifying (1.12) in the case of dependence, which is discussed on pp. 92–93 of Stein (1986). Here, one simply writes

$$\mathbb{E}(I_i g(W_i + 1)) = p_i \mathbb{E}\{g(W)|I_i = 1\},$$

leading to the equation

$$\mathbb{P}[W \in A] - \mathrm{Po}(\lambda)\{A\} = \sum_{i=1}^{n} p_i[\mathbb{E}g(W+1) - \mathbb{E}\{g(W)|I_i = 1\}]. \quad (1.31)$$

Now suppose that, for each i, one has constructed random variables U_i and V_i on the same probability space, in such a way that $U_i \stackrel{\mathcal{D}}{=} W$ and that $V_i + 1$ has the distribution of W conditional on $I_i = 1$: we refer to this as a coupling. Then the following estimate is immediate.

Theorem 1.B *With the above definitions, whatever the choice of couplings* (U_i, V_i),

$$|\mathbb{P}[W \in A] - \mathrm{Po}(\lambda)\{A\}| = |\sum_{i=1}^{n} p_i[\mathbb{E}g(U_i + 1) - \mathbb{E}g(V_i + 1)]|$$
$$\leq \Delta g \sum_{i=1}^{n} p_i \mathbb{E}|U_i - V_i|. \qquad (1.32)$$

□

Thus, using the inequality $\Delta g \leq \lambda^{-1}$, the total variation distance between $\mathcal{L}(W)$ and $\mathrm{Po}(\lambda)$ is no greater than an average of the quantities $\mathbb{E}|U_i - V_i|$ with weights p_i. Thus, if the couplings (U_i, V_i) can be constructed in such a way that $\mathbb{E}|U_i - V_i|$ is small, (1.32) yields good Poisson approximation. For instance, with independent summands, one can take the original probability space and set $U_i = W$ and $V_i = W_i$, leading to the same estimate as in (1.23).

A less trivial example is given by the classical occupancy problem. A series of r independent experiments is carried out, in each of which the possible outcomes, labelled $1, 2, \ldots, n$, each have probability $1/n$. If Z_j denotes the outcome of experiment j, let $I_i = \prod_j I[Z_j \neq i]$, so that W counts the number of outcomes which never occurred. Once again, it is possible to take $U_i = W$, but now the probability space is enlarged by, for index i, repeating each experiment j for which $Z_j = i$ (independently of all else) until an outcome $Z_j' \neq i$ is obtained. Let Y_k denote the number of further occurrences of outcome k, let $I_k' = I_k I[Y_k = 0]$, and set $V_i = \sum_{k \neq i} I_k'$. Then

$$U_i - V_i = I_i + \sum_{k \neq i} I_k I[Y_k \neq 0] \geq 0. \qquad (1.33)$$

A heuristic estimate of the expectation of (1.33) is given by

$$p + np[1 - (1 - 1/n)^{r/n}],$$

where $p = \mathbb{E}I_k$, on the grounds that the number of experiments repeated for index i is typically about r/n, and for Y_k to be zero they must all have outcomes different from k. Thus, for $r \gg n$, the error in making a Poisson approximation for W should be of order rp/n. This estimate is made precise in much greater generality in Corollary 6.E.1 below, where it is also shown to be sharp.

The structure of the occupancy problem is quite different from that in the earlier example of a stationary mixing sequence. In the latter, each of the summands I_i may be relatively heavily dependent on a few of the other summands, but is almost independent of all the others. In the occupancy problem, the relationships between the various summands are entirely symmetric: each depends in exactly the same way on all the others. Broadly speaking, the method of exploiting the Stein equation using Theorem B, which we shall refer to as the ***coupling approach,*** is more effective in dealing with such situations, where there is no natural way of defining the index sets Γ_i^s designating the summands more closely related to I_i, whereas the method via Theorem A, the ***local approach,*** may be easier to implement if a local dependence structure is naturally available. The two approaches are actually not mutually exclusive, a good example being given by dissociated random variables; see Section 2.3. The difficulty in the coupling approach is that a suitable coupling may be tricky to construct, though it is shown in Chapter 2 that there are times when this is not necessary, and that all that is needed is to demonstrate the existence of a coupling with a certain positive or negative dependence property. On the other hand, if a coupling can be found, the estimate obtained in (1.32) always has the magic factor $1/\lambda$.

Another useful result, which can also be obtained by Stein's method using coupling ideas, concerns Poisson approximation of mixtures, rather than sums. It sometimes simplifies a Poisson approximation argument to proceed conditionally on the value of some random element Z. One then typically obtains an estimate for the distance between $\mathcal{L}(W)$ and a mixed Poisson distribution $\mathrm{Po}(\Lambda)$, where Λ is a function of Z. A further step is then required to show that, if Λ is not too variable, the distribution $\mathrm{Po}(\Lambda)$ is close to $\mathrm{Po}(\lambda)$, where $\lambda = \mathbb{E}\Lambda$. The following theorem provides the necessary link: see Pfeifer (1987) for related results, proved using the operator method.

Theorem 1.C *Let W have the mixed Poisson distribution* $\mathrm{Po}(\Lambda)$, *and let λ be arbitrary; then*

(i) $d_{TV}(\mathcal{L}(W), \mathrm{Po}(\lambda)) \le \min(1, \lambda^{-1/2})\mathbb{E}|\Lambda - \lambda|$;

(ii) $d_{TV}(\mathcal{L}(W), \mathrm{Po}(\lambda)) \le \lambda^{-1}(1 - e^{-\lambda})\mathrm{Var}\,\Lambda$ *if* $\lambda = \mathbb{E}\Lambda$.

Proof We use (1.11), computing $\mathbb{E}\{\lambda g(W+1) - W g(W)\}$ by conditioning on Λ. Since, when Λ is given, W has a Poisson distribution with mean Λ, it follows from (1.9) that

$$\mathbb{E}(Wg(W)|\Lambda) = \Lambda\mathbb{E}(g(W+1)|\Lambda),$$

and hence

$$\mathbb{E}\{\lambda g(W+1) - Wg(W)\} = \mathbb{E}\{(\lambda - \Lambda)\mathbb{E}(g(W+1)|\Lambda)\}. \tag{1.34}$$

The estimate (i) is now immediate from Lemma 1.1.1.

For the second estimate, define $k(\lambda) = \mathbb{E}g(Z+1)$, where $Z \sim \mathrm{Po}(\lambda)$. Now note that if, for instance, $\Lambda > \lambda$, W can be realized as the sum of two independent Poisson random variables W_1 and W_2, having means λ and $\Lambda - \lambda$ respectively. Thus, for $\Lambda > \lambda$,

$$\begin{aligned}|\mathbb{E}(g(W+1)|\Lambda) - k(\lambda)| &= |\mathbb{E}(g(W+1)|\Lambda) - \mathbb{E}g(W_1+1)| \\ &\leq \Delta g \mathbb{E}W_2 = \Delta g(\Lambda - \lambda).\end{aligned}$$

The proof is concluded by using an analogous argument for $\Lambda \leq \lambda$, and combining the results with (1.34). □

Remark 1.1.4 Taking $\Lambda = \mu$ with probability one, for any fixed $\mu < \lambda$, Theorem C(i) gives an estimate of the total variation distance between two Poisson distributions with different means.

The essence of the Stein–Chen method is that it enables one to obtain effective bounds on the difference between $\mathbb{E}f(W)$ and $\mathrm{Po}(\lambda)\{f\}$ for a test function f, where the latter expression denotes the expectation of f with respect to the distribution $\mathrm{Po}(\lambda)$. Given the function f, one solves the *Stein equation*

$$f(j) - \mathrm{Po}(\lambda)\{f\} = \lambda g(j+1) - jg(j) \tag{1.35}$$

for the unknown function g. Arguments similar to those given above can then be used to estimate the expectation $\mathbb{E}\{\lambda g(W+1) - Wg(W)\}$ in terms of the quantities $\|g\|$ and Δg, as is illustrated in the later chapters of the book, thus by (1.35) yielding bounds for $\mathbb{E}f(W) - \mathrm{Po}(\lambda)\{f\}$.

There are a number of metrics on the space of probability distributions over $\mathbb{Z}^+$, with respect to which the distance between $\mathcal{L}(W)$ and $\mathrm{Po}(\lambda)$ can be estimated by applying the above procedure to each member f of a class $\mathcal{F}$ of test functions. If total variation distance is being considered, the set $\mathcal{F}_{TV}$ of functions $(f_A, A \subset \mathbb{Z}^+)$ of the form

$$f_A(\cdot) = I[\cdot \in A] - \mathrm{Po}(\lambda)\{A\}$$

is appropriate, since $d_{TV}(\mathcal{L}(W), \mathrm{Po}(\lambda))$ can be expressed as

$$\sup_A \left|\mathbb{E}f_A(W) - \mathrm{Po}(\lambda)\{f_A\}\right|,$$

and Lemma 1.1.1 gives corresponding estimates of the quantities $\|g\|$ and Δg, valid uniformly for all such f. However, the method is also clearly applicable in the context of any other measure of distance d satisfying

$$d(\mathcal{L}(W), \mathrm{Po}(\lambda)) = \sup_{f \in \mathcal{F}} \left|\mathbb{E}f(W) - \mathrm{Po}(\lambda)\{f\}\right| \tag{1.36}$$

for some $\mathcal{F}$, provided always that useful estimates of $\|g\|$ and Δg can be derived. One such distance is the Wasserstein or Fortet–Mourier distance d_W, discussed in Section A.1 of the appendix, for which

$$\mathcal{F}_W = \{f : \Delta f \leq 1\}.$$

For this class of test functions, an analogue of Lemma 1.1.1 can also be proved, as follows.

Lemma 1.1.5 *Suppose that g is the solution to (1.35) for some $f \in \mathcal{F}_W$. Then*

$$\|g\| \le c_1 \qquad \text{and} \qquad \Delta g \le c_2(1 \wedge \lambda^{-1/2}),$$

where $c_1 \le 1 + 2\sqrt{2/e} \le 3$ and $c_2 \le 2 + \sqrt{2/e} \le 3$.

Proof Extend the definition of f to the whole of $\mathbb{R}^+$ by linear interpolation, and let $\bar{f}$ denote $\mathrm{Po}(\lambda)\{f\}$. Then, since $f \in \mathcal{F}_W$,

$$|f(\lambda) - \bar{f}| \le \mathbb{E}|Z - \lambda| = 2\lambda \max_{j\ge 0} \mathrm{Po}(\lambda)\{j\} \le \sqrt{2\lambda/e},$$

the latter inequality being Proposition A.2.7, where Z denotes a random variable with the distribution $\mathrm{Po}(\lambda)$. Hence, when estimating $|g(m+1)|$ for $m+1 \ge \lambda$, we can write

$$\begin{aligned}
|g(m+1)| &= \left|-\frac{m!}{\lambda^{m+1}} \sum_{j\ge m+1} \frac{\lambda^j}{j!}(f(j) - \bar{f})\right| \\
&\le \frac{m!}{\lambda^{m+1}} \sum_{j\ge m+1} \frac{\lambda^j}{j!}\left\{(j-\lambda) + \sqrt{\frac{2\lambda}{e}}\right\} \\
&= 1 + \sqrt{\frac{2\lambda}{e}}\frac{m!}{\lambda^{m+1}} \sum_{j\ge m+1} \frac{\lambda^j}{j!} \le 1 + \sqrt{\frac{2\lambda}{e}} \cdot \frac{2}{\sqrt{\lambda}},
\end{aligned}$$

where the final estimate follows as in the proof of Lemma 1.1.1. The argument bounding $|g(m+1)|$ for $m \le \lambda$ is similar.

The proof of the estimate of Δg is rather more complicated. We start with the range $m+1 \ge \lambda$, and use the identity

$$\begin{aligned}
\lambda\{g(m+2) - g(m+1)\} &= (m+1-\lambda)g(m+1) + f(m+1) - \bar{f} \\
&= \left\{1 - (m+1-\lambda) \sum_{j\ge m+1} \frac{m!}{j!}\lambda^{j-m-1}\right\}(f(m+1) - \bar{f}) \\
&\quad - (m+1-\lambda) \sum_{j\ge m+1} \frac{m!}{j!}\lambda^{j-m-1}(f(j) - f(m+1)). \qquad (1.37)
\end{aligned}$$

Taking the first of these quantities, observe that

$$\begin{aligned}
&1 - (m+1-\lambda) \sum_{j\ge m+1} \frac{m!}{j!}\lambda^{j-m-1} \\
&\quad = 1 - (m+1-\lambda)\left[\frac{1}{m+1} + \frac{1}{m+1}\cdot\frac{\lambda}{m+2} \right. \\
&\qquad\qquad \left. + \frac{1}{m+1}\cdot\frac{\lambda}{m+2}\cdot\frac{\lambda}{m+3} + \cdots\right], \qquad (1.38)
\end{aligned}$$

which, by comparison with a geometric progression with common ratio $\lambda/(m+1)$, lies between 0 and 1, an estimate which is used whenever $0 \leq m+1-\lambda \leq \sqrt{\lambda}$. For larger values of m, develop (1.38) further as

$$\begin{aligned}
(m+1-\lambda)\Big[&\frac{1}{m+1}\Big(\frac{\lambda}{m+1}-\frac{\lambda}{m+2}\Big) \\
&+\frac{1}{m+1}\Big(\Big(\frac{\lambda}{m+1}\Big)^2-\frac{\lambda}{m+2}\cdot\frac{\lambda}{m+3}\Big)+\cdots\Big] \\
&\leq \Big(\frac{m+1-\lambda}{(m+1)(m+2)}\Big)\sum_{r\geq 1}\tfrac{1}{2}r(r+1)\Big(\frac{\lambda}{m+1}\Big)^r \leq \lambda(m+1-\lambda)^{-2}.
\end{aligned}$$

Thus the first quantity in (1.37) is no larger than

$$\{(m+1-\lambda)+\sqrt{2\lambda/e}\}\left(1\wedge\lambda(m+1-\lambda)^{-2}\right) \leq \sqrt{\lambda}(1+\sqrt{2/e}).$$

In order to estimate the second quantity in (1.37), the crude inequality $|f(j)-f(m+1)| \leq j-\lambda$ suffices for an upper bound of $m+1-\lambda$. Otherwise, we can write

$$\begin{aligned}
(m+1-\lambda)&\sum_{j\geq m+1}\frac{m!}{j!}\lambda^{j-m-1}|f(j)-f(m+1)| \\
&\leq \frac{\lambda(m+1-\lambda)}{(m+1)(m+2)}\Big\{1+\frac{2\lambda}{m+3}+\frac{3\lambda^2}{(m+3)(m+4)}+\cdots\Big\} \\
&\leq \frac{\lambda(m+1-\lambda)(m+3)^2}{(m+1)(m+2)(m+3-\lambda)^2} \leq \frac{\lambda}{m+2-\lambda},
\end{aligned}$$

which, taken with the first upper bound, gives an estimate of at most $\sqrt{\lambda}$ for the second quantity in (1.37).

These arguments, and analogous estimates when $m+1<\lambda$, are enough to establish the estimate $\Delta g \leq c_2\lambda^{-1/2}$. If $\lambda<1$, the arguments already given show that $|g(m+2)-g(m+1)| \leq c_2$ if $m \geq 1$, and for $m=0$ direct inspection of (1.37) completes the proof. □

Remark 1.1.6 The constants c_1 and c_2 can be improved, combining Remark 10.2.4, Equation (10.2.12) and the proof of Lemma 10.2.3: c_1 may be taken to be 1, and the estimate of Δg may be replaced by

$$\Delta g \leq 1\wedge\frac{4}{3}\sqrt{\frac{2}{e\lambda}}.$$

Remark 1.1.7 The Wasserstein distance may be used to write the basic estimate (1.32) of Theorem 1.B succinctly as

$$d_{TV}(\mathcal{L}(W),\mathrm{Po}(\lambda)) \leq (1\wedge\lambda^{-1})\sum_{i=1}^{n}p_i\, d_W\big(\mathcal{L}(W+1),\mathcal{L}(W|I_i=1)\big).$$

Similarly, Lemma 1.1.5 and Remark 1.1.6 yield

$$d_W(\mathcal{L}(W), \mathrm{Po}(\lambda)) \le \frac{4}{3}\sqrt{\frac{2}{e}}(1 \wedge \lambda^{-1/2}) \sum_{i=1}^{n} p_i \, d_W(\mathcal{L}(W+1), \mathcal{L}(W | I_i = 1)).$$

For completeness, observe that the bounds for $\|g\|$ in Lemma 1.1.1 and Remark 1.1.6 yield the companion estimates

$$d_{TV}(\mathcal{L}(W), \mathrm{Po}(\lambda)) \le 2(1 \wedge \lambda^{-1/2}) \sum_{i=1}^{n} p_i \, d_{TV}(\mathcal{L}(W+1), \mathcal{L}(W | I_i = 1))$$

and

$$d_W(\mathcal{L}(W), \mathrm{Po}(\lambda)) \le 2 \sum_{i=1}^{n} p_i d_{TV}(\mathcal{L}(W+1), \mathcal{L}(W | I_i = 1)).$$

These are however less useful, because of the loss of a factor $\lambda^{-1/2}$.

In view of Lemma 1.1.5, for each estimate of total variation distance proved by the Stein–Chen method, one can immediately write down a corresponding estimate of the Wasserstein distance. However, the estimates are less attractive than those for the total variation distance when λ is large, in so far as the magic factor $1/\lambda$ is replaced by $1/\sqrt{\lambda}$ (and $1/\sqrt{\lambda}$ by one). This is not the fault of the method, but reflects the difference between the total variation and Wasserstein distances: it is the case that

$$d_{TV}(\mathrm{Bi}\,(n,p), \mathrm{Po}\,(np)) \asymp p,$$

whereas

$$d_W(\mathrm{Bi}\,(n,p), \mathrm{Po}\,(np)) \asymp p\sqrt{np}:$$

see Deheuvels and Pfeifer (1988) and Remark 3.2.3.

The Wasserstein distance can equally well be defined for random variables with finite expectation over $\mathbb{R}$, taking $\mathcal{F}'_W = \{f : \mathbb{R} \to \mathbb{R}, \|f'\| \le 1\}$, and has the property that

$$d_W(\mathcal{L}(cX), \mathcal{L}(cY)) = c\, d_W(\mathcal{L}(X), \mathcal{L}(Y))$$

for any $c > 0$ and random variables X, Y. In the case of Poisson approximation for large λ, there are then two natural choices of scale: one the original scale, where $\mathcal{L}(W)$ is compared with $\mathrm{Po}(\lambda)$, and one appropriate to the comparison of the standardized random variables, in which $\mathcal{L}(W^*)$, where $W^* = \lambda^{-1/2}(W - \lambda)$, is compared with $\mathrm{Po}^*(\lambda)$, the correspondingly centred and normalized Poisson distribution. Obtaining bounds on

$d_W(\mathcal{L}(W^*), \mathrm{Po}^*(\lambda))$ when λ is large is equivalent to using the family of test functions

$$\mathcal{F}_\lambda = \{f : \Delta f \leq \lambda^{-1/2}\} = \{f : \lambda^{1/2} f \in \mathcal{F}_W\},$$

for which the corresponding estimates of $\|g\|$ and Δg are $c_1\lambda^{-1/2}$ and $c_2\lambda^{-1}$, of the same order in λ as for total variation distance. Thus, if the Wasserstein distance between the *standardized* random variables is of primary interest, the same estimates — up to a constant — can be used as for the total variation distance between $\mathcal{L}(W)$ and $\mathrm{Po}(\lambda)$. Note that, in view of the inequality

$$d_W(\mathrm{Po}^*(\lambda), \mathcal{N}(0,1)) \leq 1/\sqrt{\lambda}, \tag{1.39}$$

which can be proved using Stein's method for the normal distribution, good approximation of the standardized random variable W^* by $\mathrm{Po}^*(\lambda)$ when λ is large also implies good normal approximation. As a weak consequence, if W_n is a sequence of random variables for which

$$d_{TV}(\mathcal{L}(W_n), \mathrm{Po}(\lambda_n)) \to 0,$$

where $\lambda_n \to \infty$, it follows that $W_n^* \xrightarrow{\mathcal{D}} \mathcal{N}(0,1)$.

The local approach to the Stein–Chen method, which was first proposed in Chen (1975a), is very similar in spirit to Stein's method for normal approximation, and the techniques are accordingly quite well known. There have been a number of successful applications of the method, to be found, for instance, in Arratia, Goldstein and Gordon (1989,1990) and in Smith (1988), as well as in Chen's original papers. In the remainder of the book, we concentrate instead on developing and illustrating the coupling approach.

The basic theory is contained in Chapters 2 and 3. In the former, the principal inequalities used when applying the coupling approach are presented. A central feature of the approach is that it is possible to exploit any ordering between the random variables to be coupled to simplify and sharpen the estimates obtained. In particular, it may only be necessary to know that a coupling of a certain sort exists, without specifying it, in order to obtain explicit bounds for the error in Poisson approximation. Since Strassen's (1965) theorem gives readily used conditions for the existence of such couplings, often satisfied in applications because a collection of random variables involved in defining the indicators are negatively or positively associated, this can be of considerable benefit.

The error bounds obtained in this way are very often found to be of order $|\mathrm{Var}\, W/\mathbb{E}W - 1|$. It is an intuitively appealing expression, suggesting that it might in fact be the true order of approximation, as is indeed the case for sums of independent indicator random variables. In Chapter 3, this possibility is investigated in some detail. Two sets of lower bounds for the error in Poisson approximation are derived. For the first set, nothing is required of the distribution of W, apart for the existence of a higher moment if $\operatorname{Var} W > \mathbb{E}W$. None the less, the bounds obtained are good enough to show that, in most cases, the error cannot be much smaller than $|\operatorname{Var} W/\mathbb{E}W - 1|$. For the second set, the special coupling structure is presupposed, and the sharper bounds thereby obtained are often sufficient to prove that $|\operatorname{Var} W/\mathbb{E}W - 1|$ is exactly the right order of error — though there are examples to show that this need not (quite) be the case.

Chapters 4 to 8 contain illustrations of the use of the Stein–Chen method in a variety of well-known settings. The examples are deliberately chosen to emphasize the coupling approach, and the kind of dependence involved is typically one in which the value taken by any one indicator has a non-trivial influence on the values taken by all others, even if the influence may not always be the same for all of them. The settings are permutation-based statistics of Wald–Wolfowitz type, including the classical matching and ménage problems, and a two-dimensional generalization which has wide application in spatial statistics; random graphs, and in particular subgraph counts, vertex degrees and birthday problems; occupancy and urn models, with the Poisson approximation to the hypergeometric distribution, multinomial allocation, sampling with replacement, Pólya sampling, capture–recapture and further models; spacings, large and small, of order one or more, in one or more dimensions; and extreme value theory. The examples illustrate that very good bounds can be simply obtained using the Stein–Chen method, in many cases where no other approach has proved feasible. When the special coupling structure is present, as is quite often the case, the argument simplifies dramatically. A particular example is that of occupancy, where essentially one short argument suffices to establish simple but accurate bounds on the error in Poisson approximation, valid irrespective of the chosen sampling scheme: all that remains to be done is to compute the mean and variance of W in the different cases. Without this structure, couplings have to be explicitly constructed. However, there is often a natural candidate for the coupling, which is found to work well. Those cases where the natural coupling does not prove sharp enough are thus of extra interest: one such is investigated in detail in connection with the number of vertices of given degree in a random graph. Note that extreme value theory can also be treated, indeed perhaps more naturally, using the local approach, as has been done by Smith (1988). The results obtained in this book by the coupling approach differ from those of Smith, in that rather more structure is assumed here, in exchange for which sharper estimates are obtained.

The last two chapters are concerned with developments of the basic Stein–Chen method. The first of them, Chapter 9, is concerned with refining the main results. A few general extensions, to the estimation of point probabilities and to exponential bounds in the far tails, appear earlier in Section 2.4. Here, in order to see what else might be obtainable, it is first assumed that the indicators are independent. This enables one to follow a programme suggested by Chen (1975a), and develop asymptotic expansions for both point and set probabilities, with leading term given by the Poisson approximation and with an explicit error estimate of appropriate order at each stage. These expansions can also be combined with Cramér's method of conjugate distributions, leading to asymptotic expansions for tail probabilities with explicitly bounded relative error. A little progress in the same direction is also made for sums of dissociated indicators. However, the computations rapidly become very complicated, and it is by no means clear that there is even a generally applicable natural ordering of small quantities, so that no attempt is made to push the expansions further.

When considering expansions with two adjustable parameters, as opposed to the single parameter available in the Poisson approximation, it is natural to consider also what could be achieved by using the binomial or negative-binomial distribution as an approximation, rather than the Poisson–Charlier signed measure which appears in the asymptotic series: this is analogous to using a chi-squared approximation in place of a two-term Edgeworth expansion about the normal limit. Stein's method is by no means restricted to the normal and Poisson distributions, and could have direct application in such a context, circumventing the iterative nature of the argument which yields the asymptotic expansion. In Section 9.2, the necessary groundwork is laid for applying Stein's method with, as approximating distribution, the equilibrium distribution of any birth and death process on the non-negative integers. For sums of independent indicators, the binomial approximation is appropriate, and an error estimate is obtained for a binomial $\mathrm{Bi}\,(m,p)$ fit in which both m and p are free to be chosen. This complements a result of Ehm (1991), who restricts m to be the same as the number of indicators appearing in the sum.

The extension of Stein's method to other distributions on the non-negative integers leads one to consider extensions to distributions on larger spaces, and in particular to the problem of approximating the distribution of a whole process. The process which, in the context of this book, is the most natural candidate for the 'limiting' process is the Poisson process. In Chapter 10, it is shown how to set up the machinery for using Stein's method to estimate the error in approximating a process generated by a collection of indicator random variables by a Poisson point process. The method used is different from that of Arratia, Goldstein and Gordon (1989). A natural analogue of the Stein equation (1.35) is constructed, and properties of the solutions analogous to those proved in Lemma 1.1.1 are derived. An interesting feature is that purely probabilistic techniques are used to obtain these estimates. However, in process approximation, the choice of metric turns out to be more of a problem than for the one-dimensional distributions, since, for the total variation distance, there is unfortunately in general no magic factor $1/\lambda$. This deficiency can be remedied at the cost of weakening the metric, which it is in any case often natural to do, for instance when approximating a Bernoulli process on the space $n^{-1}\mathbb{Z}$ by a Poisson process of unit rate on $\mathbb{R}$. Section 10.2 shows how the magic factor can be reinstated by using a Wasserstein metric.

Another approach is to introduce marks associated with the underlying sequence of indicator random variables, and to consider total variation approximation of the marked point process by a Poisson process on the mark space. This, under appropriate conditions, also allows magic factors to be reintroduced. Broadly speaking, the less the points in the mark space identify the indices of the indicators with which they are associated, the better the magic factor that can be incorporated. One instance of this is in multivariate Poisson approximation. Whether the best magic factor can always be introduced if the marks are exchangeable with respect to the indices is not as yet known: the study of process approximation by Stein's method is still in its early stages, and the chapter only gives a taste of what might one day be realizable.

An obvious omission so far, under the heading of extensions, is the compound Poisson distribution. One way of handling problems of compound Poisson approximation is to embed them in a problem concerning marked point processes in the obvious way, and to use the techniques of Chapter 10: this is the principal theme of Section 10.4. However, only a rather restricted subset of those test functions naturally associated with process approximation is of interest in the context of compound Poisson approximation, and it can hardly be hoped that this way is likely to lead to best possible results, even though the bounds are in many cases very useful. What seems much more promising is to use Stein's method directly. There is a natural candidate for a Stein equation, known to Chen since the 1970s, and re-discovered a number of times since, and its structure is such as to suggest that it can be profitably used in conjunction with the coupling approach: indeed, the parallel with some of the techniques proposed by Aldous (1989b) is unmistakable. There is, however, an unexpected problem: there are as yet no general analogues of the bounds in Lemma 1.1.1 which look even remotely sharp. There are problems where this does not matter; if the compound Poisson distribution can be realized as the equilibrium distribution of an immigration–death process (immigration in groups allowed), and especially if the immigration rate of single individuals dominates, so that the distribution is nearly Poisson, methods deriving from those used in Chapter 10 can be brought to bear, leading to reasonable bounds: see Barbour, Chen and Loh (1992). Nevertheless, the general situation is unsatisfactory, and the time seems not yet ripe for an attempt at a coherent exposition of compound Poisson approximation using Stein's method.

2
Upper bounds

In Theorem 1.B, the total variation distance between the distribution of a sum W of indicator random variables $(I_i)_{1\leq i\leq n}$ and the Poisson distribution with the same mean is bounded by an average of quantities $\mathbb{E}|U_i - V_i|$, where U_i and V_i are random variables on some probability space such that $U_i \stackrel{\mathcal{D}}{=} W$ and $1 + V_i$ has the distribution of W, given $I_i = 1$. Thus, in order to demonstrate a given accuracy of approximation, it is enough to be able to couple the pairs (U_i, V_i) in such a way that $\mathbb{E}|U_i - V_i|$ is small enough. In this chapter, we look at the coupling approach in more detail.

The discussion is motivated by the observation in Section 2.1 that if, for instance, the couplings (U_i, V_i) of Theorem 1.B can be realized in such a way that $U_i \geq V_i$ a.s., the bound (1.1.31) simplifies dramatically: letting $p_i = \mathbb{E}I_i$ and $\lambda = \sum_{i=1}^{n} p_i$ as before, it then follows that

$$\sum_{i=1}^{n} p_i\mathbb{E}|U_i - V_i| = \sum_{i=1}^{n} p_i\mathbb{E}\{(U_i + 1) - (V_i + 1)\}$$
$$= \lambda\mathbb{E}(W + 1) - \sum_{i=1}^{n} \mathbb{E}(I_i W) = \lambda - \operatorname{Var} W. \quad (0.1)$$

Thus, if such a coupling can be realized, an upper bound $(1 - \lambda^{-1}\operatorname{Var} W)$ for the accuracy of the Poisson approximation can be immediately written down, with no more computation required than that of the mean and variance of W. A further bonus is that the coupling itself need not be specified: all that is necessary is that a coupling with $U_i \geq V_i$ should be known to exist. This is particularly useful, since it can happen that the existence of a suitable coupling can be deduced from general considerations, where an explicit construction is not at all obvious. These ideas are enlarged upon in Section 2.2.

It has already been shown in (1.1.23) in the Introduction how to use the Stein–Chen method for sums of independent random variables. In Section 2.3, the upper bounds so far derived are applied to dissociated indicator random variables. Dissociation is a rather natural generalization of independence to arrays rather than sequences, introduced by McGinley and Sibson (1975), which has widespread practical application. The structure is well suited for analysis using the local approach, enabling more or less convenient upper bounds to be easily derived. More detailed analysis, for independent and dissociated random variables, is also feasible: further results are presented in Chapter 9.

Finally, in Section 2.4, some miscellaneous results are collected, concerned with modes of approximation other than in total variation. In the Introduction, Poisson approximation in the simplest cases with independent summands was established by means of accurate estimates of point probabilities, and total variation approximation was then derived as a consequence. Of course, total variation approximation also implies approximation of point probabilities, but, in cases where λ is large, point probabilities are at most of order $\lambda^{-1/2}$, and the relative error of approximation obtained by appealing directly to total variation convergence is therefore increased by at least the factor $\lambda^{1/2}$. It is thus of interest to see whether, as in the case of independent summands, the relative error in approximating point probabilities can in fact be shown to be of the same order as the absolute error in total variation. Theorem 2.Q below addresses the problem. Although the statement is rather cumbersome, the result is sufficient to establish the improved accuracy of approximation to point probabilities in many cases, at least in the body of the distribution.

The accurate approximation of point probabilities is not only interesting in its own right, but is also necessary for estimating, for instance, the Hellinger distance between $\mathcal{L}(W)$ and $\mathrm{Po}(\lambda)$, as in Falk and Reiss (1988). Such results are also used as a basis for the large deviation expansions for tail probabilities in the case of independent summands in Section 9.1. Often, only upper estimates of the probabilities of large deviations are important, as an ingredient in the proof of more sophisticated results. Even then, they may in general be difficult to obtain, as soon as independence of the summands is abandoned. In the latter part of Section 2.4, some general estimates are derived. If the indicators are assumed to be negatively related, as for Corollary 2.C.2, there are useful upper bounds for the tail probabilities, given in Theorem 2.R. Sums of positively related random variables pose greater problems, as is to be expected, since they tend to be more dispersed than Poisson random variables. However, with some extra structure, an exponential upper bound for the lower tail probabilities can be derived, as is shown in Theorem 2.S. An example is also given to show that, at this level of generality, there can be no exponential bound for the upper tail probabilities.

The notation used in this chapter and hereafter is slightly different from that of the Introduction, in so far as, in many applications, indexing the indicator random variables by $1, 2, \ldots, n$ is unnatural and even misleading: in random graph theory, for instance, the vertices of the carrier graph are conventionally labelled $1, 2, \ldots, n$, and the indicator random variables may well be indexed by a collection of subsets of these vertices. As a result, we let Γ denote an arbitrary finite collection of indices, usually denoted by α, β and so on. The notation in the Introduction is now reserved for applications where it occurs naturally. Then we can state the following slight generalization of Theorem 1.B.

Theorem 2.A *Suppose that $W = \sum_{\alpha\in\Gamma} I_\alpha$, where the I_α are indicator random variables with expectations π_α, and suppose that, for each $\alpha \in \Gamma$, random variables U_α and V_α can be constructed on a common probability space, in such a way that*

$$\mathcal{L}(U_\alpha) = \mathcal{L}(W); \qquad \mathcal{L}(1 + V_\alpha) = \mathcal{L}(W|I_\alpha = 1). \tag{0.2}$$

Then

$$d_{TV}(\mathcal{L}(W), \mathrm{Po}(\lambda)) \le \sum_{\alpha\in\Gamma} \pi_\alpha \mathbb{E}\{\lambda^{-1}(1 - e^{-\lambda})|U_\alpha - V_\alpha| \wedge 2\min(1, \lambda^{-1/2})\},$$

where $\lambda = \sum_{\alpha\in\Gamma} \pi_\alpha$. In particular,

$$d_{TV}(\mathcal{L}(W), \mathrm{Po}(\lambda)) \le \lambda^{-1}(1 - e^{-\lambda}) \sum_{\alpha\in\Gamma} \pi_\alpha \mathbb{E}|U_\alpha - V_\alpha|.$$

Proof As for Theorem 1.B, it follows from (1.1.31) that

$$\mathbb{P}[W \in A] - \mathrm{Po}(\lambda)\{A\} = \sum_{\alpha\in\Gamma} \pi_\alpha \mathbb{E}\{g(U_\alpha + 1) - g(V_\alpha + 1)\}. \tag{0.3}$$

Now observe that $|g(U_\alpha + 1) - g(V_\alpha + 1)| \le \min(\Delta g|U_\alpha - V_\alpha|, 2\|g\|)$, and use Lemma 1.1.1. □

Frequently, U_α is taken to be the original W.

It is sometimes convenient to use a more detailed coupling than that of Theorem 2.A, conditioning not only on $I_\alpha = 1$, but also on the value of another random element X_α. Suppose that random variables $U_{\alpha,x}$ and $V_{\alpha,x}$ can be defined on the same probability space in such a way that

$$\mathcal{L}(U_{\alpha,x}) = \mathcal{L}(W); \qquad \mathcal{L}(1 + V_{\alpha,x}) = \mathcal{L}(W|I_\alpha = 1, X_\alpha = x).$$

Then the following result can be proved in much the same way as Theorem 1.B.

Theorem 2.B *Under the above hypotheses,*

$$\begin{aligned} d_{TV}(\mathcal{L}(W), \mathrm{Po}(\lambda)) &\le \lambda^{-1}(1 - e^{-\lambda}) \sum_{\alpha\in\Gamma} \pi_\alpha \mathbb{E}\{\theta_\alpha(X_\alpha)|I_\alpha = 1\} \\ &= \frac{1 - e^{-\lambda}}{\lambda} \sum_{\alpha\in\Gamma} \mathbb{E}(I_\alpha \theta_\alpha(X_\alpha)), \end{aligned}$$

where $\theta_\alpha(x) = \mathbb{E}|U_{\alpha,x} - V_{\alpha,x}|$. □

2.1 Upper bounds and monotonicity

Suppose that for each α there exists random variables $(J_{\beta\alpha}, \beta \in \Gamma)$ defined on the same probability space as $(I_\beta, \beta \in \Gamma)$ with

$$\mathcal{L}(J_{\beta\alpha}; \beta \in \Gamma) = \mathcal{L}(I_\beta; \beta \in \Gamma | I_\alpha = 1), \tag{1.1}$$

and set $V_\alpha = \sum_{\beta \neq \alpha} J_{\beta\alpha}$. For W, λ as above it follows from Theorem 2.A that

$$\begin{aligned} d_{TV}(\mathcal{L}(W), \mathrm{Po}(\lambda)) &\leq \frac{1 - e^{-\lambda}}{\lambda} \sum_{\alpha \in \Gamma} \pi_\alpha \mathbb{E}|W - V_\alpha| \\ &= \frac{1 - e^{-\lambda}}{\lambda} \sum_{\alpha \in \Gamma} \pi_\alpha \mathbb{E}|I_\alpha + \sum_{\beta \neq \alpha} (I_\beta - J_{\beta\alpha})|. \end{aligned} \tag{1.2}$$

Expression (1.2) can be significantly simplified under additional assumptions on the $J_{\beta\alpha}$. Suppose that, for each α, the set $\Gamma_\alpha = \Gamma \setminus \{\alpha\}$ is partitioned into $\Gamma_\alpha^-, \Gamma_\alpha^+, \Gamma_\alpha^0$ in such a way that

$$J_{\beta\alpha} \begin{cases} \leq I_\beta, & \text{if } \beta \in \Gamma_\alpha^-, \\ \geq I_\beta, & \text{if } \beta \in \Gamma_\alpha^+, \end{cases} \tag{1.3}$$

with no condition if $\beta \in \Gamma_\alpha^0$. For the particular cases where $\Gamma_\alpha = \Gamma_\alpha^+$ or $\Gamma_\alpha = \Gamma_\alpha^-$, when the couplings are *monotone*, we make the following definition.

Definition 2.1.1 The indicator random variables $(I_\alpha, \alpha \in \Gamma)$ are said to be *positively related* if, for each α, random variables $(J_{\beta\alpha}, \beta \in \Gamma)$ can be defined in such a way that $\Gamma_\alpha = \Gamma_\alpha^+$, and to be *negatively related* if $\Gamma_\alpha = \Gamma_\alpha^-$.

Remark 2.1.2 If the indicators $(I_\alpha, \alpha \in \Gamma)$ are positively (negatively) related, they are necessarily positively (negatively) correlated, but the converse does not hold. Correlation is only a pairwise property, whereas being positively or negatively related depends on the whole joint distribution. Counterexamples can be constructed in traditional fashion, using three indicators $(I_i, 1 \leq i \leq 3)$ which satisfy $I_3 = I_1 + I_2 (\mathrm{mod}\ 2)$. It is then sufficient to ensure that the random variables $(I_i, 1 \leq i \leq 3)$ are either all positively or all negatively correlated, and that the unconditional distribution puts positive mass on the events $I_1 = I_2 = 0$ and $I_1 = I_2 = 1$. This is because, given $I_3 = 1$, exactly one of I_1 and I_2 takes the value 1, a state which cannot be monotonely reached from both $I_1 = I_2 = 0$ and $I_1 = I_2 = 1$, so that, under these circumstances, the I_i's can be neither negatively nor positively related. Two suitable examples are given by

$$\mathbb{P}[I_1 = I_2 = 0] = 7/16; \qquad \mathbb{P}[I_1 = r, I_2 = s] = 3/16 \text{ for } (r, s) \neq (0, 0)$$

and

$$\mathbb{P}[I_1 = I_2 = 0] = 1/16; \qquad \mathbb{P}[I_1 = r, I_2 = s] = 5/16 \text{ for } (r,s) \neq (0,0),$$

and the case of I_1 and I_2 being independent Be $(1/2)$ also provides a (somewhat degenerate) example, since then the I_i's are both positively and negatively correlated.

To exploit the partition of Γ defined by (1.3), observe that, since

$$\pi_\alpha \mathbb{E} J_{\beta\alpha} = \mathbb{P}(I_\alpha = 1)\mathbb{E}(I_\beta \,|\, I_\alpha = 1) = \mathbb{E} I_\alpha I_\beta,$$

the covariance of I_α and I_β can be expressed as

$$\mathrm{Cov}\,(I_\alpha, I_\beta) = \mathbb{E} I_\alpha I_\beta - \pi_\alpha \pi_\beta = \pi_\alpha \mathbb{E}(J_{\beta\alpha} - I_\beta) \begin{cases} \leq 0, & \text{if } \beta \in \Gamma_\alpha^-, \\ \geq 0, & \text{if } \beta \in \Gamma_\alpha^+, \end{cases}$$

and therefore

$$\begin{aligned} \pi_\alpha \mathbb{E}|W - V_\alpha| &= \pi_\alpha \mathbb{E}|I_\alpha + \sum_{\beta \neq \alpha} (I_\beta - J_{\beta\alpha})| \\ &\leq \pi_\alpha \mathbb{E} I_\alpha + \pi_\alpha \mathbb{E} \sum_{\beta \in \Gamma_\alpha^-} (I_\beta - J_{\beta\alpha}) \\ &\quad + \pi_\alpha \mathbb{E} \sum_{\beta \in \Gamma_\alpha^+} (J_{\beta\alpha} - I_\beta) + \pi_\alpha \mathbb{E} \sum_{\beta \in \Gamma_\alpha^0} (I_\beta + J_{\beta\alpha}) \\ &= \pi_\alpha^2 - \sum_{\beta \in \Gamma_\alpha^-} \mathrm{Cov}\,(I_\alpha, I_\beta) \\ &\quad + \sum_{\beta \in \Gamma_\alpha^+} \mathrm{Cov}\,(I_\alpha, I_\beta) + \sum_{\beta \in \Gamma_\alpha^0} (\pi_\alpha \pi_\beta + \mathbb{E} I_\alpha I_\beta). \end{aligned} \tag{1.4}$$

Thus we have proved the following theorem.

Theorem 2.C *If there exists a coupling satisfying (1.1), (1.3), then*

$$\begin{aligned} d_{TV}(\mathcal{L}(W), \mathrm{Po}(\lambda)) \leq \frac{1 - e^{-\lambda}}{\lambda} \Big(&\sum_{\alpha \in \Gamma} \pi_\alpha^2 + \sum_{\alpha \in \Gamma} \sum_{\beta \in \Gamma_\alpha^-} |\mathrm{Cov}\,(I_\alpha, I_\beta)| \\ &+ \sum_{\alpha \in \Gamma} \sum_{\beta \in \Gamma_\alpha^+} \mathrm{Cov}\,(I_\alpha, I_\beta) + \sum_{\alpha \in \Gamma} \sum_{\beta \in \Gamma_\alpha^0} (\mathbb{E} I_\alpha I_\beta + \pi_\alpha \pi_\beta) \Big). \quad \square \end{aligned}$$

The following corollaries are now immediate.

Corollary 2.C.1 *If $\Gamma_\alpha^+ = \emptyset$, then*

$$d_{TV}(\mathcal{L}(W), \mathrm{Po}(\lambda)) \le \frac{1-e^{-\lambda}}{\lambda}\Big(\lambda - \mathrm{Var}\, W + 2\sum_\alpha \sum_{\beta\in\Gamma_\alpha^0} \mathbb{E} I_\alpha I_\beta\Big). \quad \square$$

Corollary 2.C.2 *If the random variables $(I_\alpha; \alpha \in \Gamma)$ are negatively related, that is, if $\Gamma_\alpha = \Gamma_\alpha^-$, then*

$$d_{TV}(\mathcal{L}(W), \mathrm{Po}(\lambda)) \le (1-e^{-\lambda})\Big(1 - \frac{\mathrm{Var}\, W}{\lambda}\Big). \quad \square$$

Corollary 2.C.3 *If $\Gamma_\alpha^- = \emptyset$, then*

$$d_{TV}(\mathcal{L}(W), \mathrm{Po}(\lambda)) \le \frac{1-e^{-\lambda}}{\lambda}\Big(\mathrm{Var}\, W - \lambda + 2\sum_\alpha \pi_\alpha^2 + 2\sum_\alpha \sum_{\beta\in\Gamma_\alpha^0} \pi_\alpha \pi_\beta\Big). \quad \square$$

Corollary 2.C.4 *If the random variables are positively related, that is, if $\Gamma_\alpha = \Gamma_\alpha^+$, then*

$$d_{TV}(\mathcal{L}(W), \mathrm{Po}(\lambda)) \le \frac{1-e^{-\lambda}}{\lambda}\Big(\mathrm{Var}\, W - \lambda + 2\sum_\alpha \pi_\alpha^2\Big). \quad \square$$

Corollary 2.C.5 *Suppose that Γ_α can be partitioned into $\Gamma_\alpha^0, \Gamma_\alpha^i$ such that I_α and $\{I_\beta; \beta \in \Gamma_\alpha^i\}$ are independent. Then*

$$d_{TV}(\mathcal{L}(W), \mathrm{Po}(\lambda)) \le \frac{1-e^{-\lambda}}{\lambda}\Big(\sum_{\alpha\in\Gamma} \pi_\alpha^2 + \sum_\alpha \sum_{\beta\in\Gamma_\alpha^0} (\mathbb{E} I_\alpha I_\beta + \pi_\alpha \pi_\beta)\Big). \quad \square$$

Remark 2.1.3 As $1 - e^{-\lambda} \le 1 \wedge \lambda$ for $\lambda > 0$, the factor $1 - e^{-\lambda}$ can be replaced by $1 \wedge \lambda$ in the inequalities.

Remark 2.1.4 The inequalities (1.3) can be replaced by the the weaker assumption $\sum_{\beta\in\Gamma_\alpha^-} J_{\beta\alpha} \le \sum_{\beta\in\Gamma_\alpha^-} I_\beta$ and $\sum_{\beta\in\Gamma_\alpha^+} J_{\beta\alpha} \ge \sum_{\beta\in\Gamma_\alpha^+} I_\beta$.

Remark 2.1.5 In many applications, the estimate given in Theorem 2.C is of the correct order of magnitude. There are, however, cases in which it over-estimates the true discrepancy. This occurs mainly because a crude estimate is used in (1.4), for the terms with $\beta \in \Gamma_\alpha^0$. Better bounds may then be obtainable by making a more detailed analysis of the right hand side of (1.2), using an explicit coupling: an example is given in Theorem 5.H, and another in the proof of Theorem B(b) in Barbour, Janson, Karoński and Ruciński (1990). When the indicators $(I_\alpha, \alpha \in \Gamma)$ are positively or negatively related, this problem does not arise, and it is shown in Chapter 3 that estimates of best order are typically obtained.

Remark 2.1.6 Theorem 2.C and its corollaries extend immediately to sums with an infinite index set Γ, provided $\lambda = \sum_{\alpha\in\Gamma} \pi_\alpha < \infty$.

2.2 Existence of monotone couplings

In order to apply Theorem 2.C, we need only to know that couplings satisfying (1.1) and (1.3) exist, for a given collection of partitions $\Gamma_\alpha^-, \Gamma_\alpha^+, \Gamma_\alpha^0$: evaluation of the estimate then merely requires the computation of the appropriate correlations. Such couplings can frequently be constructed explicitly. The estimates become particularly simple to evaluate if a positive or negative relation between the indicators $(I_\alpha; \alpha \in \Gamma)$ can be established, since only the mean and variance of W are needed: but what is then even more useful is that, in many cases, the existence of suitable monotone couplings can be established by appealing to the following general theorems. As a result, the construction of a particular coupling becomes unnecessary, and the application of the estimates is then rather straightforward.

Let $S = \{0,1\}^n$ be the space of sequences of 0's and 1's of length n with the natural partial ordering: $y = (y_1, \ldots, y_n) \leq x = (x_1, \ldots, x_n)$ if $y_i \leq x_i$ for all i. A function ϕ on S is increasing if $\phi(y) \leq \phi(x)$ for all $y \leq x$; ϕ is decreasing if $-\phi$ is increasing. Consider a random element (I, J) on $S \times S$ having distribution μ with given marginals $\mathcal{L}(I)$ and $\mathcal{L}(J)$. If such a probability measure μ exists and fulfills $\mu(I \geq J) = 1$, then we say that there exists a (monotone) *coupling* of I and J with $I \geq J$. The following result, going back to Strassen (1965), can be found in Kamae, Krengel and O'Brien (1977) Theorem 1 or in Liggett (1985) page 72, where it is established for random elements of very general partially ordered spaces S. For the finite dimensional case used here, Preston (1974) Proposition 2 gives a proof using the min–cut, max–flow theorem of graph theory.

Theorem 2.D *A coupling with $I \geq J$ exists if and only if, for every increasing indicator function ϕ,*

$$\mathbb{E}\phi(I) \geq \mathbb{E}\phi(J). \qquad \square$$

Corollary 2.D.1 *Suppose that ψ is an indicator function which satisfies $\mathbb{P}(\psi(I) = 1) > 0$ and $\mathcal{L}(J) = \mathcal{L}(I|\psi(I) = 1)$. Then a coupling with $I \geq J$ ($I \leq J$) exists if and only if $\psi(I)$ and $\phi(I)$ are negatively (positively) correlated, for every increasing indicator function ϕ.*

Proof We prove the assertion with negative correlation: the other case is similar. From the definition of J,

$$\mathbb{E}\phi(J) = \mathbb{E}(\phi(I)|\psi(I) = 1) = \mathbb{E}(\psi(I)\phi(I))/\mathbb{P}(\psi(I) = 1).$$

Hence the negative correlation can be written as

$$\mathbb{E}\phi(I) \geq \mathbb{E}(\psi(I)\phi(I))/\mathbb{E}\psi(I) = \mathbb{E}\phi(J)$$

from which the assertion follows from Theorem 2.D. $\square$

In typical applications, $I = (I_\alpha, \alpha \in \Gamma)$ and $\psi(I) = I_\beta$ for some $\beta \in \Gamma$.

The indicators I_α are often functions of a set of independent random variables. The following result may then be useful for proving positive (or negative) correlation; a simple proof is given in Liggett (1985) page 78.

Proposition 2.2.1 *Any collection $X_1, \ldots, X_n$ of independent random variables satisfy the FKG inequality: if $f(\cdot)$ and $g(\cdot)$ are bounded increasing functions, then*

$$\mathbb{E}f(X)g(X) \geq \mathbb{E}f(X)\mathbb{E}g(X), \tag{2.1}$$

where X denotes $(X_1, \ldots, X_n)$. □

This leads immediately to the next theorem, whose corollary is then deduced from Corollary C.4.

Theorem 2.E *Let $(I_\alpha, \alpha \in \Gamma)$ be increasing functions of independent random variables $X_1, \ldots, X_n$. Then the random variables $(I_\alpha, \alpha \in \Gamma)$ are positively related.*

Proof We take $I = (I_\beta, \beta \in \Gamma)$ in Corollary 2.D.1, and, for any given α, let $\psi(I) = I_\alpha$. Then, for any increasing function ϕ,

$$\mathbb{E}I_\alpha\phi(I) \geq \mathbb{E}I_\alpha\mathbb{E}\phi(I),$$

by the FKG inequality (2.1), since I_α and $\phi(I)$ are increasing functions of $X_1, \ldots, X_n$. The conclusion now follows from Corollary 2.D.1. □

Corollary 2.E.1 *Under the assumptions of Theorem E, if $W = \sum_{\alpha\in\Gamma} I_\alpha$,*

$$d_{TV}(\mathcal{L}(W), \mathrm{Po}(\lambda)) \leq \frac{1-e^{-\lambda}}{\lambda}\Big(\mathrm{Var}\, W - \lambda + 2\sum_{\alpha\in\Gamma}\pi_\alpha^2\Big). \qquad \square$$

A more complicated result, based on similar considerations, is at times convenient.

Theorem 2.F *Let (X_ν) be independent random variables. Suppose for each $\alpha \in \Gamma$ that*

(a) *I_α is a function of (X_ν),*

(b) *there exists $G_\alpha \subset \Gamma$ with $\alpha \in G_\alpha$ and a partition of the ν's into $E_{1\alpha} \cup E_{2\alpha} \cup E_{3\alpha}$ such that:*
   - (i) *I_α is a function of $(X_\nu, \nu \in E_{1\alpha} \cup E_{2\alpha})$ and increasing (decreasing) as a function of every X_ν, $\nu \in E_{2\alpha}$,*
   - (ii) *if $\beta \notin G_\alpha$, then I_β is a function of $(X_\nu, \nu \in E_{2\alpha} \cup E_{3\alpha})$, and increasing (decreasing) as a function of every X_ν, $\nu \in E_{2\alpha}$.*

Then there exists a coupling with $\mathcal{L}(J_{\beta\alpha}, \beta \in \Gamma) = \mathcal{L}(I_\beta, \beta \in \Gamma | I_\alpha = 1)$ and $J_{\beta\alpha} \geq I_\beta$ for $\beta \notin G_\alpha$.

Proof For any increasing function φ of $I_\beta^\alpha = (I_\beta, \beta \notin G_\alpha)$,

$$\mathbb{E}(\varphi(I_\beta^\alpha)I_\alpha | X_\nu, \nu \in E_{1\alpha} \cup E_{3\alpha}) \geq \mathbb{E}(\varphi(I_\beta^\alpha) | X_\nu, \nu \in E_{3\alpha})\mathbb{E}(I_\alpha | X_\nu, \nu \in E_{1\alpha})$$

by the FKG inequality. Hence

$$\mathbb{E}\varphi(I_\beta^\alpha)I_\alpha \geq \mathbb{E}\varphi(I_\beta^\alpha)\mathbb{E}I_\alpha,$$

whence, by Corollary 2.D.1, there exists a coupling of I_β^α and $(J_{\beta\alpha}, \beta \notin G_\alpha)$ with $J_{\beta\alpha} \geq I_\beta$ for $\beta \notin G_\alpha$. This coupling can of course be extended to all $\beta \in \Gamma$, with $J_{\beta\alpha} \geq I_\beta$ for $\beta \notin G_\alpha$. □

Thus we can take $\Gamma_\alpha^+ = G_\alpha^c$, $\Gamma_\alpha^0 = G_\alpha \setminus \{\alpha\}$ in Corollary 2.C.3, giving the following estimate.

Corollary 2.F.1 *Under the above assumptions,*

$$d_{TV}(\mathcal{L}(W), \mathrm{Po}(\lambda)) \leq \frac{1 - e^{-\lambda}}{\lambda}\left(\mathrm{Var}\, W - \lambda + 2\sum_{\alpha\in\Gamma}\sum_{\beta\in G_\alpha} \pi_\alpha\pi_\beta\right). \qquad (2.2)$$

□

Associated random variables

As is evident from the proof of Theorem 2.E, the only property of the underlying random variables $X_1, \ldots, X_n$ used in establishing that the $(I_\alpha, \alpha \in \Gamma)$ are positively related is that they satisfy the FKG inequality (2.1). In general, random variables are said to be *associated* if they satisfy the FKG inequality: that is, if (2.1) holds for all bounded increasing functions f and g. This definition was introduced in probability theory by Esary, Proschan and Walkup (1967), and independently in statistical physics, see Liggett (1985) Chapter II:2. The following result is an immediate consequence of the definition of association.

Proposition 2.2.2 *Let the random variables (X_ν) be associated (for example independent). Then any random variables (Y_ν) which are increasing functions of the (X_ν) are also associated.* □

The same proof as for Theorem 2.E leads to the following generalization, which shows in particular that associated indicator random variables are positively related.

Theorem 2.G *Let (X_ν) be associated random variables. Suppose for each $\alpha \in \Gamma$ that I_α is an increasing (decreasing) function of every X_ν. Then the random variables $(I_\alpha; \alpha \in \Gamma)$ are positively related.* □

Consequently Corollary 2.C.4 is applicable under the hypotheses of Theorem 2.G.

As observed above, Theorem 2.D and Corollary 2.D.1 hold for very general spaces S. The following results are useful for establishing monotone couplings in many applications.

Theorem 2.H *Let T be independent of the associated random variables $U, X_1, \ldots, X_n$. Then there exists a probability space with random variables $Y_1, \ldots, Y_n$, $Z_1, \ldots, Z_n$ such that*

$$\begin{aligned}\mathcal{L}(Y_1, \ldots, Y_n) &= \mathcal{L}(X_1, \ldots, X_n),\\ \mathcal{L}(Z_1, \ldots, Z_n) &= \mathcal{L}(X_1, \ldots, X_n | U > T),\\ Y_i &\le Z_i, \quad 1 \le i \le n.\end{aligned}$$

Proof By the hypothesis we have, for any bounded increasing function f,

$$\begin{aligned}\mathbb{E}(I(U > T)f(X_1, \ldots, X_n)) &= \mathbb{E}(\mathbb{E}(I(U > T)f(X_1, \ldots, X_n)|T))\\ &\ge \mathbb{E}(\mathbb{E}(I(U > T)|T)\mathbb{E}(f(X_1, \ldots, X_n)|T))\\ &= \mathbb{E}(\mathbb{E}(I(U > T)|T)\mathbb{E}(f(X_1, \ldots, X_n)))\\ &= \mathbb{E}(I(U > T))\mathbb{E}(f(X_1, \ldots, X_n)).\end{aligned}$$

Hence the assertion follows as in Corollary 2.D.1. □

Remark 2.2.3 Obviously the random variables U, T in Theorem 2.H could be vector valued.

Negatively associated random variables

A collection of random variables is defined to be ***negatively associated*** if, for all disjoint sets A_1, A_2 of indices and all bounded functions f and g increasing in every variable,

$$\mathbb{E}f(X_\alpha, \alpha \in A_1)g(X_\alpha, \alpha \in A_2) \le \mathbb{E}f(X_\alpha, \alpha \in A_1)\mathbb{E}g(X_\alpha, \alpha \in A_2), \qquad (2.3)$$

see Joag-Dev and Proschan (1983). The proofs of the following two theorems are very similar to those of Theorems 2.E and 2.H.

Theorem 2.I *Let (X_ν) be negatively associated random variables and (S_α) disjoint subsets of them. Suppose for each $\alpha \in \Gamma$ that $I_\alpha = I_\alpha(S_\alpha)$ is an increasing (decreasing) function of every $X_\nu \in S_\alpha$. Then the random variables $(I_\alpha; \alpha \in \Gamma)$ are negatively related.* □

In particular, negatively associated indicators are negatively related.

Theorem 2.J *Let T be independent of the negatively associated random variables $U, X_1, \ldots, X_n$. Then there exists a probability space with random variables $Y_1, \ldots, Y_n$, $Z_1, \ldots, Z_n$ such that*

$$\begin{aligned}\mathcal{L}(Y_1, \ldots, Y_n) &= \mathcal{L}(X_1, \ldots, X_n), \\ \mathcal{L}(Z_1, \ldots, Z_n) &= \mathcal{L}(X_1, \ldots, X_n | U > T), \\ Y_i &\geq Z_i, \quad 1 \leq i \leq n.\end{aligned}$$

□

Theorem 2.I can be directly combined with Corollary 2.C.2 in applications.

Negatively related random variables frequently occur in interesting applications, as is seen, for instance, in Chapters 6 and 7. The property of being negatively related is usually established by first verifying that a set of random variables is negatively associated. Joag-Dev and Proschan (1983) proved a number of results useful for that. We state some of them here.

Proposition 2.2.4 *The union of independent sets of negatively associated random variables is negatively associated.* □

Proposition 2.2.5 *Increasing functions of disjoint subsets of negatively associated random variables are negatively associated.* □

From the relation

$$\mathrm{Cov}\,(U, V) = \mathbb{E}(\mathrm{Cov}\,(U, V|T)) + \mathrm{Cov}\,(\mathbb{E}(U|T), \mathbb{E}(V|T)),$$

we get the following result.

Proposition 2.2.6 *Suppose that the negatively associated random variables $X_1, \ldots, X_n$ are independent of the independent random variables $T_1, \ldots, T_n$. Let all the measurable functions $F_1(x,t), \ldots, F_n(x,t)$ be increasing (decreasing) in x. Then the random variables $F_1(X_1, T_1), \ldots, F_n(X_n, T_n)$ are negatively associated.* □

These closure properties have the following consequence.

Proposition 2.2.7 *A sum of independent vectors of negatively associated random variables is negatively associated.* □

The next few results show how negative association is induced by independence.

Proposition 2.2.8 *Let $X_1, \ldots, X_n$ be independent random variables. Suppose that $\mathbb{E}(f(X_i; i \in A)| \sum_{i\in A} X_i)$ is increasing in $\sum_{i\in A} X_i$ for every increasing function f and every non-empty subset A of $\{1, \ldots, n\}$. Then the distribution of $X_1, \ldots, X_n$ conditional on $\sum_{i=1}^n X_i$ is negatively associated (a.s.).* □

Proposition 2.2.9 *Let $X_1, \ldots, X_n$ be independent with log concave densities. Then the distribution of $X_1, \ldots, X_n$ conditional on $\sum_{i=1}^n X_i$ is negatively associated (a.s.).* □

We can extend this result as follows.

Proposition 2.2.10 *Let $X_1, \ldots, X_n$ be independent with log concave densities, and let I be any interval (open, closed or halfopen; finite or infinite) with $\mathbb{P}[\sum_{i=1}^n X_i \in I] > 0$. Then the distribution of $X_1, \ldots, X_n$ conditional on $\sum_{i=1}^n X_i \in I$ is negatively associated.*

Proof Let $d > 0$ and let X_0 be uniformly distributed on $[0, d]$ and independent of $(X_i)_1^n$. Denote the sum $\sum_{i=1}^n X_i$ by S_n. Proposition 2.2.9 then implies that the conditional distribution of $(X_i, 0 \le i \le n)$ given $X_0 + S_n = s$ is negatively associated for ν-a.e. s, where ν is the distribution of $X_0 + S_n$. Consequently, for ν-a.e. s, the conditional distribution of $(X_i, 1 \le i \le n)$ given $S_n + X_0 = s$, which equals the conditional distribution given $S_n \in [s - d, s]$, is negatively associated. The restriction to ν-a.e. s is easily lifted: if $\mathbb{P}[S_n \in (a, b]] > 0$, we take $d = b - a$ and may then choose a sequence of 'good' s that decreases to b; it follows, by taking limits in (2.3), that the conclusion holds for $I = (a, b]$. Finally, the conclusion for any other interval follows by taking limits for an appropriate sequence $(a_m, b_m]$. □

Remark 2.2.11 Propositions 2.2.9 and 2.2.10 include the case of integer valued random variables, densities being with respect to counting measure.

Let $x = (x_1, \ldots, x_n)$ be a real vector. The *permutation distribution* of the vector $(X_1, \ldots, X_n)$ takes as its values all permutations of x with equal probabilities $\frac{1}{n!}$.

Proposition 2.2.12 *Permutation distributions are negatively associated.* □

Gaussian random variables

For Gaussian random variables, the conditions of association and negative association can be replaced by conditions on the covariances.

Theorem 2.K *Let the random variable T be independent of the random vector $(U, X_1, \ldots, X_n)$ having a Gaussian distribution. Suppose that $\mathrm{Cov}(U, X_i) \ge 0$ for $1 \le i \le p$ and $\mathrm{Cov}(U, X_i) \le 0$ for $p < i \le n$. Then there exists a probability space with random variables $Y_1, \ldots, Y_n$, $Z_1, \ldots Z_n$ such that*

$$\mathcal{L}(Y_1, \ldots, Y_n) = \mathcal{L}(X_1, \ldots, X_n);$$
$$\mathcal{L}(Z_1, \ldots, Z_n) = \mathcal{L}(X_1, \ldots, X_n | U > T);$$
$$Y_i \le Z_i,\ 1 \le i \le p; \qquad Y_i \ge Z_i,\ p < i \le n.$$

Proof Note that U is positively correlated with every $X_1, \ldots, X_p$ and $-X_{p+1}, \ldots, -X_n$. Thus, for any bounded function f increasing in every variable it follows from a correlation inequality, see Joag-Dev, Perlman and Pitt (1983) Corollary 3, that

$$\begin{aligned}
&\mathbb{E}(I(U > T)f(X_1, \ldots, X_p, -X_{p+1}, \ldots, -X_n)) \\
&\quad = \mathbb{E}(\mathbb{E}(I(U > T)f(X_1, \ldots, X_p, -X_{p+1}, \ldots, -X_n)|T)) \\
&\quad \geq \mathbb{E}(\mathbb{E}(I(U > T)|T)\mathbb{E}(f(X_1, \ldots, X_p, -X_{p+1}, \ldots, -X_n)|T)) \\
&\quad = \mathbb{E}(I(U > T))\mathbb{E}(f(X_1, \ldots, X_p, -X_{p+1}, \ldots, -X_n)),
\end{aligned}$$

from which the assertion follows as in Theorem 2.E. □

Remark 2.2.13 Theorem 2.K is often used when the covariances are either all positive or all negative. It can however also be shown directly that a collection of Gaussian random variables is associated if all covariances are positive, and negatively associated if all off-diagonal covariances are negative, as in Joag-Dev, Perlman and Pitt (1983), so that, in such cases, Theorem 2.H or Theorem 2.J would give the same conclusion as Theorem 2.K.

Theorem 2.K can frequently be used to obtain the couplings required in Theorem 2.C and its corollaries. The next result is a typical example.

Theorem 2.L *Suppose that $(X_\alpha, \alpha \in \Gamma)$ are jointly normally distributed random variables. Let $(t_\alpha, \alpha \in \Gamma)$ be real numbers, and set*

$$I_\alpha = I[X_\alpha > t_\alpha], \ \pi_\alpha = \mathbb{P}[X_\alpha > t_\alpha], \ W = \sum_{\alpha \in \Gamma} I_\alpha, \ \lambda = \mathbb{E}W = \sum_{\alpha \in \Gamma} \pi_\alpha.$$

Then

$$d_{TV}(\mathcal{L}(W), \mathrm{Po}(\lambda)) \leq \lambda^{-1}(1 - e^{-\lambda})\Big(\sum_{\alpha \in \Gamma} \pi_\alpha^2 + \sum\sum\nolimits_{\alpha \neq \beta \in \Gamma} |\mathrm{Cov}\,(I_\alpha, I_\beta)|\Big).$$

Proof For any fixed $\alpha \in \Gamma$, the existence of a partition of Γ_α with $\Gamma_\alpha^0 = \emptyset$ in Theorem 2.C follows from Theorem 2.K, implying the assertion. □

2.3 Independent and dissociated summands

The simplest application of the foregoing results is to a sum W of independent indicators $(I_\alpha, \alpha \in \Gamma)$. Since independent indicators are both positively and negatively related, either of Corollaries 2.C.2 and 2.C.4 could be applied: however, in this case, the bounds thus obtained are both equal to the estimate derived in (1.1.23).

Theorem 2.M *If $(I_\alpha, \alpha \in \Gamma)$ are independent indicators,*

$$d_{TV}(\mathcal{L}(W), \mathrm{Po}(\lambda)) \leq \frac{1 - e^{-\lambda}}{\lambda} \sum_{\alpha \in \Gamma} \pi_\alpha^2. \qquad \square$$

The order of the estimate turns out to be exactly correct, as is shown in Corollary 3.D.1 below. There are a number of applications of the Poisson approximation for sums of independent non-identically distributed indicators in the literature; among others, to average-case analysis of algorithms (Kemp 1984; Ross 1982), to secretary and record problems (Bruss 1984; Nevzorov 1986; Pfeifer 1989) and to chain letter systems (Gastwirth and Bhattacharya 1984).

There are many ways in which the assumption of independence may be relaxed. One of the most natural and useful generalizations is to families of dissociated indicator random variables (McGinley and Sibson 1975). Suppose that Γ is a collection of k-subsets $\alpha = \{\alpha_1, \ldots, \alpha_k\}$ of $\{1, 2, \ldots, n\}$. Then the family $(I_\alpha, \alpha \in \Gamma)$ is said to be *dissociated* if the subsets of random variables $(I_\alpha, \alpha \in A)$ and $(I_\alpha, \alpha \in B)$ are independent whenever $(\bigcup_{\alpha \in A} \alpha) \cap (\bigcup_{\alpha \in B} \alpha) = \emptyset$. If $k = 1$, dissociated and independent are equivalent, but for $k \geq 2$ there are much wider possibilities. The most common setting is that in which $I_\alpha = \phi_\alpha(Y_{\alpha_1}, \ldots, Y_{\alpha_k})$ for some symmetric 0–1 valued functions ϕ_α, where the $Y_1, \ldots, Y_n$ are independent random elements of some space $\mathcal{Y}$. An important practical instance is when the Y_i's are points in three dimensions, or on a two-dimensional manifold, and one wishes to test for departures from the null hypothesis of independent Y_i's which entail clustering. A simple statistic is obtained by taking $k = 2$, defining Γ to be the set of all 2-subsets of $\{1, 2, \ldots, n\}$ and setting $I_\alpha = I[|Y_{\alpha_1} - Y_{\alpha_2}| < \theta]$ for some $\theta > 0$; $W = \sum_{\alpha \in \Gamma} I_\alpha$ then counts the number of close pairs of points, and can be expected to be large in the presence of clustering. More generally, one might wish to take $k > 2$, and count how many groups of k points have diameter less than θ, or even perhaps consider as a statistic the size of the largest group of points with diameter less than θ. Under appropriate circumstances, the null hypothesis distribution of such statistics can be approximated by the Stein–Chen method, as follows.

For each α, define

$$\Gamma_\alpha^s = \{\beta \in \Gamma : \beta \neq \alpha, \beta \cap \alpha \neq \emptyset\}$$

and

$$\Gamma_\alpha^i = \{\beta \in \Gamma : \beta \cap \alpha = \emptyset\}.$$

Then Corollary 2.C.5, with $\Gamma_\alpha^0 = \Gamma_\alpha^s$, directly yields the following theorem (Barbour and Eagleson 1984).

Theorem 2.N *If $(I_\alpha, \alpha \in \Gamma)$ are dissociated and Γ_α^s is as defined above,*

$$d_{TV}(\mathcal{L}(W), \mathrm{Po}(\lambda)) \le \lambda^{-1}(1-e^{-\lambda}) \sum_{\alpha\in\Gamma} \Big\{ \pi_\alpha^2 + \sum_{\beta\in\Gamma_\alpha^s} (\pi_\alpha\pi_\beta + \mathbb{E}(I_\alpha I_\beta)) \Big\}. \quad (3.1)$$

□

Remark 2.3.1 If the $\mathbb{E}(I_\alpha I_\beta)$ are so large that $\sum_{\alpha\in\Gamma}\sum_{\beta\in\Gamma_\alpha^s} \mathbb{E}(I_\alpha I_\beta)$ and $\sum_{\alpha\in\Gamma}\sum_{\beta\in\Gamma_\alpha^s} \mathrm{Cov}(I_\alpha, I_\beta)$ are of the same order, the estimate in Theorem 2.N is typically of the right order of magnitude, as can be seen from Theorem 9.G below. The same theorem also indicates that the estimate is pessimistic if $\sum_{\alpha\in\Gamma}\sum_{\beta\in\Gamma_\alpha^s} \mathrm{Cov}(I_\alpha, I_\beta)$ happens to be much smaller than $\sum_{\alpha\in\Gamma}\sum_{\beta\in\Gamma_\alpha^s} \mathbb{E}(I_\alpha I_\beta)$. For example, a better estimate is obtained by using Corollary 2.C.4 or 2.C.2 in place of Theorem 2.N if the $(I_\alpha, \alpha \in \Gamma)$ are positively or negatively related.

Remark 2.3.2 Poisson approximation in this context was first considered by Silverman and Brown (1978), when considering the statistical analysis of point patterns. The approximation also arises when counting the number of occurrences of a pattern in a sequence (Chryssaphinou and Papastavridis 1988) or of failures in certain reliability models, and in connection with subgraph statistics (Nowicki 1988).

In the particular case where Γ consists of all k-subsets of $\{1, 2, \dots, n\}$, $I_\alpha = \phi(Y_{\alpha_1}, \dots, Y_{\alpha_k})$ and the $Y_1, \dots, Y_n$ are independent and identically distributed, the estimate (3.1) simplifies considerably. Let π denote the common value of the π_α's, and let σ_r be the common value of $\pi^{-1}\mathbb{E}(I_\alpha I_\beta) = \mathbb{P}[I_\beta = 1 | I_\alpha = 1]$ for all pairs α, β for which $|\beta \cap \alpha| = r$. Then it follows directly that

$$\begin{aligned} d_{TV}(\mathcal{L}(W), \mathrm{Po}(\lambda)) &\le (1-e^{-\lambda}) \Big\{ \pi \Big[\binom{n}{k} - \binom{n-k}{k} \Big] + \sum_{r=1}^{k-1} \binom{k}{r}\binom{n-k}{k-r} \sigma_r \Big\} \\ &= O(\pi n^{k-1} + \sum_{r=1}^{k-1} \sigma_r n^{k-r}). \end{aligned} \quad (3.2)$$

Note that, as observed by Silverman and Brown, $\pi \le \sigma_1 \le \ldots \le \sigma_k = 1$. Therefore, if $k \ge 2$, the order term is dominated by the sum.

When $k = 2$, it is natural to identify Γ with the edge set of a graph on $\{1, 2, \dots, n\}$. Let d_i denote the degree of vertex i in Γ, and, given any pair $\alpha = \{i, j\} \in \Gamma$, set $d_\alpha = d_i \wedge d_j$ and $D_\alpha = d_i + d_j$, so that, in particular, $|\Gamma_\alpha^s| = D_\alpha - 2$. Then, in the symmetric case in which $\pi_\alpha = \pi$ for all α and $\mathbb{E}(I_\alpha I_\beta) = \pi\sigma_1$ for all α, β such that $|\beta \cap \alpha| = 1$, the estimate of Theorem 2.N can also be substantially simplified.

Corollary 2.N.1 *Under the special assumptions above,*

$$d_{TV}(\mathcal{L}(W), \mathrm{Po}(\lambda)) \le \frac{1-e^{-\lambda}}{\lambda}(N\pi^2 + \sum_\alpha (D_\alpha - 2)(\pi\sigma_1 + \pi^2))$$
$$= (1-e^{-\lambda})\{\pi + (\sigma_1 + \pi)N^{-1}\sum_{i=1}^n d_i(d_i - 1)\}, \quad (3.3)$$

where N denotes $|\Gamma|$. □

The estimate given in Corollary 2.N.1 can be unduly pessimistic when the covariances $\mathrm{Cov}(I_\alpha, I_\beta)$, $|\beta \cap \alpha| = 1$, are small compared to $\mathbb{E}(I_\alpha I_\beta)$, as observed in Remark 2.3.1. It turns out surprisingly often in practice that, because of some extra uniformity, the covariances are actually exactly zero. Such is the case, for instance, when counting close inter-point distances, if the Y_i's are uniformly distributed on a sphere. This motivates the following definition.

Definition 2.3.3 A family $(I_\alpha : \alpha \in \Gamma)$ of indicator random variables, indexed by the edges of a graph Γ, is *strongly dissociated* if it is dissociated and if, furthermore, for all $H \subset \Gamma$,

$$I_\alpha \text{ is independent of } (I_\beta : \beta \in H) \text{ whenever } |\alpha \cap \bigcup_{\beta \in H} \beta| = 1.$$

Note that every strongly dissociated family is dissociated and pairwise independent, though the converse does not hold: see Section 5.3. Hence, if furthermore $\mathbb{E}I_\alpha = \pi$ for every α, $\mathbb{E}I_\alpha I_\beta = \pi^2$ when $\alpha \ne \beta$. We will show that the estimate in Corollary 2.N.1 then holds with D_α replaced by $d_\alpha + 1$. In many cases, D_α and d_α are of the same order of magnitude, but sometimes the difference is important; for example, if Γ is an N-star, then $D_\alpha = N + 1$ and $d_\alpha = 1$ for every α.

Theorem 2.O *Let $(I_\alpha, \alpha \in \Gamma)$ be a strongly dissociated family of* $\mathrm{Be}(\pi)$ *random variables, and set $W = \sum_{\alpha \in \Gamma} I_\alpha$. Then, with $\lambda = \mathbb{E}W = N\pi$,*

$$d_{TV}(\mathcal{L}(W), \mathrm{Po}(\lambda)) \le (1-e^{-\lambda})2\pi(N^{-1}\sum_{\alpha\in\Gamma} d_\alpha - \tfrac{1}{2})$$
$$\le \sqrt{8}(1-e^{-\lambda})\lambda N^{-1/2}. \quad (3.4)$$

Proof Fix the edge α, having endpoints k and j with $d_\alpha = d_k \wedge d_j = d_j$, say. Set

$$\Gamma_\alpha^i = \{\text{edges not having endpoint } j\} \subset \Gamma,$$

and

$$\Gamma_\alpha^s = \{\text{edges joining } j \text{ but not } k\} \subset \Gamma.$$

Because of the strong dissociation, I_α and $(I_\beta; \beta \in \Gamma_\alpha^i)$ are independent. Hence it follows from Corollary 2.C.5 that

$$d_{TV}(\mathcal{L}(W), \mathrm{Po}(\lambda)) \leq \frac{1-e^{-\lambda}}{\lambda}(N\pi^2 + \sum_{\alpha\in\Gamma}(d_\alpha - 1)2\pi^2)$$
$$= (1-e^{-\lambda})2\pi(N^{-1}\sum_{\alpha\in\Gamma} d_\alpha - \tfrac{1}{2}).$$

It remains to estimate $\sum_{\alpha\in\Gamma} d_\alpha$. Write $i \in \alpha$ if i is an endpoint of the edge α. We have

$$\sum_{i=1}^n d_i = 2N \quad \text{and} \quad \sum_{i=1}^n \sum_{\alpha\ni i} d_\alpha = 2\sum_{\alpha\in\Gamma} d_\alpha.$$

For a fixed i,

$$\sum_{\alpha\ni i} d_\alpha \leq \sum_{\alpha\ni i} d_i = d_i^2,$$

and, if j_α is the other endpoint of α,

$$\sum_{\alpha\ni i} d_\alpha \leq \sum_{\alpha\ni i} d_{j_\alpha} \leq \sum_j d_j = 2N.$$

Hence,

$$\sum_{\alpha\ni i} d_\alpha \leq d_i^2 \wedge 2N \leq \sqrt{2Nd_i^2},$$

and therefore

$$\sum_\alpha d_\alpha = \tfrac{1}{2}\sum_i \sum_{\alpha\ni i} d_\alpha \leq \tfrac{1}{2}\sum_i d_i\sqrt{2N} = N\sqrt{2N}.$$

As $\lambda = N\pi$, we get

$$d_{TV}(\mathcal{L}(W), \mathrm{Po}(\lambda)) \leq (1-e^{-\lambda})2\lambda\sqrt{2/N},$$

proving the assertion. □

In particular, if we vary Γ and (I_α) in such a way that $\lambda \to \lambda_\infty$ and $N \to \infty$, keeping (I_α) strongly dissociated and equidistributed, then $W \xrightarrow{\mathcal{D}} \mathrm{Po}(\lambda)$. This is not true in general if we only assume that (I_α) is dissociated as in Corollary 2.N.1, as the following example shows.

Example 2.3.4 Let Γ be an N-star with $N+1$ vertices $1, \ldots, N+1$ and N edges connecting vertex $N+1$ to each of the others. We colour the vertices with $2N$ different colours with the probabilities

$$\mathbb{P}(X_i = k) = \begin{cases} \frac{1}{3N} & k = 1, \ldots, N; \\ \frac{2}{3N} & k = N+1, \ldots, 2N, \end{cases}$$

and let $I_\alpha = I[X_{\alpha_1} = X_{\alpha_2}]$, $W = \sum_{\alpha\in\Gamma} I_\alpha$. Then the conditional distribution of W given X_{N+1} is $\mathrm{Bi}(N, \frac{1}{3N})$ if $X_{N+1} \le N$ and $\mathrm{Bi}(N, \frac{2}{3N})$ if $X_{N+1} > N$. Consequently, W converges in distribution to a mixture of $\mathrm{Po}(\frac{1}{3})$ and $\mathrm{Po}(\frac{2}{3})$, and not to $\mathrm{Po}(\lambda_\infty) = \mathrm{Po}(\frac{5}{9})$.

Strong dissociation is considered in more detail in Chapter 5. A further extension of the above setting is to allow the index set Γ to be randomly chosen. We examine here only the case in which $\Gamma = \Gamma(N)$ consists of all k-subsets of $\{1, 2, \ldots, N\}$ and $I_\alpha = \phi(Y_{\alpha_1}, \ldots, Y_{\alpha_k})$, where the $(Y_i)_{i\ge1}$ are independent and identically distributed, and N is a random variable independent of the Y_i's. Let π and σ_r be defined as before, and set $W = W(N) = \sum_{\alpha\in\Gamma(N)} I_\alpha$. Clearly, from (3.2), for any $A \subset \mathbb{Z}^+$,

$$\begin{aligned} |\mathbb{P}[W \in A|N] - \mathrm{Po}(\tbinom{N}{k}\pi)\{A\}| &\le (1 - e^{-\binom{N}{k}\pi}) \\ &\times \left\{\pi\left[\binom{N}{k} - \binom{N-k}{k}\right] + \sum_{r=1}^{k-1} \binom{k}{r}\binom{N-k}{k-r}\sigma_r\right\}, \end{aligned} \tag{3.5}$$

and hence we obtain the following estimate.

Proposition 2.3.5 *With $W = W(N)$ defined as above,*

$$\begin{aligned} &d_{TV}(\mathcal{L}(W), \mathrm{Po}(\Lambda)) \\ &\quad \le \mathbb{E}\left\{\pi\left[\binom{N}{k} - \binom{N-k}{k}\right] + \sum_{r=1}^{k-1} \binom{k}{r}\binom{N-k}{k-r}\sigma_r\right\}, \end{aligned} \tag{3.6}$$

where $\mathrm{Po}(\Lambda)$ *denotes the mixed Poisson distribution with mean* $\Lambda \overset{\mathcal{D}}{=} \binom{N}{k}\pi$. □

In order to connect this statement to one comparing $\mathcal{L}(W)$ with $\mathrm{Po}(\lambda)$, where $\lambda = \pi\mathbb{E}\binom{N}{k}$, it suffices to combine it with the estimates

$$d_{TV}(\mathrm{Po}(\Lambda), \mathrm{Po}(\lambda)) \le \begin{cases} \pi \min\{1, \lambda^{-1/2}\}\mathbb{E}\left|\binom{N}{k} - \mathbb{E}\binom{N}{k}\right| \\ \\ \pi(1 - e^{-\lambda})\mathrm{Var}\binom{N}{k}/\mathbb{E}\binom{N}{k} \end{cases} \tag{3.7}$$

obtained from Theorem 1.C.

In the case where $k = 1$, this yields the estimate

$$d_{TV}(\mathcal{L}(W), \mathrm{Po}(\lambda)) \leq \pi\{1 + \mathrm{Var}\, N/\mathbb{E}N\}, \tag{3.8}$$

which is typically sharp if either $1 \gg \mathrm{Var}\, N/\mathbb{E}N$ or $1 \ll \mathrm{Var}\, N/\mathbb{E}N$, as follows from Corollary 3.D.1 and Theorem 3.F. However, if N has the Poisson distribution $\mathrm{Po}(\mu)$, so that $\mathrm{Var}\, N/\mathbb{E}N = 1$, (3.8) gives an estimate of 2π, whereas the true distance is zero. The case where N has a Poisson distribution is important in applications, where statistics are derived from a configuration of points which occur as the points of a Poisson process, and one might hope that the improvement in Poisson approximation obtained by having a Poisson number of summands would also be reproduced when $k \geq 2$. However, for $k \geq 2$ and $N = n$ fixed,

$$\mathrm{Var}\, W(n)/\mathbb{E}W(n) - 1 = -\pi + \sum_{r=1}^{k-1} \binom{k}{r}\binom{n-k}{k-r}(\sigma_r - \pi). \tag{3.9}$$

If instead N is random with $\mathbb{E}N = n$,

$$\begin{aligned}
\mathrm{Var}\, W(N)/\mathbb{E}W(N) - 1 &\geq \mathbb{E}\{\mathrm{Var}\,(W(N)|N)\}/\mathbb{E}W(N) - 1 \\
&= -\pi + \sum_{r=1}^{k-1} \mathbb{E}\left\{\pi\binom{N}{k}\binom{k}{r}\binom{N-k}{k-r}(\sigma_r - \pi)\right\}\Big/\mathbb{E}\left\{\pi\binom{N}{k}\right\} \\
&\geq -\pi + \sum_{r=1}^{k-1} \binom{k}{r}\mathbb{E}\binom{N-k}{k-r}(\sigma_r - \pi) \\
&\geq -\pi + \sum_{r=1}^{k-1} \binom{k}{r}\binom{n-k}{k-r}(\sigma_r - \pi) \\
&= \mathrm{Var}\, W(n)/\mathbb{E}W(n) - 1,
\end{aligned} \tag{3.10}$$

where the last two inequalities follow from the FKG inequality and from Jensen's inequality. Since (3.9) is frequently dominated by the positive contributions from the terms involving the σ_r's, randomizing N, even using the Poisson distribution, only increases $\mathrm{Var}\, W(N)/\mathbb{E}W(N) - 1$ further away from zero, typically reducing the accuracy of approximation: see Section 3.1. Thus, for $k \geq 2$, having a Poisson distributed N need not improve the accuracy of Poisson approximation. However, since the quantities $\mathrm{Var}\, W(N)/\mathbb{E}W(N) - 1$ and $\mathrm{Var}\, W(n)/\mathbb{E}W(n) - 1$ are often of the same order, the extra randomization usually makes little difference.

Although having a Poisson distributed number N of summands may not improve the order of approximation, it does make it possible to simplify and improve upon the error estimate obtained from (3.6) and (3.7). The new estimate once again involves the quantity $(\lambda^{-1}\mathrm{Var}\, W - 1)$, and could be obtained as a limiting case of Corollary 2.C.4, though the argument given below is much simpler.

Theorem 2.P *Let $W = W(N)$ be defined as for Proposition 2.3.5. If also N has the Poisson distribution* $\mathrm{Po}(\mu)$, *then*

$$d_{TV}(\mathcal{L}(W), \mathrm{Po}(\lambda)) \leq (1-e^{-\lambda}) \sum_{r=1}^{k-1} \binom{k}{r} (k-r)!^{-1} \mu^{k-r} \sigma_r$$
$$= (1-e^{-\lambda})(\lambda^{-1} \mathrm{Var}\, W - 1), \tag{3.11}$$

where $\lambda = \mathbb{E}W = \mu^k \pi / k!$.

Proof Take first the case where $N \sim \mathrm{Bi}(n,p)$, and denote the corresponding sum by W_{np}. Let $(J_i, 1 \leq i \leq n)$ be $\mathrm{Be}(p)$ random variables, independent of each other and of $(Y_i, i \geq 1)$, and let A be the random set $\{i : 1 \leq i \leq n, J_i = 1\}$. Since $|A| \stackrel{\mathcal{D}}{=} N$,

$$W_{np} \stackrel{\mathcal{D}}{=} \sum_{\substack{\alpha \in \Gamma(n) \\ \alpha \subset A}} I_\alpha = \sum_{\alpha \in \Gamma(n)} I'_\alpha,$$

where $I'_\alpha = I_\alpha \prod_{r=1}^k J_{\alpha_r}$. As a result, defining

$$\pi' = p^k \pi; \quad \sigma'_r = p^{k-r} \sigma_r; \quad \lambda' = \binom{n}{k} \pi',$$

(3.2) yields

$$d_{TV}(W_{np}, \mathrm{Po}(\lambda'))$$
$$\leq (1-e^{-\lambda'}) \left(\pi' \left(\binom{n}{k} - \binom{n-k}{k} \right) + \sum_{r=1}^{k-1} \binom{k}{r} \binom{n-k}{k-r} \sigma'_r \right)$$
$$\leq (1-e^{-\lambda'}) \left(k n^{k-1} p^k \pi + \sum_{r=1}^{k-1} \binom{k}{r} (np)^{k-r} \sigma_r / (k-r)! \right).$$

To complete the proof, let $n \to \infty$ with $p = \mu/n$. Then $W_{np} \xrightarrow{\mathcal{D}} W$ and $\lambda' \to \mu^k \pi / k! = \mathbb{E}W$, proving the first assertion. The second expression follows by direct computation. □

As an example of the application of Theorem 2.P, suppose that $k = 2$ and that $\xi_1, \ldots, \xi_N$ are the points of a Poisson process with intensity measure ν on a set S. Let W be the sum $\sum_{1 \leq i < j \leq N} \phi(\xi_i, \xi_j)$, where ϕ is a symmetric indicator function on S^2; for instance, W might be the number of close pairs of points.

Corollary 2.P.1 *With the above notation,*

$$\begin{aligned} d_{TV}(\mathcal{L}(W), \mathrm{Po}(\lambda)) &\le (1-e^{-\lambda})\{\lambda^{-1}\mathrm{Var}\, W - 1\} \\ &= \frac{1-e^{-\lambda}}{\lambda} \iint \phi(x_1, x_2)\phi(x_1, x_3)\nu(dx_1)\nu(dx_2)\nu(dx_3), \end{aligned}$$

where $\lambda = \frac{1}{2}\iint \phi(x_1, x_2)\nu(dx_1)\nu(dx_2)$. *In particular, if* $\int \phi(x_1, x_2)\nu(dx_2)$ *does not depend on* x_1, *as when* S *is a sphere,* ϕ *depends only on the distance between the points, and* ν *is uniform,*

$$d_{TV}(\mathcal{L}(W), \mathrm{Po}(\lambda)) \le 4\lambda(1-e^{-\lambda})/\nu(S). \qquad \square$$

See Aldous (1989a) for a related application of Stein's method to point processes in the plane.

2.4 Miscellany

The first problem addressed in this section is the approximation of point probabilities, to an accuracy greater than that implied by Theorem 2.A. It is, of course, immediate from (0.3) that

$$\mathbb{P}[W=j] - \mathrm{Po}(\lambda)\{j\} = \sum_{\alpha\in\Gamma} \pi_\alpha \mathbb{E}\{g_j(U_\alpha+1) - g_j(V_\alpha+1)\}, \tag{4.1}$$

where g_j denotes $g_{\lambda,\{j\}}$ and (U_α, V_α) satisfy (0.2) as usual. Theorem 2.A then gives the estimate

$$|\mathbb{P}[W=j] - \mathrm{Po}(\lambda)\{j\}| \le \psi,$$

where $\psi = \min(\lambda^{-1}, 1)\sum_{\alpha\in\Gamma} \pi_\alpha \mathbb{E}|U_\alpha - V_\alpha|$. Better estimates than this depend on exploiting the special structure of the functions g_j, as compared to the greater variety possible for the general $g_{\lambda,A}$. The properties to be used were proved in the course of Lemma 1.1.1: that

$$\begin{aligned} g_j(i) < 0,\ i \le j;\quad g_j(i) > 0,\ i \ge j+1;\quad g_j(i+1) - g_j(i) < 0,\ i \ne j; \\ 0 < g_j(j+1) - g_j(j) \le \min(\lambda^{-1}, j^{-1}, 1). \end{aligned} \tag{4.2}$$

Now, for each $\alpha \in \Gamma$, pick k_α, $K_\alpha \ge 1$, and let

$$\begin{aligned} \varepsilon &= \max_{\alpha\in\Gamma}\{\mathbb{P}[|U_\alpha - V_\alpha| > k_\alpha,\ U_\alpha \le K_\alpha]/\mathbb{E}|U_\alpha - V_\alpha|\} \\ \eta &= \max_{\alpha\in\Gamma}\{\mathbb{P}[W > K_\alpha]/\mathbb{E}|U_\alpha - V_\alpha|\}. \end{aligned} \tag{4.3}$$

Define also

$$\mu = \max_{\alpha\in\Gamma}\{\sum_{0<|l|\le k_\alpha} |l| \max_{0\le m\le K_\alpha} \mathbb{P}[V_\alpha = l+m | U_\alpha = m]/\mathbb{E}|U_\alpha - V_\alpha|\}. \tag{4.4}$$

Theorem 2.Q *With the definitions above, for any choices of K_α and k_α, $\alpha \in \Gamma$,*

$$|\mathbb{P}[W=j] - \mathrm{Po}(\lambda)\{j\}| \le 2\psi_j(\varepsilon + \eta + \mu(2e\lambda)^{-1/2}),$$

provided only that $\mu\psi \le 1/2$, where

$$\psi_j = \min(\lambda^{-1}, j^{-1}, 1) \sum_{\alpha\in\Gamma} \pi_\alpha \mathbb{E}|U_\alpha - V_\alpha| \tag{4.5}$$

and

$$\psi = \max_j \psi_j = \psi_1.$$

Proof It follows from (4.1) and (4.2) that

$$\begin{aligned}
|\mathbb{P}[W=j] - \mathrm{Po}(\lambda)\{j\}| &\le \min(\lambda^{-1}, j^{-1}, 1) \\
&\quad \times \Big\{\sum_{\alpha\in\Gamma} \pi_\alpha(\mathbb{P}[U_\alpha > K_\alpha] + \mathbb{P}[|U_\alpha - V_\alpha| > k_\alpha,\ U_\alpha \le K_\alpha])\Big\} \\
&\quad + \Big|\sum_{\alpha\in\Gamma} \pi_\alpha \sum_{0<|l|\le k_\alpha} \sum_{m=0}^{K_\alpha} \mathbb{P}[U_\alpha = m,\ V_\alpha - U_\alpha = l] \\
&\qquad\qquad \times \{g_j(m+l+1) - g_j(m+1)\}\Big|.
\end{aligned}$$

Now, if $p_{ml} \ge 0$ are any positive numbers and p_l denotes $\max_m p_{ml}$, the properties (4.2) of the function g_j imply that, for $l > 0$,

$$\begin{aligned}
&\sum_m p_{ml}\{g_j(m+l+1) - g_j(m+1)\} \\
&\qquad \le \sum_{m=(j-l)\vee 0}^{j-1} p_l\{g_j(m+l+1) - g_j(m+1)\} \le l p_l\{g_j(j+1) - g_j(j)\},
\end{aligned}$$

and, collapsing the telescoping sums, that

$$\begin{aligned}
\sum_m p_{ml}\{g_j(m+l+1) - g_j(m+1)\} & \\
\ge \Big\{\sum_{m<j-l} + \sum_{m\ge j}\Big\} & p_l\{g_j(m+l+1) - g_j(m+1)\} \\
\ge l p_l\{g_j(j) - g_j(j+1)\}. &
\end{aligned}$$

Together with similar estimates for $l < 0$, this yields

$$\Big|\sum_m p_{ml}\{g_j(m+l+1) - g_j(m+1)\}\Big| \le |l| p_l \Delta g_j.$$

Thus, recalling that $U_\alpha \stackrel{\mathcal{D}}{=} W$, it follows that

$$\begin{aligned} &|\mathbb{P}[W = j] - \mathrm{Po}(\lambda)\{j\}| \\ &\quad \le \psi_j(\varepsilon + \eta) + \Delta g_j \sum_{\alpha\in\Gamma} \pi_\alpha \sum_{0<|l|\le k_\alpha} |l| \max_{0\le m\le K_\alpha} \mathbb{P}[V_\alpha = l+m,\ U_\alpha = m] \\ &\quad \le \psi_j\Big(\varepsilon + \eta + \mu \max_{m\ge 0} \mathbb{P}[W = m]\Big). \end{aligned} \tag{4.6}$$

Hence, if $\mu\psi \le 1/2$, and using Proposition A.2.7,

$$\begin{aligned} \max_{m\ge 0} \mathbb{P}[W = m] &\le 2 \max_{m\ge 0} \mathrm{Po}(\lambda)\{m\} + 2\psi(\varepsilon + \eta) \\ &\le (2/e\lambda)^{1/2} + 2\psi(\varepsilon + \eta), \end{aligned}$$

which, with (4.6), completes the proof. □

Theorem 2.Q has been successfully applied to a problem in number theory by Grimmett and Hall (1990), but the conditions are far from convenient. The estimate is only useful in so far as it can be shown that ε and η can be made small — preferably as small as $\lambda^{-1/2}$ — by suitable choice of the k_α and K_α, while at the same time keeping μ bounded. Controlling ε is usually a matter of being able to show that $\mathbb{P}[|U_\alpha - V_\alpha| \ge 2]$ is much smaller than $\mathbb{P}[|U_\alpha - V_\alpha| \ge 1]$, which is a natural requirement for good Poisson approximation, and η can be estimated by, for instance, using Chebyshev's inequality. However, μ can prove rather awkward, because of the appearance of the conditional distribution of V_α given $U_\alpha = m$, especially when m may be large: keeping K_α small to avoid this problem conflicts with the aim of controlling η. Thus, if the best results are to be proved, it is helpful if the probabilities $\mathbb{P}[W > K]$ for large K can be estimated more sharply than by using a second moment estimate. This is possible under the conditions of Corollary 2.C.2, as is shown in the next theorem.

Theorem 2.R *If the* $(I_\alpha, \alpha \in \Gamma)$ *are negatively related,*

(a) *for* $K \ge 3\lambda$, $\mathbb{P}[W \ge K] \le 2\mathrm{Po}(\lambda)\{[K,\infty)\}$;

(b) *for* $J \le \lambda/4$, $\mathbb{P}[W \le J] \le 2\mathrm{Po}(\lambda)\{J\}$.

Proof To prove (a), take $A = [K,\infty)$ and $g = g_{\lambda,A}$ in (0.3). Then it is easy to check that g is decreasing in $[1, K]$ and increasing in $[K,\infty)$, and that $0 \ge g(j) \ge -(K-\lambda)^{-1}$ for all $j \ge 1$ provided that $K > \lambda$. Thus, from (0.3),

$$\begin{aligned} \mathbb{P}[W \ge K] - \mathrm{Po}(\lambda)\{[K,\infty)\} &= \sum_{\alpha\in\Gamma} \pi_\alpha \mathbb{E}\{g(U_\alpha + 1) - g(V_\alpha + 1)\} \\ &\le \sum_{\alpha\in\Gamma} \pi_\alpha \mathbb{P}[U_\alpha \ge K]/(K-\lambda) = [\lambda/(K-\lambda)]\mathbb{P}[W \ge K]. \end{aligned}$$

Hence $\mathbb{P}[W \geq K] \leq [(K-\lambda)/(K-2\lambda)]\mathrm{Po}(\lambda)\{[K,\infty)\}$ in $K > 2\lambda$, from which (a) follows.

For part (b), the argument is a little more complicated. First, taking $A = \{0\}$, the function $g = g_0$ is decreasing throughout $[1,\infty)$, so that $\mathbb{P}[W=0] \leq \mathrm{Po}(\lambda)\{0\}$ follows immediately from (0.3). To extend the result to events $\{W \leq J\}$ for $J > 0$, it is necessary to take linear combinations of the functions $\{g_j,\ j \leq J\}$, with some coefficients exceeding one, in order to arrive at a function g which is decreasing. A crude choice is given by $g = \sum_{j=0}^{J} c^{J-j} g_j$, where $c = \lambda/[(\lambda-J)\mathrm{Po}(\lambda)\{[J,\infty)\}]$. For this g, the only positive contribution to $g(j+1)-g(j)$ for $j \geq 1$ is that made by the term $c^{J-j}g_j$, and this is already outweighed by the negative jump of $c^{J-j+1}g_{j-1}$, since $g_j(j+1)-g_j(j) \leq \lambda^{-1}$ as in Lemma 1.1.1, whereas

$$g_{j-1}(j)-g_{j-1}(j+1) = \lambda^{-2}(\lambda-j)\mathrm{Po}(\lambda)\{[j,\infty)\}+j\lambda^{-2}\mathrm{Po}(\lambda)\{j\} \geq 1/(c\lambda).$$

Hence, from (0.3),

$$\sum_{j=0}^{J} c^{J-j}(\mathbb{P}[W=j]-\mathrm{Po}(\lambda)\{j\}) \leq 0,$$

and thus

$$\mathbb{P}[W \leq J] \leq \sum_{j=0}^{J} c^{J-j}e^{-\lambda}\lambda^j/j! \leq \mathrm{Po}(\lambda)\{J\}\sum_{k\geq 0}(Jc/\lambda)^k.$$

Since $\mathrm{Po}(\lambda)\{[J,\infty)\} \geq 3/4$ for all $J \leq \lambda/4$, as a consequence of Chebychev's inequality, it follows that $Jc/\lambda < 1/2$, which completes the proof of (b). □

It seems to be harder to obtain similar estimates for positively related variables, since a positive relation generally increases the probability of extremely large or small values of the sum. The following result gives good bounds in the lower tail, but assumes that the variables have a special structure.

Consider a collection $(J_i, i \in Q)$ of independent random indicator variables and a (finite) family $(Q(\alpha), \alpha \in \Gamma_\alpha)$ of subsets of the index set Q, and define $I_\alpha = \prod_{i\in Q(\alpha)} J_i$ and $W = \sum_{\alpha\in\Gamma_\alpha} I_\alpha$. Thus W counts the number of sets $Q(\alpha)$ that are contained in the random set $\{i \in Q : J_i = 1\}$ with independently occurring elements. It is easily seen that the I_α are positively related. Partition Γ_α into $\Gamma_\alpha^+ \cup \Gamma_\alpha^i$, where $\Gamma_\alpha^+ = \{\beta \neq \alpha : Q(\alpha)\cap Q(\beta) \neq \emptyset\}$; thus I_α is independent of $(I_\beta)_{\beta\in\Gamma_\alpha^i}$. Define

$$p_\alpha = \mathbb{E}I_\alpha, \quad \lambda = \mathbb{E}W = \sum_\alpha p_\alpha, \quad \delta = \frac{1}{\lambda}\sum_\alpha \sum_{\beta\in\Gamma_\alpha^+} \mathbb{E}I_\alpha I_\beta.$$

Theorem 2.S *With notation as above, if* $0 \le \varepsilon \le 1$, *then*

$$\begin{aligned}\mathbb{P}(W \le (1-\varepsilon)\lambda) &\le \exp\Big(-\frac{\lambda}{1+\delta}(\varepsilon + (1-\varepsilon)\log(1-\varepsilon))\Big)\\ &\le \exp\Big(-\frac{1}{2(1+\delta)}\varepsilon^2\lambda\Big) = \exp\Big(-\frac{1}{2}\frac{(\varepsilon\lambda)^2}{\lambda + \sum_\alpha \sum_{\beta\in\Gamma_\alpha^+} EI_\alpha I_\beta}\Big), \quad (4.7)\end{aligned}$$

and if k *is an integer with* $0 \le k \le \lambda$, *then*

$$\mathbb{P}(W \le k) \le \Big(\sqrt{2\pi(k+1)}\frac{\lambda^k}{k!}e^{-\lambda}\Big)^{1/(1+\delta)}. \qquad (4.8)$$

Proof We begin by estimating the Laplace transform of W. First set $\psi(t) = \mathbb{E}e^{-tW}$, $t \ge 0$. Then

$$-\frac{d\psi(t)}{dt} = \mathbb{E}We^{-tW} = \sum_\alpha \mathbb{E}I_\alpha e^{-tW}.$$

Now, for each α, write $W'_\alpha = I_\alpha + \sum_{\beta\in\Gamma_\alpha^+} I_\beta$ and $W''_\alpha = W - W'_\alpha$, so that W''_α is independent of I_α. By definition,

$$\mathbb{E}(I_\alpha e^{-tW}) = p_\alpha \mathbb{E}(e^{-tW'_\alpha}e^{-tW''_\alpha} \mid I_\alpha = 1).$$

The condition $I_\alpha = 1$ fixes J_i, $i \in Q(\alpha)$. Since $e^{-tW'_\alpha}$ and $e^{-tW''_\alpha}$ both are decreasing functions of the remaining J_i, the FKG inequality yields

$$\begin{aligned}\mathbb{E}(I_\alpha e^{-tW}) &\ge p_\alpha \mathbb{E}(e^{-tW'_\alpha} \mid I_\alpha = 1)\mathbb{E}(e^{-tW''_\alpha} \mid I_\alpha = 1)\\ &= p_\alpha \mathbb{E}(e^{-tW'_\alpha} \mid I_\alpha = 1)\mathbb{E}e^{-tW''_\alpha} \ge p_\alpha \mathbb{E}(e^{-tW'_\alpha} \mid I_\alpha = 1)\psi(t).\end{aligned}$$

We now sum over α and apply Jensen's inequality twice to obtain

$$\begin{aligned}-\frac{d}{dt}(\log\psi(t)) &= \frac{1}{\psi(t)}\sum_\alpha \mathbb{E}(I_\alpha e^{-tW}) \ge \sum_\alpha p_\alpha \mathbb{E}(e^{-tW'_\alpha} \mid I_\alpha = 1)\\ &\ge \sum_\alpha p_\alpha \exp(-\mathbb{E}(tW'_\alpha \mid I_\alpha = 1)) = \lambda \sum_\alpha \frac{p_\alpha}{\lambda}\exp(-t\mathbb{E}(W'_\alpha \mid I_\alpha = 1))\\ &\ge \lambda\exp\Big(-\sum_\alpha \frac{p_\alpha}{\lambda}t\mathbb{E}(W'_\alpha \mid I_\alpha = 1)\Big) = \lambda\exp\Big(-t\lambda^{-1}\sum_\alpha \mathbb{E}(W'_\alpha I_\alpha)\Big)\\ &= \lambda\exp\Big(-t\lambda^{-1}(\sum_\alpha \mathbb{E}I_\alpha^2 + \sum_\alpha \sum_{\beta\in\Gamma_\alpha^+} \mathbb{E}I_\alpha I_\beta)\Big) = \lambda\exp(-(1+\delta)t).\end{aligned}$$

Consequently, since $\psi(0) = 1$,

$$-\log\psi(t) \ge \int_0^t \lambda e^{-(1+\delta)u}\,du = \frac{\lambda}{1+\delta}(1 - e^{-(1+\delta)t}).$$

The proof is now easily completed. Let $a = 1 - \varepsilon \geq 0$. Then, for any $t \geq 0$, $\psi(t) = \mathbb{E}e^{-tW} \geq e^{-ta\lambda}\mathbb{P}(W \leq a\lambda)$, and thus

$$\log \mathbb{P}(W \leq a\lambda) \leq \log \psi(t) + ta\lambda \leq -\frac{\lambda}{1+\delta}(1 - e^{-(1+\delta)t}) + ta\lambda.$$

The right hand side is minimized by choosing $t = -(1+\delta)^{-1}\log a$, which yields

$$\begin{aligned}\log \mathbb{P}(W \leq a\lambda) &\leq -\frac{\lambda}{1+\delta}(1-a) - \frac{\lambda}{1+\delta}a\log a \\ &= -\frac{\lambda}{1+\delta}(\varepsilon + (1-\varepsilon)\log(1-\varepsilon)).\end{aligned}$$

This gives the first inequality in (4.7), and the second follows because

$$\varepsilon + (1-\varepsilon)\log(1-\varepsilon) \geq \varepsilon^2/2, \quad 0 \leq \varepsilon \leq 1.$$

In order to obtain (4.8), set $1 - \varepsilon = k/\lambda$. Then (4.7) yields

$$\mathbb{P}(W \leq k) \leq \exp\Big(-\frac{1}{1+\delta}\big(k\log\frac{k}{\lambda} + \lambda - k\big)\Big) = \Big(\big(\frac{\lambda}{k}\big)^k e^k e^{-\lambda}\Big)^{1/1+\delta},$$

and (4.8) follows because $k! \leq \sqrt{2\pi(k+1)}k^k e^{-k}$. □

Of the two versions of the bound given in Theorem 2.S, (4.7) is most convenient for applications, while (4.8) provides a direct comparision with the Poisson probabilities. Choosing $\varepsilon = 1$, we obtain in particular

$$\mathbb{P}(W = 0) \leq e^{-\lambda/(1+\delta)} \leq \exp\Big\{-\lambda + \sum_\alpha \sum_{\beta\in\Gamma_\alpha^+} \mathbb{E}I_\alpha I_\beta\Big\}. \tag{4.9}$$

Remark 2.4.1 It is also possible to derive similar, although more complicated, *lower* bounds to $\mathbb{P}(W \leq (1-\varepsilon)\lambda)$ by this method. In fact, if

$$\eta = \frac{1}{\lambda}\sum_\alpha(-\log(1-p_\alpha)) - 1 \leq \max_\alpha p_\alpha/(2(1-p_\alpha)),$$

then

$$\mathbb{P}(W < (1-\varepsilon)\lambda) \geq \exp\Big(-\frac{4\lambda(\varepsilon+\eta)(\varepsilon + 2\delta\varepsilon + \varepsilon^2 + 2(1+\delta)\eta)}{(1-\varepsilon)^2} - \frac{4(1+\delta)}{\lambda\varepsilon^2}\Big)$$

for any $0 < \varepsilon < 1$. If $\varepsilon, \delta, \eta/\varepsilon$ are small and $\lambda\varepsilon^2$ is big, then the exponent in this estimate is about 8 times the exponent in the upper bound (4.7). A proof of the inequality can be found in Janson (1990).

Note also that the FKG inequality immediately gives a lower bound

$$\mathbb{P}(W=0) \geq \prod_{\alpha}(1-p_\alpha) = \exp(-\lambda(1+\eta)), \tag{4.10}$$

corresponding to (4.9).

In Corollary 2.C.4, the error is estimated using

$$\operatorname{Var} W - \lambda + 2\sum_{\alpha}\pi_\alpha^2 = \sum_{\alpha}\sum_{\beta\in\Gamma_\alpha^+}\operatorname{Cov}(I_\alpha, I_\beta) + \sum_{\alpha}\pi_\alpha^2.$$

This suggests that $\delta = \frac{1}{\lambda}\sum_\alpha\sum_{\beta\in\Gamma_\alpha^+}\mathbb{E}I_\alpha I_\beta$ could perhaps be replaced by $\frac{1}{\lambda}\sum_\alpha\sum_{\beta\in\Gamma_\alpha^+}\operatorname{Cov}(I_\alpha, I_\beta)$ in Theorem 2.S, but it cannot, even for (4.9), as the following example shows.

Example 2.4.2 Let $1<\mu<n$ and put $p=\mu/n$, $p_0 = 1-\mu^{-2}$. Let $J_0 \sim \operatorname{Be}(p_0)$ and $J_i \sim \operatorname{Be}(p)$, $i=1,\dots,n$, be independent; define $I_i = J_0 J_i$, $i=1,\dots,n$. Then $W = J_0\sum_1^n J_i$ and $\mathbb{P}(W=0) \geq \mathbb{P}(J_0=0) = \mu^{-2}$. On the other hand,

$$\lambda = np_0 p = \mu - \mu^{-1}$$

and

$$\sum_i\sum_j \operatorname{Cov}(I_i, I_j) = n(n-1)p^2 p_0(1-p_0) = (1-\frac{1}{n})(1-\mu^{-2}) < 1.$$

Thus, if μ is large,

$$e^{-\lambda/(1+\delta)} \leq \exp\Big\{-\lambda + \sum_i\sum_j \operatorname{Cov}(I_i, I_j)\Big\} < e^{-\mu+2} \ll \mu^{-2} < \mathbb{P}(W=0).$$

Theorem 2.S deals only with the lower tail of the distribution of W. The next example shows that there is no corresponding estimate in the upper tail.

Example 2.4.3 Let $\lambda>1$ be an integer and let $\{I_\alpha\}$ consist of λ^2 independent variables with $p_\alpha = \lambda^{-1}-\lambda^{-4}$ together with λ^2 coinciding variables with $p_\alpha = \lambda^{-4}$. Then

$$\mathbb{E}W = \lambda; \quad \sum_{\alpha}\sum_{\beta\in\Gamma_\alpha^+}\mathbb{E}I_\alpha I_\beta = \lambda^2(\lambda^2-1)\lambda^{-4} < 1,$$

and thus $\delta < 1/\lambda$. Nevertheless, for any $a<\infty$ and $c>0$, if λ is large enough,

$$\mathbb{P}(W > a\lambda) \geq \lambda^{-4} > \exp(-c\lambda).$$

Suen (1990) has proved an estimate for $\mathbb{P}(W = 0)$ which is similar to (4.9) and (4.10), without any assumption that the variables are positively or negatively related. We do not know whether his result can be extended to give bounds for $\mathbb{P}(W \leq k)$, $k > 0$. Large deviation approximations with small relative error for $\mathbb{P}(W = 0)$ have recently been obtained in some specific problems by Godsil and McKay (1990) and McKay and Wormald (1990), using recursive computation of ratios of probabilities. Stein has also developed an analogous technique based on coupling ideas, starting from the equation

$$\mathbb{E}\{I_\alpha g(W)\} = \pi_\alpha \mathbb{E}\{g(W) \mid I_\alpha = 1\},$$

which can be used in a variety of problems to give similar estimates: see also Stein (1978).

3
Lower bounds

We now turn to the problem of finding lower bounds for Poisson approximation. Although the Stein–Chen method is widely applicable, it would be much less interesting if it did not give accurate estimates, at least in the sense that the upper bounds it gave were of the right order of magnitude. Of course, bad choices of the couplings (U_i, V_i) can always lead to poor estimates, and it may not necessarily be obvious how to recognize the best possible choices, so that some sort of qualification is needed. However, in many applications of the results of Section 2.1, the bounds proved are essentially of order

$$\varepsilon = (1 \wedge \mathbb{E}W)\left|\frac{\operatorname{Var} W}{\mathbb{E}W} - 1\right|, \tag{0.1}$$

which looks very appealing, and often turns out to be sharp. Unfortunately, even when the bound given by the Stein–Chen method is of order ε, it need not be best possible, as is seen in Example 3.4.1 below. Nonetheless, there is much that can now be said.

First, in Section 3.1, it is proved under rather general assumptions that, for some constant c,

$$d_{TV}(\mathcal{L}(W), \operatorname{Po}(\lambda)) \geq c\varepsilon|\log \varepsilon|^{-2}.$$

Thus, if an upper bound of order ε can also be established, as under the conditions of Corollary 2.C.2, ε must be the correct order of magnitude, at least up to logarithmic factors, and Poisson convergence takes place if and only if $\varepsilon \to 0$. Example 3.4.1 shows that logarithmic factors are at times necessary, but experience suggests that the true order of magnitude of the approximation error is often precisely ε. In Section 3.2, assuming the monotone coupling structure of either Corollary 2.C.2 or Corollary 2.C.4, lower bounds involving the fourth cumulant of W are derived, which are sharp enough to show that the order is indeed ε under very reasonable assumptions. Similar bounds are established in Section 3.3 for the mixed Poisson case, and the chapter concludes with examples illustrating the limits of what is obtainable with the Stein–Chen method.

3.1 General lower bounds

In this section, we show that the value of ε defined in (0.1) implies a lower bound for the error in Poisson approximation to the distribution of W. One of the simplest and best known facts about the Poisson distribution is that its variance is equal to its mean, and ε is a natural measure of the extent to which this property fails to be true for W. It is thus encouraging to find that ε is also closely related to the discrepancy in total variation.

The approach used to prove the lower bounds of this section does not involve the Stein–Chen method, though the results obtained confirm the accuracy of the Stein–Chen upper bounds. Two cases are distinguished, according to the sign of $\operatorname{Var} W - \mathbb{E}W$. When $\operatorname{Var} W < \mathbb{E}W$, no further assumptions about the distribution of W are required to obtain a bound, but when $\operatorname{Var} W > \mathbb{E}W$ the bounds obtained also involve at least a higher central moment of W. Some such extra ingredient is, however, inevitable in this case. For suppose that W_M has as distribution the mixture

$$\{(1 - M^{-1}), \mathrm{Po}(1)\} + \{M^{-1}, \delta_{M^{3/4}}\}, \tag{1.1}$$

where δ_k denotes unit mass at the point k, and the notation $\{\theta, \mathcal{L}_1\} + \{1 - \theta, \mathcal{L}_2\}$ is used to denote the mixture of the distributions $\mathcal{L}_1$ and $\mathcal{L}_2$ with weights θ and $1 - \theta$. Then $\varepsilon \asymp M^{1/2} \to \infty$ as $M \to \infty$, whereas $d_{TV}(\mathcal{L}(W_M), \mathrm{Po}(1)) \to 0$.

We begin with the case $\operatorname{Var} W < \mathbb{E}W = \lambda$, which occurs, for example, when the summands are negatively correlated. The notation

$$\delta = d_{TV}(\mathcal{L}(W), \mathrm{Po}(\lambda)) \tag{1.2}$$

is used throughout the section. The first step is to establish a technical lemma, used in the proof of the first lower bound.

Lemma 3.1.1 *Let $Z \sim \mathrm{Po}(\lambda)$ and W be such that $\mathbb{P}(W \neq Z) = \delta$, and let $\tilde{Z}$ have the distribution $\mathcal{L}(Z - \lambda | W \neq Z)$. Then*

$$\mathbb{E}\tilde{Z}^2 \leq 4\lambda \log\left(\frac{2e}{\delta}\right)\left\{1 \vee \frac{1}{\lambda}\log\left(\frac{2e}{\delta}\right)\right\}.$$

Proof Since $e^{\sqrt{x}}$ is a convex function on $[1, \infty)$, we obtain, by a straightforward argument using Jensen's inequality, that, for any $t > 0$,

$$\begin{aligned}
\exp\{t\sqrt{\mathbb{E}\tilde{Z}^2}\} &\leq \exp\{\sqrt{\mathbb{E}(1 + t^2\tilde{Z}^2)}\} \leq \mathbb{E}\exp\{\sqrt{1 + t^2\tilde{Z}^2}\} \\
&\leq \mathbb{E}\exp\{1 + t|\tilde{Z}|\} \leq e\mathbb{E}e^{t\tilde{Z}} + e\mathbb{E}e^{-t\tilde{Z}} \\
&= e\mathbb{E}(e^{t(Z-\lambda)}|Z \neq W) + e\mathbb{E}(e^{-t(Z-\lambda)}|Z \neq W) \\
&\leq \tfrac{e}{\delta}(\mathbb{E}e^{t(Z-\lambda)} + \mathbb{E}e^{-t(Z-\lambda)}) = \tfrac{e}{\delta}(e^{\lambda(e^t - 1 - t)} + e^{\lambda(e^{-t} - 1 + t)}).
\end{aligned} \tag{1.3}$$

Hence, for any $0 < t \leq 1$,

$$\exp\{t\sqrt{\mathbb{E}\tilde{Z}^2}\} \leq \frac{2e}{\delta}e^{\lambda t^2},$$

and thus

$$\mathbb{E}\tilde{Z}^2 \leq \frac{1}{t^2}\left\{\log\left(\frac{2e}{\delta}\right) + \lambda t^2\right\}^2. \tag{1.4}$$

If $\lambda \geq \log(\frac{2e}{\delta})$, take $t^2 = \frac{1}{\lambda}\log(\frac{2e}{\delta})$, giving

$$\mathbb{E}\tilde{Z}^2 \leq 4\lambda \log\left(\frac{2e}{\delta}\right).$$

Otherwise, take $t = 1$ to give

$$\mathbb{E}\tilde{Z}^2 \leq 4\left\{\log\left(\frac{2e}{\delta}\right)\right\}^2. \qquad \square$$

Theorem 3.A *The following inequalities hold for any integer valued random variable W with* $\mathrm{Var}\, W < \mathbb{E}W = \lambda$*:*

$$\text{(a)} \quad \varepsilon \leq \begin{cases} 4\delta\{\log(\frac{2e}{\delta})\}^2 & \text{if } \lambda \leq 1; \\ 4\delta \log(\frac{2e}{\delta})\{1 \vee \frac{1}{\lambda}\log(\frac{2e}{\delta})\} & \text{if } \lambda \geq 1: \end{cases}$$

$$\text{(b)} \quad \delta \geq c\varepsilon(1 + \log(\tfrac{1}{\varepsilon}))^{-1}\{1 \wedge (\lambda \vee 1)(1 + \log(\tfrac{1}{\varepsilon}))^{-1}\},$$

for some universal constant $c > 0$.

Proof It is easy to see that (b) follows from (a). As observed in (A.1.4), enlarging the probability space if necessary, we can define a random variable $Z \sim \mathrm{Po}(\lambda)$ with $\mathbb{P}(Z \neq W) = \delta$: that is, the pair (W, Z) is a maximal coupling. Let $\tilde{W}$ and $\tilde{Z}$ be random variables which have the distributions $\mathcal{L}(W - \lambda | W \neq Z)$ and $\mathcal{L}(Z - \lambda | W \neq Z)$, respectively. Then it is immediate that

$$\begin{aligned} \mathbb{E}(W - \lambda)^2 &= \mathbb{E}(\{(W - \lambda)^2 - (Z - \lambda)^2\} I[W \neq Z]) + \mathbb{E}(Z - \lambda)^2 \\ &= \delta(\mathbb{E}\tilde{W}^2 - \mathbb{E}\tilde{Z}^2) + \lambda. \end{aligned}$$

Hence it follows that

$$\mathbb{E}W - \mathrm{Var}\, W = \lambda - \mathbb{E}(W - \lambda)^2 = -\delta\mathbb{E}\tilde{W}^2 + \delta\mathbb{E}\tilde{Z}^2 \leq \delta\mathbb{E}\tilde{Z}^2.$$

For $\lambda \leq 1$, it is immediate that $\log(\frac{2e}{\delta}) \geq 1 \geq \lambda$, so that, from Lemma 3.1.1,

$$\varepsilon = \lambda - \mathrm{Var}\, W \leq 4\delta\left\{\log\left(\frac{2e}{\delta}\right)\right\}^2.$$

For $\lambda \geq 1$, from Lemma 3.1.1,

$$\varepsilon = 1 - \lambda^{-1}\mathrm{Var}\, W \leq 4\delta \log\left(\frac{2e}{\delta}\right)\left\{1 \vee \frac{1}{\lambda}\log\left(\frac{2e}{\delta}\right)\right\}. \qquad \square$$

It is natural, when approximating the distribution of a random variable W by a Poisson distribution $\mathrm{Po}(\lambda)$, to take $\lambda = \mathbb{E}W$. However, there are times when other choices of λ may be better: see Serfling (1975) and Deheuvels and Pfeifer (1988), and also Example 3.4.5. The following theorem strengthens Theorem 3.A, in that a lower bound of the same order is obtained for the approximation of $\mathcal{L}(W)$ by $\mathrm{Po}(\mu)$, for any μ. We use the notation

$$\delta^* = \inf_{\mu \geq 0} d_{TV}(\mathcal{L}(W), \mathrm{Po}(\mu)). \tag{1.5}$$

By continuity, the infimum is attained for some $\mu \geq 0$.

Theorem 3.A* *Under the conditions of Theorem 3.A,*

$$\varepsilon \leq \delta^*\left[\frac{7}{2} + 6\log\left(\frac{2e}{\delta^*}\right)\max\left\{3, \frac{1}{\lambda \vee 1}\log\left(\frac{2e}{\delta^*}\right)\right\}\right].$$

Thus there exists a universal constant $c^* > 0$ such that

$$\delta^* \geq c^*\varepsilon(1 + \log(\tfrac{1}{\varepsilon}))^{-1}\{1 \wedge (\lambda \vee 1)(1 + \log(\tfrac{1}{\varepsilon}))^{-1}\}.$$

Proof Suppose first that $\delta^* \leq 1/4$, and that, much as before, W and $Z \sim \mathrm{Po}(\mu)$ are constructed in such a way that $\mathbb{P}[W \neq Z] = \delta^*$. Let $\tilde{W}$ and $\tilde{Z}$ be random variables with distributions $\mathcal{L}(W - \mu | W \neq Z)$ and $\mathcal{L}(Z - \mu | W \neq Z)$. By direct calculation, we have

$$\mathbb{E}W = \mathbb{E}\{(W - Z)I[W \neq Z]\} + \mathbb{E}Z = \delta^*(\mathbb{E}\tilde{W} - \mathbb{E}\tilde{Z}) + \mu$$

and

$$\begin{aligned}\mathbb{E}(W-\mu)^2 &= \mathbb{E}\{[(W-\mu)^2 - (Z-\mu)^2]I[W \neq Z]\} + \mathbb{E}(Z-\mu)^2 \\ &= \delta^*(\mathbb{E}\tilde{W}^2 - \mathbb{E}\tilde{Z}^2) + \mu.\end{aligned}$$

Hence it follows that

$$\begin{aligned}\mathbb{E}W - \mathrm{Var}\, W &= \mathbb{E}W + (\mathbb{E}W - \mu)^2 - \mathbb{E}(W-\mu)^2 \\ &= \delta^*(\mathbb{E}\tilde{W} - \mathbb{E}\tilde{Z}) + (\delta^*)^2(\mathbb{E}\tilde{W} - \mathbb{E}\tilde{Z})^2 - \delta^*(\mathbb{E}\tilde{W}^2 - \mathbb{E}\tilde{Z}^2) \\ &\leq \delta^*(\mathbb{E}\tilde{W} - \mathbb{E}\tilde{Z}) + 2(\delta^*)^2\{(\mathbb{E}\tilde{W})^2 + (\mathbb{E}\tilde{Z})^2\} - \delta^*(\mathbb{E}\tilde{W}^2 - \mathbb{E}\tilde{Z}^2).\end{aligned}$$

Now, since $\delta^* \leq 1/4$ and $\tilde{Z} \geq -\mu$, this implies that

$$\begin{aligned}\mathbb{E}W - \mathrm{Var}\, W &\leq \delta^*(\mathbb{E}\tilde{W} - \tfrac{1}{2}(\mathbb{E}\tilde{W})^2) + \tfrac{3}{2}\delta^*\mathbb{E}\tilde{Z}^2 + \delta^*\mu \\ &\leq \tfrac{1}{2}\delta^* + \tfrac{3}{2}\delta^*\mathbb{E}\tilde{Z}^2 + \delta^*\mu,\end{aligned}$$

and hence, using Lemma 3.1.1 to give the estimate

$$\mathbb{E}\tilde{Z}^2 \le 4\mu \log\left(\frac{2e}{\delta^*}\right)\left\{1 \vee \frac{1}{\mu}\log\left(\frac{2e}{\delta^*}\right)\right\},$$

it follows that

$$\mathbb{E}W - \mathrm{Var}\,W \le \delta^*\left[\frac{1}{2} + \mu + 6\mu\log\left(\frac{2e}{\delta^*}\right)\left(1 \vee \frac{1}{\mu}\log\left(\frac{2e}{\delta^*}\right)\right)\right].$$

The proof for $\delta^* \le 1/4$ now requires only the estimate $\mu \le 3(1 \vee \lambda)$. To prove this, use the Cauchy–Schwarz inequality to give

$$\begin{aligned}(\mathbb{E}W - \mu)^2 &= (\mathbb{E}\{I[W \ne Z](W - Z)\})^2 \le \delta^*\mathbb{E}(W - Z)^2 \\ &= \delta^*\{[\mathbb{E}(W - Z)]^2 + \mathrm{Var}\,(W - Z)\} \\ &\le \delta^*\{(\mathbb{E}W - \mu)^2 + 2\mathrm{Var}\,W + 2\mathrm{Var}\,Z\} \\ &\le \tfrac{1}{4}\{(\mathbb{E}W - \mu)^2 + 2\mathbb{E}W + 2\mu\},\end{aligned}$$

since $\delta^* \le 1/4$ and $\mathrm{Var}\,W \le \mathbb{E}W$. Hence

$$(\mathbb{E}W - \mu)^2 \le 2(\mathbb{E}W + \mu)/3,$$

and the result follows easily.

If $\delta^* > 1/4$, the inequality $\varepsilon = (\lambda - \mathrm{Var}\,W)/(\lambda \vee 1) \le 1$ shows that $\varepsilon \le 4\delta^*$. □

Remark 3.1.2 Although an inequality of the form $\delta \ge c\varepsilon$ or $\delta^* \ge c\varepsilon$ has not quite been achieved, Theorems 3.A(b) and 3.A* come very close. The results can be improved very slightly, for $\log(\frac{2e}{\delta}) \ge e\lambda$, by taking

$$t = \log\left\{\frac{1}{\lambda}\log\left(\frac{2e}{\delta}\right)\right\}$$

in (1.3), instead of taking $t = 1$ in (1.4), when estimating $\mathbb{E}\tilde{Z}^2$. This, in the proofs above, implies the inequalities

$$\delta \ge \delta^* \ge c\varepsilon(1+\log(\tfrac{1}{\varepsilon}))^{-1}\{1\wedge(\lambda\vee 1)(1+\log(\tfrac{1}{\varepsilon}))^{-1}\log^2[e+\lambda^{-1}(1+\log(\tfrac{1}{\varepsilon}))]\}.$$

Example 3.4.1 shows that it is possible for δ^* to attain the order of this lower bound; we do not know whether it is possible for δ to attain it, or if there exists a slightly greater lower bound for δ.

The following result is a simple consequence of Theorems 3.A and 3.A*.

Corollary 3.A*.1 *Suppose that* $(W_n, n \ge 1)$ *are integer valued random variables such that* $\mathrm{Var}\,W_n \le \mathbb{E}W_n < \infty$ *for each* n *and* $W_n \xrightarrow{\mathcal{D}} \mathrm{Po}\,(\lambda)$ *as* $n \to \infty$, *for some* $\lambda \ge 0$. *Then* $\lim \mathbb{E}W_n = \lim \mathrm{Var}\,W_n = \lambda$.

Proof Let $\lambda_n = \mathbb{E}W_n$. It follows by Chebyshev's inequality that

$$0 < e^{-\lambda} = \lim \mathbb{P}[W_n = 0] \leq \liminf \frac{\operatorname{Var} W_n}{\lambda_n^2} \leq \liminf \frac{1}{\lambda_n}.$$

Hence the expectations λ_n are bounded. Moreover, the second moments $\mathbb{E}W_n^2 \leq \lambda_n + \lambda_n^2$ are also bounded, and thus the W_n are uniformly integrable, which together with the convergence in distribution yields $\mathbb{E}W_n \to \lambda$. Finally, Theorem 3.A* yields $\varepsilon \to 0$ and thus $\mathbb{E}W_n - \operatorname{Var} W_n \to 0$. □

We now turn to the case $\operatorname{Var}W > \mathbb{E}W$. Here, as observed earlier, we cannot expect any lower bounds for the order of approximation without imposing extra conditions. The estimates we give involve at least a higher moment of W, in addition to the mean and variance.

We give two related estimates. The first, Theorem 3.B, is particulary useful if we consider a sequence of random variables with uniformly bounded rth standardized absolute central moments, for some $r > 2$. This includes any situation where convergence to a Poisson distribution can be proved by the method of moments. In such cases, we obtain from Theorem 3.B that, with ε given by (0.1) and δ given by (1.2),

$$\delta \geq C\varepsilon^{r/(r-2)}. \tag{1.6}$$

The second estimate, Theorem 3.C, which requires the existence of the Laplace transform of W for some positive argument, yields the sharper bound $C\varepsilon(1 + \log 1/\varepsilon)^{-2}$.

Theorem 3.B *Suppose that W satisfying* $\operatorname{Var} W > \mathbb{E}W = \lambda$ *is such that* $\|W - \lambda\|_r = (\mathbb{E}|W - \lambda|^r)^{1/r} < \infty$ *for some* $r > 2$, *and set*

$$\mu_r = (\lambda^{-1/2} \wedge 1)\|W - \lambda\|_r.$$

Then

$$\delta \geq \varepsilon^{r/(r-2)}\mu_r^{-2r/(r-2)}.$$

Proof Let $Z \sim \operatorname{Po}(\lambda)$ be maximally coupled to W as in the proof of Theorem 3.A. Then, by Hölder's inequality,

$$\begin{aligned} \operatorname{Var} W - \mathbb{E}W &= \mathbb{E}(W - \lambda)^2 - \mathbb{E}(Z - \lambda)^2 \\ &= \mathbb{E}\{[(W - \lambda)^2 - (Z - \lambda)^2]I[W \neq Z]\} \leq \mathbb{E}\{(W - \lambda)^2 I[W \neq Z]\} \\ &\leq \delta^{1-2/r}\|W - \lambda\|_r^2 = \delta^{1-2/r}(1 \vee \lambda^{1/2})^2\mu_r^2. \end{aligned}$$

The result now follows. □

Again, there is a corresponding refinement of the argument, leading to a lower bound for δ^*. As a first step, we need the following technical lemma.

Lemma 3.1.3 *Suppose that W and Z are two random variables such that $Z \sim \mathrm{Po}(\mu)$ for some μ and that $\mathbb{E}W^2 < \infty$. Let $d = \mathbb{P}[W \neq Z]$ and write $\lambda = \mathbb{E}W$. If $d \le 1/2$ and either $\mu \le \lambda$ or $\mu \ge \lambda + 8d$, then*

$$\mathrm{Var}\, W - \lambda \le 3\mathbb{E}\{(W-\lambda)^2 I[W \neq Z]\}. \qquad (1.7)$$

Proof Using the fact that $\mathbb{P}[W \neq Z] \le 1/2$, an elementary calculation yields

$$\begin{aligned}
\mathrm{Var}\, W - \lambda &= \mathbb{E}(W-\mu)^2 - (\mu-\lambda)^2 - \mu + (\mu - \lambda) \\
&= \mathbb{E}\{(W-\mu)^2 - (Z-\mu)^2\} - (\mu-\lambda)^2 + (\mu-\lambda) \\
&= \mathbb{E}\{[(W-\mu)^2 - (Z-\mu)^2] I[W \neq Z]\} - (\mu-\lambda)^2 + (\mu-\lambda) \\
&\le \mathbb{E}\{[2(W-\lambda)^2 + 2(\lambda-\mu)^2] I[W \neq Z]\} \\
&\qquad - \mathbb{E}\{(Z-\mu)^2 I[W \neq Z]\} - (\mu-\lambda)^2 + (\mu-\lambda) \\
&\le 2\mathbb{E}\{(W-\lambda)^2 I[W \neq Z]\} \\
&\qquad - \mathbb{E}\{(Z-\mu)^2 I[W \neq Z]\} + (\mu-\lambda). \qquad (1.8)
\end{aligned}$$

The assertion of the lemma now follows immediately if $\mu \le \lambda$.

If $\mu \ge \lambda + 8d$, we estimate $\mu - \lambda$ using the inequality

$$\begin{aligned}
\tfrac{1}{2}(\mu-\lambda) \le (\mu-\lambda)\mathbb{P}[Z = W] &= \mathbb{E}(Z - W) - (\mu-\lambda)\mathbb{P}[W \neq Z] \\
&= \mathbb{E}\{[Z - \mu - (W-\lambda)] I[W \neq Z]\}.
\end{aligned}$$

Now, because $x \le 1 + x^2/4$, this in turn gives

$$\begin{aligned}
\mu - \lambda &\le 2\mathbb{E}\{[1 + (Z-\mu)^2/4 + 1 + (W-\lambda)^2/4] I[W \neq Z]\} \\
&= 4d + \tfrac{1}{2}\mathbb{E}\{[(W-\lambda)^2 + (Z-\mu)^2] I[W \neq Z]\}.
\end{aligned}$$

The assumption $\mu - \lambda \ge 8d$ now implies that

$$\mu - \lambda \le \mathbb{E}\{[(W-\lambda)^2 + (Z-\mu)^2] I[W \neq Z]\}, \qquad (1.9)$$

and the result follows from (1.8) and (1.9). □

Remark 3.1.4 Some condition on $\mu - \lambda$ is required for (1.7), since, for $\lambda = \mathbb{E}W$ integer and small d, an example may be constructed with $\mu = \lambda + d$ and $\mathbb{P}[W \neq Z] = \mathbb{P}[W = \lambda \text{ and } Z = \lambda + 1] = d$. For this example, $W - \lambda = d^2 > 0$, whereas $\mathbb{E}\{(W-\lambda)^2 I[W \neq Z]\} = 0$.

Theorem 3.B* *Under the conditions of Theorem 3.B,*

$$\delta^* \ge C_r \varepsilon^{r/(r-2)} \mu_r^{-2r/(r-2)},$$

where $C_r = 3^{-r/(r-2)}$, $2 < r < 4$, and $C_r = 1/9$ for $r \ge 4$.

Proof There exists a random variable $Z \sim \mathrm{Po}(\mu)$ for some μ such that $\mathbb{P}[W \neq Z] = \delta^*$. If $\delta^* \leq 1/2$ and $\mu \leq \lambda = \mathbb{E}W$ or $\mu \geq \lambda + 8\delta^*$, then Lemma 3.1.3 and Hölder's inequality yield $\varepsilon \leq 3(\delta^*)^{1-2/r}\mu_r^2$, and the result follows. If $0 < \mu - \lambda < 8\delta^*$, note that

$$\begin{aligned}\delta = d_{TV}(\mathcal{L}(W), \mathrm{Po}(\lambda)) &\leq d_{TV}(\mathcal{L}(W), \mathrm{Po}(\mu)) + d_{TV}(\mathrm{Po}(\mu), \mathrm{Po}(\lambda)) \\ &\leq \delta^* + |\mu - \lambda| < 9\delta^*,\end{aligned}$$

from Theorem 1.C, and the result follows by Theorem 3.B. The case $\delta^* > 1/2$ is trivial, because $\mu_r^2 \geq (\lambda \wedge 1)\mathbb{E}(W - \lambda)^2 \geq \varepsilon$. □

Note that δ^* may in certain circumstances be very much smaller than δ, in which case the lower bound in Theorem 3.B will not be sharp: see Example 3.4.5 below.

Theorem 3.C *If W satisfying* $\mathrm{Var}\, W > \mathbb{E}W = \lambda$ *is such that* $\mathbb{E}e^{t|W_0|} < \infty$ *for some* $t > 0$, *where* $W_0 = (\lambda^{-1/2} \wedge 1)(W - \lambda)$, *then*

$$\varepsilon \leq \{t^{-1}(1 + \log \tfrac{1}{\delta} + \nu_t)\}^2\delta,$$

where ν_t *denotes* $\log \mathbb{E}e^{t|W_0|}$.

Proof Again let $Z \sim \mathrm{Po}(\lambda)$ be such that $\mathbb{P}(Z \neq W) = \delta$. Then, arguing as for (1.3), we have

$$\begin{aligned}\mathbb{E}\{W_0^2|W \neq Z\} &\leq \{t^{-1}\log \mathbb{E}(\exp\{1 + t|W_0|\}|W \neq Z)\}^2 \\ &\leq \{\tfrac{1}{t}(1 + \log \tfrac{1}{\delta} + \nu_t)\}^2. \end{aligned} \tag{1.10}$$

Consequently, by the same calculation as for Theorem 3.B,

$$\begin{aligned}\varepsilon = (\lambda^{-1} \wedge 1)(\mathrm{Var}\, W - \mathbb{E}W) &\leq (\lambda^{-1} \wedge 1)\mathbb{E}\{(W - \lambda)^2 I[W \neq Z]\} \\ &= \delta\mathbb{E}\{W_0^2 \mid W \neq Z\} \leq \delta\{\tfrac{1}{t}(1 + \log \tfrac{1}{\delta} + \nu_t)\}^2,\end{aligned}$$

proving the theorem. □

There is also a corresponding result for δ^*.

Theorem 3.C* *Under the conditions of Theorem 3.C,*

$$\varepsilon \leq 9\{\tfrac{1}{t}(1 + \log \tfrac{1}{\delta^*} + \nu_t)\}^2\delta^*.$$

Proof The proof is as for Theorem 3.B*. When Lemma 3.1.3 is applicable, the analogue of (1.10) with $Z \sim \mathrm{Po}(\mu)$ yields

$$\varepsilon \leq 3\mathbb{E}\{W_0^2 I[W \neq Z]\} \leq 3\delta^*\{\tfrac{1}{t}(1 + \log \tfrac{1}{\delta^*} + \nu_t)\}^2,$$

and, if $\delta^* > 1/2$, the inequality $\varepsilon \leq \mathbb{E}W_0^2 \leq \{\frac{1}{t}(1 + \nu_t)\}^2$ suffices. □

The following consequence of Theorems 3.C and 3.C* is immediate.

Corollary 3.C.1 *Under the conditions of Theorem 3.C, there exists a constant $c = c(t, \nu_t)$ such that*

$$\delta \geq \delta^* \geq c\varepsilon(1 + \log^+(\tfrac{1}{\varepsilon}))^{-2}. \qquad \square$$

Remark 3.1.5 The value of t can naturally be chosen to maximize $c(t, \nu_t)$.

In order to apply Corollary 3.C.1, the exponential moment $\mathbb{E}e^{t|W_0|}$ must be estimated. The following proposition shows how to do so for an m-dependent sequence of indicators.

Proposition 3.1.6 *Suppose that $W = \sum_{k=1}^n I_k$, where $\{I_k\}_1^n$ is an m-dependent sequence of indicators, and define $\lambda = \mathbb{E}W$. Then, for $t \geq 0$,*

$$\mathbb{E}\{e^{t|W-\lambda|}\} \leq 2\exp\{\lambda((e^t - 1)e^{mt} - t)\}.$$

Proof Let $W' = \sum_{k=1}^{n-1} I_k$ and $W'' = \sum_{k=1}^{n-m-1} I_k$. Then, for $t \geq 0$,

$$\begin{aligned}
\mathbb{E}e^{tW} &= \mathbb{E}(e^{tW'}(1 + (e^t - 1)I_n)) \leq \mathbb{E}e^{tW'} + \mathbb{E}(e^{t(W''+m)}(e^t - 1)I_n) \\
&= \mathbb{E}e^{tW'} + \mathbb{E}e^{tW''}e^{tm}(e^t - 1)\mathbb{E}I_n \leq \mathbb{E}e^{tW'}(1 + e^{tm}(e^t - 1)\mathbb{E}I_n) \\
&\leq \mathbb{E}e^{tW'}\exp((e^t - 1)e^{mt}\mathbb{E}I_n).
\end{aligned}$$

Hence, by induction, for $t \geq 0$,

$$\mathbb{E}\{e^{t(W-\lambda)}\} \leq \exp\{\lambda((e^t - 1)e^{mt} - t)\}.$$

A similar argument with $-t$ for t gives

$$\mathbb{E}\{e^{-t(W-\lambda)}\} \leq \exp\{\lambda[(e^{-t} - 1)e^{-mt} + t]\} \leq \exp\{\lambda((e^t - 1)e^{mt} - t)\},$$

and the proposition now follows because, for $t \geq 0$, $e^{t|x|} \leq e^{tx} + e^{-tx}$. $\square$

Theorem 3.C* and Proposition 3.1.6 can now be combined to yield a concrete estimate of ε.

Proposition 3.1.7 *Under the conditions of Proposition 3.1.6,*

$$\varepsilon \leq 300(1 \wedge \lambda)\Big((m+1)(1 + \log\tfrac{1}{\delta^*}) \vee \lambda^{-1}(m+1)^2(1 + \log\tfrac{1}{\delta^*})^2\Big)\delta^*. \quad (1.11)$$

Proof Let $0 \leq s \leq (m+1)^{-1}$ and $t = s(1 \vee \lambda^{1/2})$. From Proposition 3.1.6, we obtain the inequalities

$$\begin{aligned}
\nu_t &= \log \mathbb{E}e^{t|W_0|} = \log \mathbb{E}e^{s|W-\lambda|} \\
&\leq \log 2 + \lambda\{e^{ms}(e^s - 1) - s\} \leq \log 2 + \tfrac{3e}{2}(m+1)s^2\lambda,
\end{aligned}$$

where the last expression follows because, whenever $0 \le x \le (m+1)^{-1}$,

$$\frac{d^2}{dx^2}\{e^{mx}(e^x - 1) - x\} = (2m+1)e^{(m+1)x} + m^2 e^{(m+1)x}(1 - e^{-x}) \\ \le (2m + 1 + m^2 x)e^{(m+1)x} \le (3m+1)e.$$

Taking

$$s = (1 + \log \tfrac{1}{\delta^*})^{1/2}(m+1)^{-1/2}\lambda^{-1/2} \wedge (m+1)^{-1}$$

then gives

$$\nu_t \le \log 2 + \tfrac{3e}{2}(1 + \log \tfrac{1}{\delta^*}).$$

Hence Theorem 3.C* yields

$$\varepsilon \le 9t^{-2}(1 + \log 2 + \tfrac{3e}{2})^2(1 + \log \tfrac{1}{\delta^*})^2 \delta^* \\ \le 300(1 \wedge \lambda)\Big((m+1)(1 + \log \tfrac{1}{\delta^*}) \vee \lambda^{-1}(m+1)^2(1 + \log \tfrac{1}{\delta^*})^2\Big)\delta^*,$$

as required. □

3.2 Lower bounds under monotonicity

In cases where it can be proved that $\delta \le C\varepsilon$, the results of the previous section are still not quite enough to show that $\delta \asymp \varepsilon$, because the relation need not in general be true: δ can be of smaller order than ε, although only through the introduction of logarithmic factors. In this section, sharper lower bounds for δ are established under additional assumptions, strong enough to show that $\delta \asymp \varepsilon$ under quite reasonable conditions.

We start with the Stein equation

$$\mathbb{E}\{\lambda g(W+1) - Wg(W)\} = \sum_{\alpha \in \Gamma} \pi_\alpha [\mathbb{E} g(W+1) - \mathbb{E}(g(W) | I_\alpha = 1)],$$

where, as before, $W = \sum_{\alpha\in\Gamma} I_\alpha$. That finding lower bounds from the Stein equation has any chance of success derives from the observation made in the Introduction, that a non-negative integer valued random variable Z has the distribution $\mathrm{Po}(\lambda)$ if and only if

$$\mathbb{E}(\lambda g(Z+1) - Zg(Z)) = 0$$

for every real bounded function g. For a sum of independent indicators, Barbour and Hall (1984) were able to turn this into an argument which yields a lower bound for $d_{TV}(\mathcal{L}(W), \mathrm{Po}(\lambda))$. Their method is extended in Theorems 3.D and 3.E to many situations where Corollaries 2.C.2 and 2.C.4 are applicable.

We begin with a technical result.

Lemma 3.2.1 *For $\theta, \lambda > 0$ define*

$$f(z) = (z-\lambda)e^{-(z-\lambda)^2/\theta\lambda}. \tag{2.1}$$

Then, for $j = 0, 1, 2, \ldots$ and $y \geq x$,

$$\text{(a)} \quad -2e^{-3/2} \leq f(j+1) - f(j) \leq 1, \tag{2.2}$$

$$\text{(b)} \quad f(y+\lambda) - f(x+\lambda) \geq y - x - \frac{y^3 - x^3}{\theta\lambda}, \tag{2.3}$$

$$\text{(c)} \quad |\lambda f(j+1) - jf(j)| \leq \lambda \max(1, 2e^{-3/2} + \theta e^{-1}). \tag{2.4}$$

Proof Inequality (a) follows from the estimate

$$-2e^{-3/2} \leq \frac{d}{dt} te^{-t^2/\theta\lambda} = (1 - \frac{2t^2}{\theta\lambda})e^{-t^2/\theta\lambda} \leq 1.$$

Since also

$$(1 - \frac{2t^2}{\theta\lambda})e^{-t^2/\theta\lambda} \geq 1 - \frac{3t^2}{\theta\lambda},$$

we have for $y \geq x$

$$\begin{aligned} ye^{-y^2/\theta\lambda} - xe^{-x^2/\theta\lambda} &= \int_x^y (1 - \frac{2t^2}{\theta\lambda})e^{-t^2/\theta\lambda} dt \\ &\geq \int_x^y (1 - \frac{3t^2}{\theta\lambda}) dt = y - x - \frac{y^3 - x^3}{\theta\lambda}, \end{aligned}$$

proving (b). Finally, (a) implies that

$$\begin{aligned} |\lambda f(j+1) - jf(j)| &= |\lambda(f(j+1) - f(j)) - (j-\lambda)^2 e^{-(j-\lambda)^2/\theta\lambda}| \\ &\leq \lambda \max(1, 2e^{-3/2} + \theta e^{-1}), \end{aligned}$$

which proves (c). □

For any real bounded function g and $Z \sim \text{Po}(\lambda)$, using (1.1.9) and writing $h(j) = \lambda g(j+1) - jg(j)$, we obtain the straightforward upper estimate

$$\begin{aligned} |\mathbb{E}(\lambda g(W+1) - Wg(W))| &= |\mathbb{E}\{h(W) - h(Z)\}| \\ &\leq 2d_{TV}(\mathcal{L}(W), \mathcal{L}(Z)) \cdot \|h\| \\ &= 2d_{TV}(\mathcal{L}(W), \text{Po}(\lambda)) \sup_j |\lambda g(j+1) - jg(j)|. \end{aligned}$$

Hence we can look for lower bounds by evaluating the right hand side of the inequality

$$d_{TV}(\mathcal{L}(W), \text{Po}(\lambda)) \geq \frac{|\mathbb{E}(\lambda g(W+1) - Wg(W))|}{2\sup_j |\lambda g(j+1) - jg(j)|} \tag{2.5}$$

for suitable functions g. This procedure turns out to be successful if the $(I_\alpha, \alpha \in \Gamma)$ are negatively or positively related. We use the notation $\kappa_4(W)$ to denote the fourth cumulant of W.

Theorem 3.D *Suppose that the $(I_\alpha, \alpha \in \Gamma)$ are negatively related. Let $\lambda = \mathbb{E}W$ and set*

$$\varepsilon_- = 1 - \frac{\operatorname{Var} W}{\lambda}, \qquad \gamma_- = 1 - \frac{\kappa_4(W)}{\lambda}.$$

Then

$$(1 - e^{-\lambda})\varepsilon_- \geq d_{TV}(\mathcal{L}(W), \operatorname{Po}(\lambda)) \geq \frac{\varepsilon_-}{11 + 3\max(0, \gamma_-/\lambda\varepsilon_-)}. \tag{2.6}$$

Proof The upper bound in (2.6) is the assertion of Corollary 2.C.2. Taking $g = f$ as in (2.1), (2.4) and (2.5) imply for $\theta \geq e$ that

$$d_{TV}(\mathcal{L}(W), \operatorname{Po}(\lambda)) \geq \frac{|\mathbb{E}(\lambda g(W+1) - Wg(W))|}{2\lambda(2e^{-3/2} + \theta e^{-1})}. \tag{2.7}$$

Writing $V_\alpha = \sum_{\beta \neq \alpha} J_{\beta\alpha}$, where the $J_{\beta\alpha}$ are defined as in (2.1.1), it follows that

$$\begin{aligned}
\mathbb{E}(\lambda g(W+1) - Wg(W)) &= \sum_\alpha (\pi_\alpha \mathbb{E}g(W+1) - \mathbb{E}I_\alpha g(W)) \\
&= \sum_\alpha \pi_\alpha (\mathbb{E}g(W+1) - \mathbb{E}(g(W) | I_\alpha = 1)) \\
&= \sum_\alpha \pi_\alpha \mathbb{E}(g(W+1) - g(V_\alpha + 1)).
\end{aligned} \tag{2.8}$$

Using (2.3), and since $W \geq V_\alpha$ by hypothesis,

$$\begin{aligned}
&\sum_\alpha \pi_\alpha \mathbb{E}(g(W+1) - g(V_\alpha+1)) \\
&\quad \geq \sum_\alpha \pi_\alpha \mathbb{E}\Big((W+1-\lambda) - (V_\alpha + 1 - \lambda) \\
&\qquad\qquad - \frac{(W+1-\lambda)^3 - (V_\alpha + 1 - \lambda)^3}{\theta\lambda}\Big) \\
&\quad = \lambda - \sum_\alpha \mathbb{E}I_\alpha(W - \lambda) \\
&\qquad\qquad - \frac{1}{\theta\lambda}(\lambda\mathbb{E}((W+1-\lambda)^3) - \sum_\alpha \mathbb{E}(I_\alpha(W-\lambda)^3)) \\
&\quad = \lambda - \operatorname{Var} W - \frac{1}{\theta\lambda}\{\lambda\mathbb{E}(W+1-\lambda)^3 - \mathbb{E}(W(W-\lambda)^3)\} \\
&\quad = \lambda - \operatorname{Var} W \\
&\qquad - \frac{1}{\theta\lambda}\{\lambda - (\mathbb{E}(W-\lambda)^4 - 3(\operatorname{Var} W)^2) + 3(\lambda - \operatorname{Var} W)\operatorname{Var} W\} \\
&\quad = \lambda\varepsilon_-\Big(1 - \frac{\gamma_-/\lambda\varepsilon_- + 3(1-\varepsilon_-)}{\theta}\Big).
\end{aligned} \tag{2.9}$$

Combining (2.7), (2.8) and (2.9) gives

$$d_{TV}(\mathcal{L}(W), \mathrm{Po}(\lambda)) \geq \varepsilon_- \frac{1 + \frac{3\varepsilon_-}{\theta} - \frac{3+\gamma_-/\lambda\varepsilon_-}{\theta}}{2(2e^{-3/2} + \theta e^{-1})}.$$

for $\theta \geq e$. Choosing $\theta = 6 + 2\max(0, \gamma_-/\lambda\varepsilon_-)$, we obtain after a small computation that

$$d_{TV}(\mathcal{L}(W), \mathrm{Po}(\lambda)) \geq \varepsilon_- \frac{1}{11 + 3\max(0, \gamma_-/\lambda\varepsilon_-)},$$

which proves the assertion. □

In particular, if γ_-/ε_- is bounded, Theorem 3.D shows that the bound $(1 - e^{-\lambda})\varepsilon_-$ given by Corollary 2.C.2 is sharp to within a constant. One example of this is the following result.

Corollary 3.D.1 [Barbour and Hall 1984] *Let the* $(I_\alpha, \alpha \in \Gamma)$ *be independent. Then*

$$(1 - e^{-\lambda})\frac{1}{\lambda}\sum \pi_\alpha^2 \geq d_{TV}(\mathcal{L}(W), \mathrm{Po}(\lambda)) \geq \frac{1}{32}\min(1, \frac{1}{\lambda})\sum \pi_\alpha^2.$$

Proof As the I's are independent

$$\gamma_- = 1 - \frac{1}{\lambda}\sum \pi_\alpha(1 - \pi_\alpha)(1 - 6\pi_\alpha(1 - \pi_\alpha))$$

and

$$\varepsilon_- = 1 - \frac{1}{\lambda}\sum \pi_\alpha(1 - \pi_\alpha) = \frac{1}{\lambda}\sum \pi_\alpha^2.$$

Thus

$$\gamma_- = \varepsilon_- + \frac{6}{\lambda}\sum \pi_\alpha^2(1 - \pi_\alpha)^2 = 7\varepsilon_- - \frac{6}{\lambda}\sum \pi_\alpha^3(2 - \pi_\alpha) \leq 7\varepsilon_-.$$

Hence, from (2.6),

$$d_{TV}(\mathcal{L}(W), \mathrm{Po}(\lambda)) \geq \frac{\varepsilon_-}{11 + 21/\lambda} \geq \frac{\varepsilon_-}{32}\min(1, \lambda),$$

giving the lower bound. The upper bound is that of Theorem 2.M. □

Remark 3.2.2 The constant $\frac{1}{32}$ in the lower bound in Corollary 3.D.1 could be improved to $\frac{1}{14}$ using essentially the same argument, but now taking $g(j) = f(j - 1)$ instead of $g(j) = f(j)$ as above.

Remark 3.2.3 For the Wasserstein distance, there are analogous estimates based on the inequality

$$|\mathbb{E}\{\lambda g(W+1) - Wg(W)\}| \le d_W(\mathcal{L}(W), \mathrm{Po}(\lambda)) \sup_j |h(j+1) - h(j)|,$$

where, once again,

$$h(j) = \lambda g(j+1) - jg(j) = \lambda\{g(j+1) - g(j)\} - (j-\lambda)g(j).$$

Taking $g = f$, as defined in (2.1), this gives

$$\begin{aligned}\sup_j |h(j+1) - h(j)| &\le \lambda \sup_t \left|\frac{d^2}{dt^2} te^{-t^2/\theta\lambda}\right| + \sup_t \left|\frac{d}{dt} t^2 e^{-t^2/\theta\lambda}\right| \\ &\le \lambda^{1/2}\{1.952\theta^{-1/2} + 0.5873\theta^{1/2}\}.\end{aligned}$$

Hence, under the conditions of Theorem 3.D,

$$d_W(\mathcal{L}(W), \mathrm{Po}(\lambda)) \ge (\theta\lambda)^{1/2}\varepsilon_- \frac{1 + \frac{3\varepsilon_-}{\theta} - \frac{3+\gamma_-/\lambda\varepsilon_-}{\theta}}{1.952 + 0.5873\theta}.$$

The choice $\theta = 3.5\{3 + \max(\gamma_-/\lambda\varepsilon_-, 7)\}$ yields

$$\begin{aligned}d_W(\mathcal{L}(W), \mathrm{Po}(\lambda)) &\ge (\theta\lambda)^{1/2}\varepsilon_- \frac{1 - 1/3.5}{0.6433\theta} \\ &> 0.593\{3 + \max(\gamma_-/\lambda\varepsilon_-, 7)\}^{-1/2}\lambda^{1/2}\varepsilon_-,\end{aligned}$$

since $1.952\theta < 1.96/35 = 0.056$. For independent summands, this implies that

$$d_W(\mathcal{L}(W), \mathrm{Po}(\lambda)) \ge \frac{3}{16}(1 \wedge \lambda^{1/2})\lambda^{1/2}\varepsilon_- = \frac{3}{16}(1 \wedge \lambda^{-1/2}) \sum_{\alpha\in\Gamma} \pi_\alpha^2.$$

The constant 3/16 could in fact be improved to $0.264 > 1/4$ by the argument of Remark 3.2.2, using the value $16(1 \vee \lambda^{-1})$ for θ.

Theorem 3.E *Suppose that the $(I_\alpha, \alpha \in \Gamma)$ are positively related. Let $\lambda = \mathbb{E}W$ and set*

$$\varepsilon_+ = \frac{\mathrm{Var}\, W}{\lambda} - 1, \qquad \gamma_+ = \frac{\kappa_4(W)}{\lambda} - 1,$$

$$\psi = \max(0, \frac{\gamma_+}{\lambda\varepsilon_+}) + 3\varepsilon_+ + \frac{\sum \pi_\alpha^2}{\lambda^2\varepsilon_+} + 3\frac{\mathbb{E}\{((W-\lambda)^2 + W - \lambda)\sum \pi_\alpha I_\alpha\}}{\lambda^2\varepsilon_+}. \tag{2.10}$$

Suppose that $\varepsilon_+ > 0$. *Then*

$$(1-e^{-\lambda})\Big(\varepsilon_+ + 2\frac{\sum \pi_\alpha^2}{\lambda}\Big) \geq d_{TV}(\mathcal{L}(W), \mathrm{Po}(\lambda)) \geq \frac{\varepsilon_+}{11+3\psi}. \tag{2.11}$$

Furthermore, using $\pi^* = \max_\alpha \pi_\alpha$, ψ *may be more simply estimated by*

$$\psi \leq (1+\frac{3}{2}\pi^*)\max(0, \frac{\gamma_+}{\lambda\varepsilon_+}) + 3\varepsilon_+ + (\frac{15}{2}+\frac{7}{\lambda})(1+\varepsilon_+)^2\frac{\pi^*}{\varepsilon_+}. \tag{2.12}$$

Remark 3.2.4 Note that, using (2.12), if π^*/ε_+ and ε_+ are small, the inequalities (2.11) are almost precise analogues of the inequalities (2.6). In particular, if $\varepsilon_+ \to 0, \gamma_+ = O(\varepsilon_+)$ and $\pi^* = O(\varepsilon_+)$, then Theorem 3.E shows that $d_{TV}(\mathcal{L}(W), \mathrm{Po}(\lambda)) \asymp (1 \wedge \lambda)\varepsilon_+$, i.e. the bound given by Corollary 2.C.4 is sharp to within a constant. Note also that the supposition that $\varepsilon_+ > 0$ is not vacuous. Independent indicators are positively (and negatively) related, but $\varepsilon_+ < 0$, and further examples are to be found in Sections 4.2, 4.3 and 5.4.

Proof The upper bound in (2.11) is the assertion of Corollary 2.C.4. For the lower bound, set $V_\alpha = \sum_{\beta\neq\alpha} J_{\beta\alpha}$, where $J_{\beta\alpha}$ are as defined in (2.1.1), and set $W_\alpha = \sum_{\beta\neq\alpha} I_\beta$. Note that by hypothesis $V_\alpha \geq W_\alpha$. Hence, taking $g = f$ as in (2.1) and using (2.8), (2.2) and (2.3),

$$\begin{aligned}
&\mathbb{E}(Wg(W) - \lambda g(W+1)) \\
&= \sum \pi_\alpha \mathbb{E}(g(V_\alpha+1) - g(W_\alpha+1)) + \sum \pi_\alpha \mathbb{E}(g(W_\alpha+1) - g(W+1)) \\
&\geq \sum \pi_\alpha \mathbb{E}\Big((V_\alpha+1-\lambda) - (W_\alpha+1-\lambda) \\
&\qquad - \frac{1}{\theta\lambda}\{(V_\alpha+1-\lambda)^3 - (W_\alpha+1-\lambda)^3\}\Big) - \sum \pi_\alpha^2 \\
&= \mathrm{Var}\, W - \lambda - \frac{1}{\theta\lambda}\sum \pi_\alpha \mathbb{E}\big((V_\alpha+1-\lambda)^3 - (W_\alpha+1-\lambda)^3\big).
\end{aligned}$$

Using $\sigma^2 = \mathrm{Var} W$, we can write

$$\begin{aligned}
&\sum \pi_\alpha \mathbb{E}(V_\alpha+1-\lambda)^3 = \sum \mathbb{E} I_\alpha (W-\lambda)^3 \\
&\quad = \mathbb{E}(W-\lambda)^4 + \lambda\mathbb{E}(W-\lambda)^3 = \kappa_4(W) + 3\sigma^4 + \lambda\mathbb{E}(W-\lambda)^3,
\end{aligned} \tag{2.13}$$

and

$$\begin{aligned}
\sum \pi_\alpha \mathbb{E}(W_\alpha+1-\lambda)^3 &= \sum \pi_\alpha \mathbb{E}(W-\lambda+1-I_\alpha)^3 \\
&= \lambda\mathbb{E}(W-\lambda)^3 + 3\sum \pi_\alpha \mathbb{E}((W-\lambda))^2(1-I_\alpha)) \\
&\quad + 3\sum \pi_\alpha \mathbb{E}((W-\lambda)(1-I_\alpha)) + \sum \pi_\alpha \mathbb{E}(1-I_\alpha) \\
&= \lambda\mathbb{E}(W-\lambda)^3 + 3\lambda\sigma^2 - 3\mathbb{E}((W-\lambda)^2 \sum \pi_\alpha I_\alpha) \\
&\quad - 3\mathbb{E}((W-\lambda)\sum \pi_\alpha I_\alpha) + \lambda - \sum \pi_\alpha^2.
\end{aligned} \tag{2.14}$$

Combining (2.13) and (2.14) gives

$$\sum \pi_\alpha \mathbb{E}((V_\alpha + 1 - \lambda)^3 - (W_\alpha + 1 - \lambda)^3)$$
$$= \lambda\gamma_+ + 3\lambda\sigma^2\varepsilon_+ + \sum \pi_\alpha^2 + 3\mathbb{E}\Big\{((W-\lambda)^2 + W - \lambda)\sum \pi_\alpha I_\alpha\Big\}. \quad (2.15)$$

Using (2.7), (2.10) and (2.15), we finally arrive at the inequality

$$\begin{aligned} d_{TV}(\mathcal{L}(W), \mathrm{Po}(\lambda)) &\\ \geq \{2\lambda(2e^{-3/2} + \theta e^{-1})\}^{-1} &\\ \times \Big[\lambda\varepsilon_+ - \frac{1}{\theta\lambda}\Big(\lambda\gamma_+ + 3\lambda\sigma^2\varepsilon_+ + \sum \pi_\alpha^2 &\\ + 3\mathbb{E}\big(\{(W-\lambda)^2 + W - \lambda\}\sum \pi_\alpha I_\alpha\big)\Big)\Big] &\\ \geq \varepsilon_+ \frac{1 - (3+\psi)/\theta}{2(2e^{-3/2} + \theta e^{-1})}. & \end{aligned}$$

Choosing $\theta = 6 + 2\psi$ gives the lower bound in (2.11).

It remains to establish (2.12). By hypothesis, $J_{\beta\alpha} \geq I_\beta$, which implies non-negative correlations among the I's. Hence

$$\begin{aligned} \mathbb{E}(W-\lambda)\sum_\alpha \pi_\alpha I_\alpha &= \sum_\alpha \pi_\alpha \mathbb{E}(W-\lambda)I_\alpha = \sum_\alpha \pi_\alpha \sum_\beta \mathrm{Cov}\,(I_\beta, I_\alpha) \\ &\leq \pi^* \sum_\alpha \sum_\beta \mathrm{Cov}\,(I_\beta, I_\alpha) = \pi^* \mathbb{E}(W-\lambda)W. \end{aligned}$$

This then yields the estimate

$$\begin{aligned} \mathbb{E}\Big\{((W-\lambda)^2 + W - \lambda)\sum_\alpha \pi_\alpha I_\alpha\Big\} &\leq \pi^*\mathbb{E}\{((W-\lambda)^2 + W - \lambda)W\} \\ &= \pi^*(\mathbb{E}(W-\lambda)^3 + (\lambda+1)\mathrm{Var}\, W). \end{aligned} \quad (2.16)$$

Furthermore, the Cauchy–Schwarz inequality implies that

$$\begin{aligned} &\mathbb{E}(W-\lambda)^3 \leq (\mathbb{E}(W-\lambda)^4 \mathbb{E}(W-\lambda)^2)^{1/2} \\ &= \big\{(\kappa_4(W) + 3\sigma^4)\sigma^2\big\}^{1/2} = (\lambda\sigma^2 + \lambda\gamma_+\sigma^2 + 3\sigma^6)^{1/2} \\ &\leq \sigma^2 + \tfrac{1}{2}\lambda(\gamma_+ \vee 0) + \tfrac{3}{2}\sigma^4 \leq \tfrac{1}{2}\lambda(\gamma_+ \vee 0) + \lambda(1 + \tfrac{3}{2}\lambda)(1+\varepsilon_+)^2. \quad (2.17) \end{aligned}$$

The inequality (2.12) now follows from (2.10), (2.16), (2.17) and the estimate $\sum \pi_\alpha^2 \leq \lambda\pi^*$. □

Remark 3.2.5 Lower bounds for d_W could likewise be derived.

In order to use Theorems 3.D and 3.E, it is necessary to estimate the fourth cumulant of W and possibly $\pi^* = \max_\alpha \pi_\alpha$. The following results are sometimes helpful. Although phrased in terms of γ_+ and ε_+, they can equally well be used to estimate $\gamma_- = -\gamma_+$ in terms of $\varepsilon_- = -\varepsilon_+$.

Proposition 3.2.6 *Suppose that $W = \sum_1^n I_j$, where (I_j) is a sequence of indicator random variables. Let $p_J = \mathbb{E}(\prod_{j\in J} I_j)$ for any subset J of $\{1,2,\ldots,n\}$, and set $\lambda = \sum_1^n p_j$ and $p^* = \max_j p_j$. Assume that $\alpha(m)$ is such that*

$$|p_{jk} - p_j p_k| \le \alpha(k-j) \quad \text{for all} \quad j < k; \tag{2.18}$$

and

$$|p_{J\cup K} - p_J p_K| \le n^{-1}\lambda\alpha(m) \quad \text{for all} \quad \emptyset \ne J, K \subset \{1,2,\ldots,n\}$$
$$\text{such that} \min_{j\in J, k\in K}(k-j) = m > 0 \text{ and } |J| + |K| = 3 \text{ or } 4. \tag{2.19}$$

Let $\delta := \sum_{m=1}^n m^2\alpha(m)$. Then

$$\gamma_+ = \lambda^{-1}\kappa_4(W) - 1 = 7\varepsilon_+ + O(\delta + (p^*)^2), \tag{2.20}$$

where $\varepsilon_+ = \lambda^{-1}\mathrm{Var}\, W - 1$.

Proof We assume for simplicity that $\alpha(m)$ is decreasing: the general case follows by a slightly more careful version of the argument. Expand $\kappa_4(W)$ into a sum of mixed cumulants

$$\kappa_4(W) = \sum \kappa(I_{i_1}, I_{i_2}, I_{i_3}, I_{i_4}),$$

as, for example, in Leonov and Shiryaev (1959) or Janson (1987) Section 4. By expressing the mixed cumulants in terms of mixed moments (Leonov and Shiryaev 1959, Equation (II.c)), and by appropriately pairing terms of opposite sign, it is easily seen from (2.18) and (2.19) that all terms with three or four distinct indices $i_1 \le \ldots \le i_4$ are $O\{\alpha(m)(\frac{\lambda}{n} + \sum_{j=1}^4 p_{i_j})\}$, where m denotes $\max_j(i_j - i_{j-1})$, and that there are $O(nm^2)$ such terms in all in the sum, of which $O(m^2)$ contain a given index i. Hence the contribution of these terms is of order

$$O\Big(\sum_m m^2\alpha(m)\lambda + \sum_i p_i \sum_m m^2\alpha(m)\Big) = O(\lambda\delta).$$

Furthermore,

$$\kappa(I_i, I_i, I_i, I_i) = \kappa_4(I_i) = p_i - 7p_i^2 + O(p_i^3),$$

and the 14 terms that are mixed cumulants of I_i and I_j are all

$$p_{ij} - p_i p_j + O((p_i + p_j)\alpha(|j-i|)).$$

Consequently,

$$\begin{aligned}\kappa_4(W) &= \lambda - 7\sum_i p_i^2 + 14\sum_{i<j}\mathrm{Cov}\,(I_i, I_j) + O\Big(\lambda\delta + \sum_i p_i^3\Big) \\ &= 7\,\mathrm{Var}\,W - 6\lambda + O(\lambda\delta + \lambda(p^*)^2),\end{aligned}$$

from which the proposition follows. □

Remark 3.2.7 For a stationary m-dependent sequence,

$$\varepsilon_+ = -p_1 + \frac{2}{p_1}\sum_{k=1}^{m}(1-\frac{k}{n})(p_{1,k+1}-p_1^2).$$

The same result, without the factor $(1-k/n)$, holds for cyclic m-dependent summands.

Remark 3.2.8 In the case of m-dependent summands, if $|p_{jk}-p_jp_k| \le \alpha$ for all $j<k$, and if $p_ip_{jk}, p_jp_{ik}, p_kp_{ij}, p_{ijk} \le n^{-1}\lambda\alpha$ for all $i<j<k$, it follows that $\gamma_\pm = 7\varepsilon_\pm + O(m^3\alpha + (p^*)^2)$.

Remark 3.2.9 The conditions of the proposition are framed in terms of indices in $\mathbb{Z}$. However, replacing $k-j$ by $|k-j|$ in (2.18) and (2.19), and using $|\cdot|$ to denote the Manhattan metric, a similar estimate can be established for indices $j \in \mathbb{Z}^d$: now $\delta = \sum_m m^{3d-1}\alpha(m)$.

In order to use Theorem 3.E to show that $\varepsilon_+(1\wedge\lambda)$ is a precise asymptotic rate of convergence, it is necessary to bound $\gamma_+/\lambda\varepsilon_+$ in $\lambda \ge 1$ and γ_+/ε_+ in $\lambda<1$. Proposition 3.2.6 shows that this is possible if p^* is small and α in (2.18) and (2.19) can be so chosen that

$$\delta = \sum_{m\ge 1} m^2\alpha(m) = O((\lambda\vee 1)\varepsilon_+).$$

For example, for a stationary m-dependent sequence, the following crude but useful estimate can be derived.

Proposition 3.2.10 *Let* $W=\sum_{j=1}^n I_j$, *where* $(I_j, j\ge 0)$ *is a stationary sequence of positively correlated m-dependent indicator random variables,* $m\ge 1$. *Suppose that* $n \ge 3m$ *and, in the notation of Proposition 3.2.6, that* $\max_{1\le j\le m} p_{0j} \ge 2p_1^2$. *Then*

$$\gamma_+ = 7\varepsilon_+ + O\Big(p_1^{-1}\sum_{l=1}^{m} l^2 p_{0l}\Big) = O(m^3\varepsilon_+).$$

Proof If $j\in J$ and $k \in K$, then $p_{J\cup K} \le p_{jk} = p_{0,k-j}$ and

$$p_Jp_K \le p_jp_k \le p_{jk} = p_{0,k-j}.$$

Hence we can apply Proposition 3.2.6 with $\alpha(l) = p_1^{-1}p_{0l} = \lambda^{-1}np_{0l}$ for $1\le l\le m$, and, because of m-dependence, $\alpha(l)=0$ for $l>m$. Since also $(p^*)^2 = p_1^2 \le p_{01} \le \alpha(1)$, (2.20) yields

$$\gamma_+ = 7\varepsilon_+ + O\Big(\sum_{l=1}^{m} l^2\alpha(l)\Big),$$

which proves the first assertion.

Now let

$$q = \sum_{j=1}^{m} \mathrm{Cov}\,(I_0, I_j) \geq \max_{1 \leq j \leq m} \mathrm{Cov}\,(I_0, I_j) \geq p_1^2.$$

Then, by direct calculation,

$$\begin{aligned} np_1\varepsilon_+ = \lambda\varepsilon_+ &= -np_1^2 + 2\sum_{j=1}^{m}(n-j)\mathrm{Cov}\,(I_0, I_j) \\ &\geq 2(n-m)q - np_1^2 \geq \tfrac{4}{3}nq - nq = \tfrac{1}{3}nq, \end{aligned}$$

and thus

$$p_1 \leq qp_1^{-1} \leq 3\varepsilon_+.$$

Hence

$$p_1^{-1}\sum_{l=1}^{m} l^2 p_{0l} \leq m^2(q + mp_1^2)p_1^{-1} = m^2 qp_1^{-1} + m^3 p_1 \leq 3(m^2 + m^3)\varepsilon_+,$$

which implies the final estimate. □

Remark 3.2.11 If, furthermore, the I_j are positively related, Theorem 3.E applies. If ε_+ is bounded, (2.12) and the estimate $\pi^* = p_1 \leq 3\varepsilon_+$ proved above show that $\psi = O(1 \vee m^3\lambda^{-1})$; in particular, if m is fixed and $n \to \infty$, Theorem 3.E yields $d_{TV}(\mathcal{L}(W), \mathrm{Po}\,(\lambda)) \asymp (1 \wedge \lambda)\varepsilon_+$. The same is true if the I_j are cyclic m-dependent.

Proposition 3.2.12 *Suppose that* $W = \sum_1^n I_i$, *where* $\{I_i\}_1^n$ *is an exchangeable family of rv's with* $\mathbb{E}I_1 = p$,

$$\mathbb{E}I_1 \ldots I_k = p^k(1 + \delta_k),\ k = 2, 3, 4,$$

and $\delta = \max(|\delta_2|, |\delta_3|, |\delta_4|)$, $\lambda = \mathbb{E}W = np$. *Then*

$$\varepsilon_+ = \frac{\mathrm{Var}\,W}{\lambda} - 1 = \lambda(\delta_2 - \frac{1}{n} - \frac{1}{n}\delta_2) = np\delta_2 - p - p\delta_2,$$

$$\gamma_+ = \frac{\kappa_4(W)}{\lambda} - 1 = 7\varepsilon_+ + O\{(\lambda^2 + \lambda^3)(|\delta_3 - 3\delta_2| + |\delta_4 - 6\delta_2| + n^{-2} + \delta^2)\}.$$

Proof If W is any random variable with $\mathbb{E}W = \lambda$ and factorial moments $\mathbb{E}W^{(k)} = \lambda^k(1 + \eta_k), k = 2, 3, 4$, an elementary calculation yields the following expressions for the cumulants of W,

$$\kappa_2 = \lambda + \eta_2\lambda^2, \quad \kappa_3 = \lambda + 3\eta_2\lambda^2 + (\eta_3 - 3\eta_2)\lambda^3,$$
$$\kappa_4 = \lambda + 7\eta_2\lambda^2 + 6(\eta_3 - 3\eta_2)\lambda^3 + (\eta_4 - 4\eta_3 + 6\eta_2 - 3\eta_2^2)\lambda^4.$$

Hence $\varepsilon_+ = \eta_2\lambda$. In this case

$$\lambda = \mathbb{E}W = np, \quad \mathbb{E}W(W-1) = n(n-1)p^2(1 + \delta_2),$$

and thus $\eta_2 = \delta_2 - \frac{1}{n} - \frac{1}{n}\delta_2$, and similarly

$$\eta_3 = \delta_3 - \frac{3}{n} + O(\frac{1}{n^2} + \frac{\delta}{n}), \quad \eta_4 = \delta_4 - \frac{6}{n} + O(\frac{1}{n^2} + \frac{\delta}{n}).$$

The result thus follows. □

3.3 Lower bounds for mixed Poisson approximation

In Theorem 1.C(ii), an upper bound for the accuracy of approximation of a mixed Poisson distribution by a Poisson distribution with the same mean was established by using the Stein–Chen method. Lower bounds can also be proved. One way to do so is as a consequence of Theorem 3.E. This is because the mixed Poisson distribution can be regarded as the limit as $n \to \infty$ of a sequence of mixed binomial random variables $\mathrm{Bi}\,(n, \Lambda/n)$, where Λ is the randomized mean of the mixed Poisson distribution, and the mixed binomial, when realized in the obvious way, is a sum of positively related indicators. However, it is easier to adapt the method of Barbour and Hall (1984) directly, as was done for Theorems 3.D and 3.E: we give the argument only for total variation approximation.

Theorem 3.F *Let W be a mixed Poisson random variable with $\mathbb{E}\Lambda = \lambda$. Then, for any $\theta \geq e$,*

$$d_{TV}(\mathcal{L}(W), \mathrm{Po}\,(\lambda)) \geq \frac{1}{2(2e^{-3/2} + \theta e^{-1})}\left[\frac{\mathrm{Var}\,\Lambda}{\lambda}(1 - 3\theta^{-1}) - \frac{1}{\theta\lambda^2}(7\mathrm{Var}\,\Lambda + 6\mathbb{E}(\Lambda - \lambda)^3 + \mathbb{E}(\Lambda - \lambda)^4)\right].$$

Proof As in the proof of Theorem 3.D, we take for g the function f of Lemma 3.2.1 and use the inequality (2.7). Now, as in the proof of Theorem 1.C,

$$\mathbb{E}\{\lambda f(W+1) - Wf(W)\} = -\mathbb{E}\{(\Lambda - \lambda)(\mathbb{E}[f(W+1)|\Lambda] - \mathbb{E}f(Z+1))\},$$

where $Z \sim \mathrm{Po}\,(\lambda)$. Using the coupling of Theorem 1.C and Lemma 3.2.1(b) then yields, for $\Lambda \geq \lambda$,

$$\begin{aligned}
&\mathbb{E}[f(W+1)|\Lambda] - \mathbb{E}f(Z+1) \\
&\quad\geq (\Lambda - \lambda) - \frac{1}{\theta\lambda}[\mathbb{E}\{(W + 1 - \lambda)^3|\Lambda\} - \mathbb{E}(Z + 1 - \lambda)^3] \\
&\quad= (\Lambda - \lambda) - \frac{1}{\theta\lambda}[(3\lambda + 7)(\Lambda - \lambda) + 6(\Lambda - \lambda)^2 + (\Lambda - \lambda)^3],
\end{aligned}$$

giving

$$\begin{aligned}
&(\Lambda - \lambda)(\mathbb{E}[f(W+1)|\Lambda] - \mathbb{E}f(Z+1)) \\
&\quad\geq (\Lambda - \lambda)^2\{1 - \frac{1}{\theta\lambda}[(3\lambda + 7) + 6(\Lambda - \lambda) + (\Lambda - \lambda)^2]\}.
\end{aligned}$$

A similar argument yields precisely the same inequality for $\Lambda < \lambda$. Hence

$$\begin{aligned}
&-\mathbb{E}\{\lambda f(W+1) - Wf(W)\} \\
&\quad\geq \mathrm{Var}\,\Lambda(1 - 3\theta^{-1}) - \frac{1}{\theta\lambda}[7\,\mathrm{Var}\,\Lambda + 6\mathbb{E}(\Lambda - \lambda)^3 + \mathbb{E}(\Lambda - \lambda)^4],
\end{aligned}$$

and the theorem follows from (2.7). □

Corollary 3.F.1 *If also* $\mathbb{E}(\Lambda-\lambda)^4 \le A(1\vee\lambda)\mathrm{Var}\,\Lambda$ *for some* $A\ge 1$, *then*

$$\frac{\mathrm{Var}\,\Lambda}{1\vee\lambda} \ge d_{TV}(\mathcal{L}(W),\mathrm{Po}(\lambda)) \ge \left\{\frac{1}{32+21A}\right\}\frac{\mathrm{Var}\,\Lambda}{1\vee\lambda}.$$

Proof It follows from Hölder's inequality that

$$\mathbb{E}(\Lambda-\lambda)^3 \le \{\mathbb{E}(\Lambda-\lambda)^4\mathbb{E}(\Lambda-\lambda)^2\}^{1/2} \le A(1\vee\lambda)\mathrm{Var}\,\Lambda.$$

Taking $\theta = 6+14\lambda^{-1}\{1+A(1\vee\lambda)\}$ in Theorem 3.F, and using Theorem 1.C for the upper bound, now completes the proof. □

Example 3.3.1 If $\mathbb{P}[\Lambda=\lambda+\delta] = \mathbb{P}[\Lambda=\lambda-\delta] = 1/2$, Corollary 3.F.1 applies with $A = (\frac{\delta^2}{1\vee\lambda})\vee 1$, which yields

$$\begin{aligned}\frac{\delta^2}{1\vee\lambda} \ge d_{TV}(\mathcal{L}(W),\mathrm{Po}(\lambda)) &\ge \frac{\delta^2}{53(1\vee\lambda\vee\delta^2)}\\ &= \tfrac{1}{53}\min(\lambda^{-1}\delta^2,\delta^2,1),\end{aligned}$$

thus implying that $d_{TV}(\mathcal{L}(W),\mathrm{Po}(\lambda)) \asymp \min(\lambda^{-1}\delta^2,\delta^2,1)$ for all $\delta,\lambda>0$. More precisely, if $\delta^2\le\lambda/4$ and $\lambda\ge 1$, Theorem 3.F with the choice $\theta=21$ gives the lower bound

$$\frac{(1-10.25\theta^{-1})}{2(2e^{-3/2}+\theta e^{-1})}\cdot\frac{\delta^2}{\lambda} > \frac{1}{32}\frac{\delta^2}{\lambda}.$$

3.4 Illustrative examples

The section consists of examples illustrating features of the estimates obtained by the Stein–Chen method.

Example 3.4.1 *showing that, even for negatively related summands, a better accuracy of approximation than* ε *of (0.1) may be possible. The example also shows that Theorem 3.A* is almost sharp.*

Let $\{X_i\}_1^N$ be independent $\mathrm{Be}\,(\mu/N)$ random variables, and let $Y = \sum_1^N X_i \sim \mathrm{Bi}\,(N,\mu/N)$. We suppose that N is so large that the difference between the distributions of Y and $Z\sim\mathrm{Po}\,(\mu)$ is negligible in all estimates below. Let $\{I_i\}_1^N$ have the conditional distribution of $\{X_i\}_1^N$ given $Y<M$ for some $M\ge 1$, so that $W=\sum_1^N I_i$ has the distribution $\mathcal{L}(Y|Y<M)\approx\mathcal{L}(Z|Z<M)$. Then W is a sum of negatively related summands, by Proposition 2.2.10. (Alternatively, this may be shown by constructing an explicit coupling; we leave this as an exercise.)

We assume that $M\ge\mu+\mu^{1/2}$ and define $\Delta=(M-\mu)/\mu^{1/2}\ge 1$; for simplicity we also assume that $\mu\ge 1$. Let $\eta=\mathbb{P}[Z\ge M]\le 1/2$. Then

$$\delta^* \le d_{TV}(\mathcal{L}(W),\mathrm{Po}\,(\mu)) \approx d_{TV}(\mathcal{L}(Z|Z\ge M),\mathrm{Po}\,(\mu)) = \mathbb{P}[Z\ge M] = \eta. \tag{4.1}$$

Furthermore,

$$\lambda = \mathbb{E}W = \mathbb{E}(Y\,|Y < M)$$
$$= \frac{\mathbb{E}Y - \mathbb{E}(Y I(Y \geq M))}{1 - \mathbb{P}[Y \geq M]} = \mu - \frac{\mathbb{E}((Y-\mu)I(Y \geq M))}{1 - \mathbb{P}[Y \geq M]},$$

and thus, by Proposition A.2.6(i),

$$\mu - \lambda \asymp \mathbb{E}((Z-\mu)I(Z \geq M)) \asymp (M-\mu)\eta. \tag{4.2}$$

Similarly,

$$\operatorname{Var} W < \mathbb{E}(W-\mu)^2 = \mathbb{E}((Y-\mu)^2|Y < M)$$
$$= \operatorname{Var} Y - \frac{\mathbb{E}((Y-\mu)^2 I(Y \geq M))}{1 - \mathbb{P}(Y \geq M)}.$$

Hence

$$\begin{aligned}
&\mathbb{E}W - \operatorname{Var} W \\
&\quad > \mu - \operatorname{Var} Y + (1 - \mathbb{P}[Y \geq M])^{-1}\mathbb{E}\{((Y-\mu)^2 - (Y-\mu))I[Y \geq M]\} \\
&\quad > (M-\mu)(M-\mu-1)\mathbb{P}[Y \geq M] \geq \tfrac{1}{2}(M-\mu)^2\mathbb{P}[Y \geq M] \\
&\quad \asymp \tfrac{1}{2}(M-\mu)^2\eta
\end{aligned}$$

and

$$\varepsilon = (1 \vee \lambda)^{-1}(\mathbb{E}W - \operatorname{Var} W) \geq \tfrac{1}{2}\mu^{-1}(M-\mu)^2\eta = \tfrac{1}{2}\Delta^2\eta. \tag{4.3}$$

(Proposition A.2.6(ii) implies that this estimate is of the right order.) Furthermore, using Theorem 1.C,

$$\begin{aligned}
\delta &= d_{TV}(\mathcal{L}(W), \mathrm{Po}\,(\lambda)) \\
&\leq d_{TV}(\mathcal{L}(W), \mathrm{Po}\,(\mu)) + |\mu - \lambda|\mu^{-1/2} = O(\Delta\eta),
\end{aligned} \tag{4.4}$$

and hence, for some $c > 0$,

$$\varepsilon \geq c\Delta\delta. \tag{4.5}$$

Since Δ may be chosen arbitrarily large, this shows that δ/ε may be arbitrarily small, and that δ may tend to zero faster than ε, even for a sequence $W^{(n)}$ of sums of negatively related indicators such that $\mathbb{E}W^{(n)}$ converges.

In order to see what rates we can obtain, we use Proposition A.2.3(iv) and Remark A.2.4. These yield

$$\eta = \mathbb{P}[Z \geq m] \asymp \frac{M}{M-\mu}\mathbb{P}[Z = m] \geq c\Delta^{-1}e^{-\Delta^2} \geq ce^{-2\Delta^2},$$

and thus

$$\Delta \geq c(\log \tfrac{1}{\eta})^{1/2}, \tag{4.6}$$

where c is used throughout the example as a generic positive constant, not necessarily the same at each appearance. When $\mu^{-1}\log(1/\eta)$ is sufficiently large, (4.6) implies $\Delta \geq \mu^{1/2}$ and thus $M \geq 2\mu$, and Proposition A.2.3(v) yields the sharper estimate

$$\eta \geq \mathbb{P}[Z = m] \geq c\exp\{-M\log(M/\mu)\}$$

and hence $M \geq c\log(1/\eta)/\log\{\mu^{-1}\log(1/\eta)\}$, and

$$\Delta \geq \tfrac{1}{2}M\mu^{-1/2} \geq cm^{-1/2}\log(1/\eta)/\log\{\mu^{-1}\log(1/\eta)\}.$$

Combining this with (4.6) for smaller $\mu^{-1}\log(1/\eta)$, we obtain

$$\Delta \geq c(\log(1/\eta))^{1/2}\Big(1 \vee \mu^{-1/2}(\log(1/\eta))^{1/2}/\log\{e+\mu^{-1}\log(1/\eta)\}\Big). \quad (4.7)$$

Corollary 2.C.2 and Theorem 3.A* yield $\delta^* \leq \delta \leq \varepsilon \leq \delta^*(1+\log(1/\delta^*))$, which together with (4.1) and (4.3) implies

$$1+\log\frac{1}{\delta^*} \asymp 1+\log\frac{1}{\delta} \asymp 1+\log\frac{1}{\varepsilon} \asymp \log\frac{1}{\eta}.$$

Furthermore, it is easily seen that $\lambda \asymp \mu$. Consequently, by (4.1), (4.3), (4.7),

$$\begin{aligned}\delta^* &\leq c\Delta^{-2}\varepsilon \leq c\varepsilon(1+\log(\tfrac{1}{\varepsilon}))^{-1} \\ &\quad \times \{1 \wedge (\lambda \vee 1)(1+\log(\tfrac{1}{\varepsilon}))^{-1}\log^2[e+\lambda^{-1}(1+\log(\tfrac{1}{\varepsilon}))]\}, \quad (4.8)\end{aligned}$$

which shows that the estimate in Remark 3.1.2 is sharp for δ^*. It is sharp also in the case $0 < \mu < 1$, by a similar argument. For δ, however, we attain only

$$\begin{aligned}\delta &\leq c\Delta^{-1}\varepsilon \leq c\varepsilon\min\{(1+\log(1/\varepsilon))^{-1/2}, \\ &\quad (\lambda \vee 1)^{1/2}(1+\log(\tfrac{1}{\varepsilon}))^{-1}\log[e+\lambda^{-1}(1+\log(\tfrac{1}{\varepsilon}))]\}. \quad (4.9)\end{aligned}$$

Note that if we randomize at the endpoint $Y = M$, as below, we may here obtain any small δ^* and any $\lambda > 0$.

If μ is fixed and $M \to \infty$ (and $N \to \infty$ rapidly), the bound given by Theorem 3.D can be simply evaluated, since it is immediate that for the factorial moments

$$\mathbb{E}(W)_k = \mathbb{E}((Y)_j | Y < M) = \mu^k - M^k\mathbb{P}(Z = M)(1+o(1)).$$

This in turn implies as in the proof of Proposition 3.2.12 that

$$\gamma_- = 1 - \frac{\kappa_4(W)}{\mathbb{E}W} \sim \lambda^{-1}M^4\mathbb{P}(Z = M) \sim M^2\varepsilon_-.$$

Thus Theorem 3.D gives a lower bound to δ of order

$$\frac{\lambda\varepsilon_-}{M^2} \asymp \varepsilon_-(\log\frac{1}{\varepsilon_-})^{-2}(\log\log\frac{1}{\varepsilon_-})^2 \asymp \delta^*,$$

which is not quite sharp.

We may improve the attained level of δ somewhat by eliminating both small and large values of W. We will only sketch the details. Let

$$0 \le m \le \mu - \mu^{1/2} < \mu + \mu^{1/2} \le M$$

and $0 \le p_1, p_2 \le 1$, let $U \sim U(0,1)$ be a random variable independent of everything else, and define now $\{I_i\}_1^N$ to have the conditional distribution of $\{X_i\}_1^N$ given $m + p_1 < Y + U < M + p_2$, and $W = \sum_1^N I_i$. The effect of the auxiliary variable U is a randomization at the endpoints of the allowed interval $[m, M]$. Again, W is a sum of negatively related indicators; this follows by an extension of Proposition 2.2.10, proved by the same proof, taking

$$\begin{aligned} &\mathbb{P}(X_0 = 0) = cp_2; \quad \mathbb{P}(X_0 = M - m) = c(1 - p_1); \\ &\mathbb{P}(X_0 = 1) = \quad \dots \quad = \mathbb{P}(X_0 = M - m - 1) = c. \end{aligned}$$

We choose m, p_1, M, p_2 such that

$$\mathbb{E}\{(\mu - Y)I[Y + U \le m + p_1]\} = \mathbb{E}\{(Y - \mu)I[Y + U \ge M + p_2]\}$$

and thus $\lambda = \mu$. Let $\eta_- = \mathbb{P}[Z+U \le m+p_1]$ and $\eta_+ = \mathbb{P}[Z+U \ge M+p_2]$. Then $\delta \le \eta_- + \eta_+$ and, by Proposition A.2.6(i) and (iii),

$$\eta_-(\mu - m) \asymp \mathbb{E}\{(\mu - Y)I[Y + U \le m + p_1]\} \asymp \eta_+(M - \mu),$$

and, assuming $\lambda = \mu \ge 1$,

$$\varepsilon \ge \mu^{-1}((M - \mu)^2\eta_+ + (\mu - m)^2\eta_-).$$

For $\log(1/\delta) \le c\mu$, it follows that $M \le c\mu$ and

$$(M - \mu)/\mu^{1/2} \asymp (\log(1/\eta_+)) \asymp (\log(1/\eta_-)) \asymp (\mu - m)/\mu^{1/2},$$

and thus

$$\varepsilon \ge c\delta\log\frac{1}{\delta}.$$

Hence Theorem 3.A gives estimates of the right order in this case. For $\mu^{-1}\log(1/\delta)$ large, however, we have $m = 0$ and $M \ge 2\mu$, and thus $\eta_+ \asymp \frac{\mu}{M-\mu}\eta_- = O(\eta_-)$, $\delta \asymp \eta_-$, and

$$\begin{aligned} \varepsilon &\asymp \mu^{-1}((M - \mu)\mu\eta_- + \mu^2\eta_-) \asymp (M - \mu)\eta_- \\ &\asymp M\delta \asymp \delta\log(1/\delta)(\log(e + \lambda^{-1}\log(1/\delta))^{-1}, \end{aligned}$$

which leads to an improvement of (4.9) only in the power of λ. We do not know whether Theorem 3.A is sharp for such small δ.

Example 3.4.2 (The classical rencontre or matching problem.) *The example demonstrates that Poisson approximation may be very much better than the bound given by the Stein–Chen method, even for positively related summands.*

Let W be the number of objects fixed by a random permutation σ of n objects. Then $W = \sum_1^n I_\alpha$, with the indicator $I_\alpha = I[\text{object } \alpha \text{ is fixed}]$. We have $\mathbb{E}I_\alpha = 1/n$ and $\mathbb{E}I_\alpha I_\beta = \frac{1}{n(n-1)}$, $\alpha \neq \beta$, which gives $\mathbb{E}W = \operatorname{Var} W = 1$. It is easy to construct a coupling with $J_{\beta\alpha} \geq I_\beta$ by swapping σ_α and α, so that the indicators are positively related, and Corollary 2.C.4 yields

$$d_{TV}(\mathcal{L}(W), \mathrm{Po}(1)) \leq 2(1 - e^{-1})/n :$$

see also Section 4.3. In this example, $\mathbb{E}W = \operatorname{Var} W$ and thus $\varepsilon = 0$, so that the methods of this chapter are useless for producing lower bounds. It is easy, however, to show directly that

$$d_{TV}(\mathcal{L}(W), \mathrm{Po}(1)) \sim \frac{2^n}{(n+1)!},$$

which means that the convergence is extremely rapid, and much faster than the bound produced by the Stein–Chen method.

Example 3.4.3 [Barbour and Brown 1992] *This example shows that the factor $\lambda^{-1/2}$ multiplying $\sum_{i=1}^n \eta_i$ in Theorem 1.A cannot be greatly improved.*

Let $\Gamma = \{1, 2, \ldots, 2m\}$ and suppose that $(I_i, 1 \leq i \leq m)$ are independent and identically distributed with $\pi_i = p$. Letting W_1 denote $\sum_{i=1}^m I_i$ and $W_2 = \sum_{i=m+1}^{2m} I_i$, suppose that, given W_1, the indicators $(I_i, m+1 \leq i \leq 2m)$ are independent with means

$$\mathbb{E}(I_i | W_1) = \{p - \theta(W_1 - mp)\} \vee 0.$$

Take m to be large, and set $p = m^{-1/2}$ and $\theta = m^{-9/8}$, so that, in particular, $1 \gg m\theta \gg p \gg m\theta^2$ and $mp \gg 1$. Then direct calculation shows that

$$\varepsilon = 1 - \operatorname{Var} W/\mathbb{E}W \sim p + m\theta \sim m\theta, \tag{4.10}$$

and so Theorem 3.A implies the lower bound

$$d_{TV}(\mathcal{L}(W), \mathrm{Po}(\lambda)) \geq c\varepsilon(1 + \log(1/\varepsilon))^{-2}. \tag{4.11}$$

(It is in fact shown by Barbour and Brown that the lower bound can be increased to $c\varepsilon$.) On the other hand, for Theorem 1.A, take Γ_i^s to be empty for each i — there is no better choice — so that the first sum in (1.1.28) yields a contribution of almost p. Now, for $m+1 \leq i \leq 2m$, η_i can be simply estimated as

$$\mathbb{E}|\theta(W_1 - mp)| \sim \theta\sqrt{\frac{2mp}{\pi}} \asymp \theta\sqrt{mp}.$$

For $1 \leq i \leq m$, the argument is more complicated. By evaluating the conditional expectation of I_i, given the values of the other I's, one obtains the formula

$$\eta_i = p(1-p)\mathbb{E}\left|\left(\frac{p_{R+1}}{p}\right)^S\left(\frac{1-p_{R+1}}{1-p}\right)^{m-S} - \left(\frac{p_R}{p}\right)^S\left(\frac{1-p_R}{1-p}\right)^{m-S}\right|,$$

where $R \sim \mathrm{Bi}\,(m-1,p)$ and $S \sim \mathrm{Bi}\,(m,p)$ are independent, and $p_r = \{p - \theta(r-mp)\} \vee 0$. For $R - mp$ and $S - mp$ of the typical order $\sqrt{mp}$, the quantity in the expectation is close to $\theta|S-mp|/p$, and, by careful consideration of what happens in the tails of R and S, it is possible to show that

$$\eta_i \sim \theta(1-p)\mathbb{E}|S-mp| \sim \theta\sqrt{\frac{2mp}{\pi}}.$$

Hence (1.1.28) gives an upper bound for $d_{TV}(\mathcal{L}(W), \mathrm{Po}\,(\lambda))$ of order

$$p + \lambda^{-1/2}m\theta\sqrt{mp} \asymp m\theta \asymp \varepsilon. \tag{4.12}$$

Any higher negative power of λ than $\lambda^{-1/2}$ multiplying $\sum_{i=1}^n \eta_i$ in Theorem 1.A would give an upper bound at variance with (4.11), so that $\lambda^{-1/2}$ is the best possible order attainable.

Example 3.4.4 *showing that small correlation between the* $(I_\alpha, \alpha \in \Gamma)$ *is insufficient without further conditions to imply good Poisson approximation.*

The upper bounds for $\delta = d_{TV}(\mathcal{L}(W), \mathrm{Po}\,(\lambda))$ given, for example, in Theorem 2.C make it tempting to conjecture that

$$\delta' = \sum_{\alpha\in\Gamma} \pi_\alpha^2 + \sum_{\alpha\in\Gamma}\sum_{\beta\neq\alpha} |\mathrm{Cov}\,(I_\alpha, I_\beta)|$$

might also be an upper bound. Without further assumptions this is not true, as observed in Arratia, Goldstein and Gordon (1989). For instance, let W be such that $\mathbb{P}[W=0] = \mathbb{P}[W=2] = 1/2$, and let $I_1, \ldots, I_n$ be exchangeable $\mathrm{Be}\,(\frac{1}{n})$ random variables with sum W: for example, sample W, and then choose W of the indices $\{1, 2, \ldots, n\}$ at random to have $I_i = 1$. It is easy to evaluate $\delta' = 2/n$, but $\delta = 1 - 3/2e$ is independent of n. The same construction works for any W such that $\mathbb{E}W = \mathrm{Var}\,W < \infty$. A related example where the covariances vanish identically is to be found as Example 5.3.5.

Example 3.4.5 *showing that $\delta = d_{TV}(\mathcal{L}(W), \mathrm{Po}(\lambda))$ may be much bigger than $\delta^* = \inf_{\mu>0} d_{TV}(\mathcal{L}(W), \mathrm{Po}(\mu))$, so that Theorem 3.B can at times give an inaccurate lower bound.*

Let W have the contaminated Poisson distribution given by the mixture

$$\{(1 - M^{-\theta}), \mathrm{Po}(1)\} + \{M^{-\theta}, \delta_{2M}\}$$

for $0 < \theta \leq 1$ and M large, similar to (1.1) in the case $\theta = 1$, and let $\lambda = \mathbb{E}W \asymp M^{1-\theta}$. Then it follows that $\lambda - 1 \geq 1$, and hence that $\delta \geq \delta_0 > 0$ uniformly in M and θ, whereas δ^* is seen to be at most $M^{-\theta}$, if μ is taken as 1. On the other hand, $\varepsilon \asymp M$ and $\mu_r \asymp M^{\nu}$ with $\nu = \frac{1}{2} - \frac{\theta}{r} + \frac{\theta}{2}$, so that both Theorems 3.B and 3.B* give lower bounds of order $M^{-\theta}$, whatever the choice of r. Thus the lower bound for δ given by Theorem 3.B is in this case much too small, whereas the bound for δ^* is sharp.

Example 3.4.6 *showing that Theorem 3.B may at times give a sharp lower bound, in which case Theorem 3.B* does also.*

This example is a symmetric version of the previous one. Let W have the mixture distribution

$$\{(1 - 2M^{-\theta}), \mathrm{Po}(M)\} + \{2M^{-\theta}, 2M\mathrm{Be}(\tfrac{1}{2})\}, \quad \theta > 0,$$

which has $\lambda = \mathbb{E}W = M$ and $\mathrm{Var}\, W - \mathbb{E}W \sim 2M^{2-\theta}$. Then, provided that $M \gg 1$ and that $r > 2\theta$, direct calculation yields

$$\mu_r = \lambda^{-1/2}\{\mathbb{E}|W - \lambda|^r\}^{1/r} \asymp M^{(r-2\theta)/2r} \gg 1.$$

Thus Theorem 3.B gives the estimate

$$\delta \geq c\{M^{1-\theta}/M^{1-2\theta/r}\}^{r/(r-2)} \asymp M^{-\theta},$$

as compared with the true value of δ, which is almost exactly $2M^{-\theta}$. Note that, in this case, $\varepsilon \sim 2M^{1-\theta}$ is very much larger.

4
Random permutations

In this chapter, we investigate some quantities based on random permutations $(\sigma_1, \sigma_2, \ldots, \sigma_n)$ drawn from the uniform distribution over all permutations of $1, 2, \ldots, n$. As a simple typical example of the type of random variable we study, consider the number of fixed points W in a random permutation. W is then the random variable which occurs in the classical rencontre or matching problem, as in Example 3.4.2, and it can be written as

$$W = \sum_{i=1}^{n} I(\sigma_i = i) = \sum_{i=1}^{n} e(i, \sigma_i),$$

where $\{e(i,j)\}$ is the $n \times n$ identity matrix. Random variables W with different distributions, but where Poisson approximation may none the less still be appropriate, can evidently be obtained by replacing the identity matrix e by *any* matrix c of *zeros* and *ones*. This problem is analysed in Section 4.1. The matching and ménage problems, studied in Sections 4.2 and 4.3, correspond to particular choices of the matrix c.

The approach used is to rewrite W in the form

$$W = \sum_{i=1}^{n} c(i, \sigma_i) = \sum_{i,j=1}^{n} c(i,j) I_{ij},$$

where $I_{ij} = I[\sigma_i = j]$. The indicators I_{ij} have a joint distribution which fits in well with Theorem 2.C. If $I_{ij} = 1$, then necessarily $I_{il} = 0$ for $l \neq j$ and $I_{kj} = 0$ for $k \neq i$, so that

$$\Gamma_{ij}^{-} \supset G_{ij} := \{(i,l),\, l \neq j\} \cup \{(k,j),\, k \neq i\};$$

furthermore, a simple coupling can be constructed in such a way that

$$\Gamma_{ij}^{+} = H_{ij} := \{(k,l);\, k \neq i, l \neq j\},$$

and hence so that $\Gamma_{ij}^{0} = \emptyset$. This structure can also be exploited in more complicated permutation statistics: a generalization to two-dimensional arrays is investigated in Section 4.4.

4.1 A general result

Let c be an $n \times n$ matrix of *zeros* and *ones*, whose elements are denoted by c_{ij} or $c(i,j)$. Set

$$\pi_i = \frac{c_{i.}}{n} = \frac{1}{n}\sum_{j=1}^{n} c_{ij}, \quad \rho_j = \frac{c_{.j}}{n} = \frac{1}{n}\sum_{i=1}^{n} c_{ij}, \quad c_{..} = \sum_{i=1}^{n}\sum_{j=1}^{n} c_{ij}.$$

Proposition 4.1.1 *If W denotes the sum $\sum_{i=1}^n c(i,\sigma_i)$,*

(a) $\mathbb{E}W = \lambda = c_{..}/n$,

(b) $\operatorname{Var} W = \lambda - \frac{n}{n-1}(\sum_{i=1}^n \pi_i^2 + \sum_{j=1}^n \rho_j^2 - \frac{\lambda^2+\lambda}{n})$,

(c) $\operatorname{Var} W \le \lambda$, *with equality if and only if c is a permutation matrix or identically zero.*

Proof Defining $\Gamma = \{(i,j),\ 1 \le i,j \le n\}$ and with the indicators $I_{ij} = I(\sigma_i = j)$ as before, we have the representation

$$W = \sum_{\alpha\in\Gamma} c_\alpha I_\alpha = \sum_{\alpha\in\Gamma} I'_\alpha.$$

As $I_\alpha \sim \operatorname{Be}(1/n)$, we get $\mathbb{E}W = c_{..}/n$, proving (a).

For $\alpha = (i,j)$, write $G_\alpha = G_{ij}$ and $H_\alpha = H_{ij}$, and note that

$$\Gamma = \{\alpha\} \cup G_\alpha \cup H_\alpha.$$

Then $I_\alpha I_\beta \equiv 0$ for $\beta \in G_\alpha$ and $\mathbb{E}I_\alpha I_\beta = 1/n(n-1)$ for $\beta \in H_\alpha$, implying that

$$\operatorname{Cov}(I_\alpha, I_\beta) = \begin{cases} -1/n^2, & \beta \in G_\alpha, \\ 1/n^2(n-1), & \beta \in H_\alpha. \end{cases}$$

This in turn implies that

$$\begin{aligned}
\operatorname{Var} W &= \sum_\alpha c_\alpha \operatorname{Var} I_\alpha + \sum\sum_{\alpha\ne\beta} c_\alpha c_\beta \operatorname{Cov}(I_\alpha, I_\beta) \\
&= \sum_\alpha c_\alpha \frac{1}{n}(1-\frac{1}{n}) - \sum_\alpha c_\alpha \sum_{\beta\in G_\alpha} c_\beta \frac{1}{n^2} + \sum_\alpha c_\alpha \sum_{\beta\in H_\alpha} c_\beta \frac{1}{n^2(n-1)} \\
&= c_{..}\frac{1}{n}(1-\frac{1}{n}) - \sum_{i,j} c_{ij}(c_{i.} + c_{.j} - 2)\frac{1}{n^2} \\
&\quad + \sum_{i,j} c_{ij}(c_{..} - c_{i.} - c_{.j} + 1)\frac{1}{n^2(n-1)} \\
&= \lambda - \frac{1}{n(n-1)}\Big(\sum_i c_{i.}^2 + \sum_j c_{.j}^2 - n(\lambda + \lambda^2)\Big),
\end{aligned}$$

which proves (b).

Finally, $n\lambda = \sum_i c_{i.} \le \sum_i c_{i.}^2$, and by Schwarz's inequality we have

$$n^2\lambda^2 = (\sum_j c_{.j})^2 \le n \sum_j c_{.j}^2,$$

with equality in both inequalities if and only if $c_{i.} = c_{.j} = 1$ for all i, j or $c_{i.} = 0$ for all i. In view of this, (c) follows from (b). □

Theorem 4.A *If W denotes the sum $\sum_{i=1}^n c(i, \sigma_i)$,*

$$d_{TV}(\mathcal{L}(W), \mathrm{Po}(\lambda)) \le (1 - e^{-\lambda})\Big(\frac{n-2}{n}\Big(1 - \frac{\mathrm{Var}\, W}{\lambda}\Big) + \frac{2\lambda}{n}\Big)$$
$$\le \frac{3}{2}\frac{1-e^{-\lambda}}{\lambda}\Big(\sum_{i=1}^n \pi_i^2 + \sum_{j=1}^n \rho_j^2 - \frac{2\lambda}{3n}\Big).$$

Remark 4.1.2 Since $\pi_i = \mathbb{P}(c(i, \sigma_i) = 1)$, the first part of the last estimate is comparable to the estimates for sums of independent indicators. The second part can be understood in exactly the same way, since W can equally well be written as $\sum_{j=1}^n c(\sigma^{-1}(j), j)$.

Remark 4.1.3 Chen (1975b), Corollary 3.2, gives a similar estimate, but with the constant $45.25/2 = 22.625$ in place of $3/2$. The proof below can also be adapted to prove (2.2) of his more general Theorem 2.1, again with 3 in place of 45.25 in the constant multiplier.

Proof First we give a proof using Theorem 2.C with the following coupling. If $I_\alpha = 1$ set $J_{\beta\alpha} = I_\beta$. If $I_\alpha = I_{ij} = 0$, modify σ to $\sigma^* = \tau \circ \sigma$ by a single transposition τ, in such a way that $\sigma_i^* = j$, and set $J_{st\alpha} = I(\sigma_s^* = t)$. It now follows by construction that $J_{\beta\alpha} \ge I_\beta$ for $\beta \in \Gamma_\alpha^+ := H_\alpha$ and $J_{\beta\alpha} \le I_\beta$ for $\beta \in \Gamma_\alpha^- := G_\alpha$, and that the corresponding inequalities hold between $I'_\beta = c_\beta I_\beta$ and $J'_{\beta\alpha} = c_\beta J_{\beta\alpha}$. Thus we can apply Theorem 2.C with $\Gamma_\alpha^0 = \emptyset$. Calculations similar to those in the proof of Proposition 4.1.1 yield

$$\sum_\alpha (\mathbb{E}I'_\alpha)^2 + \sum\sum_{\alpha\ne\beta} c_\alpha c_\beta |\mathrm{Cov}\,(I_\alpha, I_\beta)|$$
$$= \frac{n-2}{n-1}\Big(\sum_i \pi_i^2 + \sum_j \rho_j^2 - \frac{\lambda}{n} + \frac{\lambda^2}{n-2}\Big) = \frac{n-2}{n}(\lambda - \mathrm{Var}\, W) + \frac{2\lambda^2}{n},$$

and hence, from Theorem 2.C, the first inequality follows. Since also

$$\Big(\sum_i c_{i.}^2 + \sum_j c_{.j}^2 - n\lambda + n\lambda^2\Big) - \frac{n-2}{n-1}\Big(\sum_i c_{i.}^2 + \sum_j c_{.j}^2 - n\lambda + \frac{n^2\lambda^2}{n-2}\Big)$$
$$= \frac{1}{n-1}\Big(\sum_i c_{i.}^2 + \sum_j c_{.j}^2 - n\lambda - n\lambda^2\Big) \ge 0,$$

as in the proof of Proposition 4.1.1(c), it follows from the Schwarz inequality that

$$\sum_\alpha (\mathbb{E}I'_\alpha)^2 + \sum\sum_{\alpha\neq\beta} c_\alpha c_\beta |\mathrm{Cov}\,(I_\alpha, I_\beta)|$$
$$\leq \frac{1}{n^2}\Big(\sum_i c_{i.}^2 + \sum_j c_{.j}^2 - n\lambda + n\lambda^2\Big) \leq \frac{3}{2}\big(\sum_i \pi_i^2 + \sum_j \rho_j^2\big) - \frac{\lambda}{n},$$

completing the proof.

An alternative proof can be given using Theorem 2.B and the representation $W = \sum_i I_i$ with $I_i = c(i, \sigma(i))$. Note that now $\Gamma = \{1, \ldots, n\}$. We couple W with random variables

$$V_{ij} = \sum_{k\neq i} c(k, \sigma^*(k)),$$

where σ^* is as above, that is $\sigma^*(i) = j$, $\sigma^*(\sigma^{-1}(j)) = \sigma(i)$, and $\sigma^*(k) = \sigma(k)$ otherwise, and where j is any index such that $c_{ij} = 1$. It is easy to see that

$$\mathcal{L}(1 + V_{ij}) = \mathcal{L}(W \mid I_i = 1, \sigma(i) = j).$$

Now

$$\begin{aligned}
\theta_i(j) = \mathbb{E}|W - V_{ij}| &= \mathbb{E}|c(i,\sigma(i)) + c(\sigma^{-1}(j), j) - c(\sigma^{-1}(j), \sigma(i))| \\
&= \frac{1}{n}c_{ij} + \frac{1}{n(n-1)}\sum_{l\neq j}\sum_{m\neq i} |c_{il} + c_{mj} - c_{ml}| \\
&\leq \frac{1}{n}c_{ij} + \frac{1}{n}(c_{i.} - c_{ij}) + \frac{1}{n}(c_{.j} - c_{ij}) + \frac{1}{n(n-1)}(c_{..} - c_{.j} - c_{i.} + c_{ij}) \\
&= \frac{1}{n-1}(\lambda + (n-2)(\pi_i + \rho_j - n^{-1}c_{ij})),
\end{aligned}$$

and thus it follows from Theorem 2.B that

$$\begin{aligned}
d_{TV}(\mathcal{L}(W), \mathrm{Po}(\lambda)) &\leq \frac{1-e^{-\lambda}}{\lambda}\sum_{i=1}^n \mathbb{E}(I_i\theta_i(\sigma(i))) \\
&= \frac{1-e^{-\lambda}}{\lambda}\sum_{i=1}^n\sum_{j=1}^n \frac{1}{n}c_{ij}\theta_i(j) \\
&\leq \frac{1-e^{-\lambda}}{\lambda(n-1)}\Big(\lambda^2 + (n-2)\big(\sum_{i=1}^n \pi_i^2 + \sum_{j=1}^n \rho_j^2 - \frac{\lambda}{n}\big)\Big). \quad \square
\end{aligned}$$

The bound for $d_{TV}(\mathcal{L}(W), \mathrm{Po}(\lambda))$ given above can in special cases be improved. For example let $I_i = I(\sigma_i \leq a_i)$, where the a_i are given numbers, and $W = I_1 + \cdots + I_n$: this is equivalent to taking $c_{ij} = I[j \leq a_i]$. As the $I[\cdot \leq a_i]$ are decreasing functions and $\sigma_1, \ldots, \sigma_n$ are negatively associated, it follows that $I_1, \ldots, I_n$ are negatively related: see Proposition 2.2.12 and Theorem 2.I. Hence it follows from Corollary 2.C.2 that

$$d_{TV}(\mathcal{L}(W), \mathrm{Po}(\lambda)) \leq (1 - e^{-\lambda})\Big(1 - \frac{\operatorname{Var} W}{\lambda}\Big),$$

improving the bound in Theorem 4.A, because

$$\begin{aligned}\frac{n-2}{n}\Big(1 - \frac{\operatorname{Var} W}{\lambda}\Big) + \frac{2\lambda}{n} - \Big(1 - \frac{\operatorname{Var} W}{\lambda}\Big)& \\ = \frac{2}{n\lambda}(\lambda^2 - \lambda + \operatorname{Var} W) = \frac{2}{n\lambda}\mathbb{E}W(W-1) \geq 0.&\end{aligned}$$

Note that the inequality is strict if $\mathbb{P}(W \geq 2) > 0$. An interesting special case is that in which $a_1 = \cdots = a_r = m$ and $a_i = 0$ otherwise, when W is hypergeometrically distributed. This case is further discussed in Section 6.1.

Consider a sequence of matrices $c_n = \{c_n(i,j)\}$ and a corresponding sequence of random variables $W_n = \sum_{i=1}^n c_n(i, \sigma_i)$ such that

$$\frac{\mathbb{E}W_n}{n} = \frac{\sum\sum_{i,j} c_n(i,j)}{n^2} \to 0, \quad \text{as} \quad n \to \infty.$$

Corollary 4.A.1 *With W_n as above:*

(a) *If $\liminf \mathbb{E}W_n > 0$, then*

$$d_{TV}(\mathcal{L}(W_n), \mathrm{Po}(\mathbb{E}W_n)) \to 0 \quad \textit{if and only if} \quad \frac{\operatorname{Var} W_n}{\mathbb{E}W_n} \to 1.$$

(b) *If $\lambda_\infty \geq 0$, then*

$$W_n \xrightarrow{\mathcal{D}} \mathrm{Po}(\lambda_\infty) \quad \textit{if and only if} \quad \mathbb{E}W_n \to \lambda_\infty \textit{ and } \operatorname{Var} W_n \to \lambda_\infty.$$

Proof Sufficiency follows from Theorem 4.A. Since $\mathbb{E}W_n \geq \operatorname{Var} W_n$, necessity is implied by Theorem 3.A and Corollary 3.A*.1. □

Remark 4.1.4 Theorem 3.A* implies that the conditions in (a) hold if $d_{TV}(\mathcal{L}(W_n), \mathrm{Po}(\mu_n)) \to 0$ for some sequence μ_n with $\liminf \mu_n > 0$.

4.2 Matching problems

Apart from problems associated with the hypergeometric distribution, random variables of the type $W_n = \sum_{i=1}^n c(i, \sigma_i)$ were probably first studied in connection with matching. To our knowledge, the first such published investigation appeared in 1708 in de Montmort's book on games of chance for the game 'Treize', see de Montmort (1713). In this game 13 cards are drawn at random one at a time from an ordinary deck of 52 cards having face values 1, ... ,13. A match occurs if a card with face value i appears in the ith drawing. The original problem was to find the probability that there was no match.

Example 4.2.1 Consider the setting with $n = rd$ cards, d of which have face value j, $j = 1, \ldots, r$, and draw r of them. With

$$c_{ij} = \begin{cases} 1, & \text{if } d(i-1) < j \leq di \text{ and } i = 1, \ldots, r, \\ 0, & \text{elsewhere,} \end{cases}$$

the number of matches W can be written as $\sum_{i=1}^{n} c(i, \sigma_i)$. From Proposition 4.1.1 we obtain

$$\lambda = \mathbb{E}W = 1, \quad \operatorname{Var} W = 1 - \frac{d-1}{n-1},$$

and by Theorem 4.A,

$$d_{TV}(\mathcal{L}(W), \operatorname{Po}(1)) \leq (1 - e^{-1}) \frac{(d+1)n - 2d}{n(n-1)} \leq (1 - e^{-1}) \frac{d+1}{n}.$$

We could also use an alternative representation, writing $W = \sum_1^r I_i$, where $I_i = I(\text{match at drawing } i)$, $i = 1, \ldots, r$. It is easy to construct a coupling with $J_{ik} \geq I_i$ by swapping the card obtained in the kth drawing with a card with face value k chosen at random. Hence $I_1, \ldots, I_r$ are positively related. Thus we can apply Corollary 2.C.4, giving the same bound on $d_{TV}(\mathcal{L}(W), \operatorname{Po}(1))$ as that found above. Note also that although the I's are positively related, and hence positively correlated,

$$\operatorname{Var}(I_1 + \cdots + I_r) = 1 - \frac{d-1}{n-1} \leq \mathbb{E}(I_1 + \cdots + I_r) = 1.$$

In this example it is possible to get an explicit closed form expression for $\mathcal{L}(W)$. From this one can prove that $d_{TV}(\mathcal{L}(W), \operatorname{Po}(1)) \asymp 1/n$ as $n \to \infty$ if $d > 1$ is held fixed. Considerably better approximation of $\mathcal{L}(W)$ is obtained by using the $\operatorname{Bi}(m, 1/m)$ distribution with variance

$$1 - 1/m \approx \operatorname{Var} W = 1 - (d-1)/(n-1);$$

it then turns out that $d_{TV}(\mathcal{L}(W), \operatorname{Bi}(m, 1/m)) \asymp 1/n^2$ if $d > 1$ is fixed. As an illustration, consider the game 'Treize', where $n = 52$ and $d = 4$. Numerical computations give $d_{TV}(\mathcal{L}(W), \operatorname{Po}(1)) = 0.01659$, whereas, for binomial approximation, $d_{TV}(\mathcal{L}(W), \operatorname{Bi}(17, 1/17)) = 0.00034$. The exact value of $1 - 2\mathbb{P}[W = 0]$ was computed in 1711 by Nicolas Bernoulli to be $\frac{7672980411874433}{26816424180170625}$, see de Montmort (1713), p. 324.

Note that we adapt the two parameters of the binomial distribution so that both the mean and the variance are (approximately) correct. The approximation with $\operatorname{Bi}(n, 1/n)$ is inferior, with an error of order $1/n$.

Example 4.2.2 The setting is the same as in the previous example, except that all $n = rd$ cards are drawn. A card of face value i gives a match in the drawings $i, i+r, \ldots, i+(d-1)r$. Using the matrix c defined by

$$c_{ij} = \begin{cases} 1, & \text{if } d(i - lr - 1) < j \leq d(i - lr), \quad l = 0, \ldots, d-1; \\ 0, & \text{elsewhere,} \end{cases}$$

the total number if matches is given by W. Using Proposition 4.1.1 and Theorem 4.A, one finds that

$$\mathbb{E}W = d, \quad \operatorname{Var} W = d - \frac{d(d-1)}{n-1}$$

and

$$d_{TV}(\mathcal{L}(W), \operatorname{Po}(d)) \leq (1 - e^{-d})\Big(\frac{3d-1}{n} - \frac{d-1}{n(n-1)}\Big).$$

Much as in Example 4.2.1 above, it is also possible to find a representation of W and a coupling such that Theorem 2.C is applicable with $\Gamma_\alpha^0 = \emptyset$. This gives the slightly better bound

$$d_{TV}(\mathcal{L}(W), \operatorname{Po}(d)) \leq (1 - e^{-d})\Big(\frac{3d-1}{n} - \frac{(d-1)(2d-1)}{n(n-1)}\Big).$$

We are not aware of any explicit expression for $\mathcal{L}(W)$, but it is possible to obtain asymptotic expressions for the factorial moments which imply that for $d \geq 2$ fixed, $d_{TV}(\mathcal{L}(W), \operatorname{Bi}(m, d/m)) = O(n^{-2})$ if $m = dn/(d-1) + O(1)$ and thus $\operatorname{Var} W \approx \operatorname{Var} \operatorname{Bi}(m, d/m)$. It follows by the triangle inequality that $d_{TV}(\mathcal{L}(W), \operatorname{Po}(d)) \asymp d_{TV}(\operatorname{Bi}(m, d/m), \operatorname{Po}(d)) \asymp n^{-1}$ and similarly $d_{TV}(\mathcal{L}(W), \operatorname{Bi}(n, d/n)) \asymp n^{-1}$.

The general matching problem can be formulated as follows. Let A and B be two decks of cards, let A have a_j cards of type j, $1 \leq j \leq r$, and let $b_1, \ldots, b_r$ be the corresponding numbers for B. Furthermore, let the total number of cards in each pack be n:

$$n = \sum_{k=1}^{r} a_k = \sum_{k=1}^{r} b_k.$$

The two decks are now matched at random, in that each card from A is paired with a card from B in such a way that each of the $n!$ ways of doing this has probability $1/n!$. The n pairs so obtained are examined for matches, where a pair is said to match if the two cards are of the same type. The problem is to find distribution of W, the total number of matches. This has been studied in several papers: see Lanke (1973) and the references therein.

The matrix c for this problem can be constructed as follows. Define $\alpha(i) = s$ if $\sum_{k=1}^{s-1} a_k < i \le \sum_{k=1}^{s} a_k$ and $\beta(j) = s$ if $\sum_{k=1}^{s-1} b_k < j \le \sum_{k=1}^{s} b_k$, and set

$$c_{ij} = 1 \quad \text{if} \quad \alpha(i) = \beta(j); \qquad c_{ij} = 0 \;\; \text{otherwise.}$$

Then

$$c_{..} = \sum_{k=1}^{r} a_k b_k, \quad \sum_{i=1}^{n} \pi_i^2 = \frac{1}{n^2} \sum_{k=1}^{r} a_k b_k^2 \quad \text{and} \quad \sum_{j=1}^{n} \rho_j^2 = \frac{1}{n^2} \sum_{k=1}^{r} a_k^2 b_k,$$

and Proposition 4.1.1 gives

$$\lambda = \mathbb{E}W = \frac{1}{n} \sum_{k=1}^{r} a_k b_k$$

and

$$\operatorname{Var} W = \lambda - \frac{1}{n-1}\Big(\frac{1}{n}\sum_{k=1}^{r} a_k b_k (a_k + b_k) - \lambda^2 - \lambda\Big) \le \lambda,$$

with equality if and only if $a_k = b_k = 1$ for all k, or $c \equiv 0$. Theorem 4.A thus gives the bound

$$d_{TV}(\mathcal{L}(W), \operatorname{Po}(\lambda))$$
$$\le (1 - e^{-\lambda})\bigg(\frac{n-2}{\lambda n(n-1)}\Big(\frac{1}{n}\sum_{k=1}^{r} a_k b_k (a_k + b_k) - \lambda^2 - \lambda\Big) + \frac{2\lambda}{n}\bigg).$$

Now consider a triangular array $a_k = a_{kn}$, $b_k = b_{kn}$, $r = r_n$ such that $\mathbb{E}W_n \to \lambda_\infty$ as $n \to \infty$, where $0 < \lambda_\infty < \infty$. Using Corollary 4.A.1, it is easily seen that $W_n \xrightarrow{\mathcal{D}} \operatorname{Po}(\lambda_\infty)$ if and only if

$$n^{-1} \max_k (a_k + b_k) I(a_k b_k > 0) \to 0 \quad \text{as} \quad n \to \infty,$$

a result first proved by Lanke (1973) using the method of moments.

4.3 The ménage problem

The classical ménage problem asks for the number of seatings of n man–women couples at a circular table, with men and women alternating, so that no one sits next to his or her partner. Consider the following generalization. Let $1 \le d \le n$ be a given integer and set $I_k = 1$, if k belongs to $\{\sigma_k, \sigma_{k+1}, \ldots, \sigma_{k+d-1}\}$, $I_k = 0$ otherwise, where $\sigma_{n+j} = \sigma_j$. We say that there is a match at place k if $I_k = 1$. We wish to study the distribution of the total number of matches: that is, $W = I_1 + \cdots + I_n$.

As $\mathbb{P}(I_k = 1) = d/n$, we have $\mathbb{E}W = d$. Note that, when $d = 1$, determining $\mathbb{P}(W = 0)$ is the classical rencontre or matching problem, and that $n!\mathbb{P}(W = 0)$, when $d = 2$, is the number asked for in the ménage problem.

The number of matches W is obtained using the matrix c defined by

$$c_{ij} = \begin{cases} 1, & \text{if } j \leq i < d + j, \\ 0, & \text{elsewhere,} \end{cases}$$

with the obvious interpretation for $d+j > n+1$. Note that all column and row sums are equal to d, as in Example 4.2.2. Thus, Proposition 4.1.1 and Theorem 4.A give the same mean, variance and bound as in that example, that is

$$\mathbb{E}W = d, \quad \operatorname{Var} W = d - \frac{d(d-1)}{n-1}$$

and

$$d_{TV}(\mathcal{L}(W), \mathrm{Po}(d)) \leq (1 - e^{-d})\left(\frac{3d-1}{n} - \frac{d-1}{n(n-1)}\right).$$

For the classical matching or rencontre problem, where $d = 1$, it is easily seen, from the explicit expression for the distribution of W, that the convergence rate to Po(1) is much faster than the bound $2(1-e^{-1})/n$ given above: see Example 3.4.2. However, an explicit form for $\mathcal{L}(W)$ is also known for $d = 2$, from which one can show that the rate of convergence to Po(2) in this case really is of order $1/n$.

In the ménage problem (with $d = 2$) we can also argue as follows. Number the places around the table from 1 to $2n$, and let $I_i = 1$ if a couple occupies seats i and $i + 1$. Then W, the number of couples sitting next to each other, can be expressed as $I_1 + \cdots + I_{2n}$. A coupling may be constructed by interchanging the partner of the person on seat α with the person sitting on seat $\alpha + 1$. This coupling satisfies Theorem 2.C with $\Gamma_\alpha^0 = \emptyset$, $\Gamma_\alpha^+ = \{1, \ldots, \alpha - 2, \alpha + 2, \ldots, 2n\}$ and $\Gamma_\alpha^- = \{\alpha - 1, \alpha + 1\}$, and gives the same bound as that above.

As $\mathbb{E}W = 2$ and $\operatorname{Var} W = 2 - 2/(n-1)$, one could also contemplate using a Bi$(2(n-1), 1/(n-1))$ approximation of $\mathcal{L}(W)$. By explicit calculation, one can prove that $d_{TV}(\mathcal{L}(W), \mathrm{Bi}(2(n-1), 1/(n-1)) \asymp n^{-2}$. Actually, Bi$(2n,1/n)$, also with error of order n^{-2}, gives a somewhat better approximation. Thus, considering the I's as independent Be$(1/n)$ random variables gives good approximation: for example, for $n = 10$, numerical computation gives $d_{TV}(\mathcal{L}(W), \mathrm{Po}(2)) = 0.02654$ and $d_{TV}(\mathcal{L}(W), \mathrm{Bi}(20, 1/10)) = 0.00280$: see Holst (1991).

For results on the exact distributions, for both the rencontre and the ménage problem, see Takács (1981) and Bogart and Doyle (1986). These papers also give many references and contain historical remarks on the two problems.

4.4 Higher order arrays

Results similar in spirit to Theorem 4.A can also be established for arrays with more than two indices, though the computations become heavier. The theorem below gives an example with statistical applications. W is taken to have the form

$$W = \sum_{i<j} a(i,j)b(\sigma_i, \sigma_j),$$

where a and b are two *symmetric* $n \times n$ matrices of zeros and ones. The matrices a and b may be thought of as the incidence matrices of two graphs, and W counts the number of edges common to both when the vertices are randomly associated. For symmetric $n \times n$ matrices a and b we use the notation

$$a_{i.} = a_{.i} = \sum_{j:j\neq i} a_{ij}, \qquad a_{..} = \sum\sum_{i,j:i<j} a_{ij},$$

$$A_1 = \sum_i a_{i.}(a_{i.} - 1), \quad A_2 = a_{..}(a_{..} - 1) - A_1,$$

and similarly for $b_{i.}, \ldots, B_2$, and further

$$\begin{aligned} AB_+ &= \Big(\frac{1}{(n)_4} - \frac{1}{(n)_2^2}\Big) 4A_2B_2 + \Big(\frac{1}{(n)_3} - \frac{1}{(n)_2^2}\Big) A_1B_1 \\ &\leq \frac{16}{(n)_5} A_2B_2 + \frac{1}{(n)_3} A_1B_1, \\ AB_- &= \frac{1}{(n)_2^2}\Big(4A_2b_{..} + 4a_{..}B_2 + 4A_2B_1 + 4A_1B_2 \\ &\qquad + 3A_1B_1 + 4a_{..}B_1 + 4A_1b_{..} + 2a_{..}b_{..}\Big), \end{aligned}$$

where $(n)_k = n(n-1)\ldots(n-k+1)$.

Theorem 4.B *Let a and b be symmetric $n \times n$ matrices with elements zero or one. Set $W = \sum\sum_{i<j} a_{ij} b_{\sigma_i \sigma_j}$. Then*

(a) $\lambda = \mathbb{E}W = a_{..}b_{..}/\binom{n}{2}$,

(b) $\operatorname{Var} W = \lambda - \lambda/(n)_2 + AB_+ - AB_-$
$\qquad = \lambda - \lambda^2 + 4A_2B_2/(n)_4 + A_1B_1/(n)_3,$

(c) $d_{TV}(\mathcal{L}(W), \operatorname{Po}(\lambda)) \leq (1 - e^{-\lambda})(1 - \lambda^{-1}\operatorname{Var} W + 2\lambda^{-1}AB_+)$
$\qquad = (1 - e^{-\lambda})\big(\lambda^{-1}[AB_+ + AB_-] + (n)_2^{-1}\big).$

Proof The random variable W can equivalently be written as

$$W = \sum\sum_{i<j}\sum\sum_{s\neq t} a_{ij} b_{st} I_{is} I_{jt} = \sum_{\alpha\in\Gamma} c_\alpha I_\alpha,$$

where the index set is

$$\Gamma = \Big\{\{(i,s),(j,t)\} : \ 1 \le i \ne j \le n, \ 1 \le s \ne t \le n\Big\},$$

$c_{\{(i,s),(j,t)\}} = a_{ij}b_{st}$ and $I_\alpha \sim \mathrm{Be}\,(1/(n)_2)$. ($\Gamma$ consists of $\frac{1}{2}(n)_2^2$ unordered pairs of ordered pairs.) Hence (a) follows from

$$\lambda = \mathbb{E}W = \sum_{\alpha\in\Gamma} c_\alpha \frac{1}{(n)_2} = \frac{a_{..}b_{..}}{\binom{n}{2}}.$$

For the variance we have

$$\begin{aligned}\mathrm{Var}\,W &= \sum_\alpha c_\alpha \mathrm{Var}\,I_\alpha + \sum\sum_{\alpha\ne\beta} c_\alpha c_\beta \mathrm{Cov}\,(I_\alpha, I_\beta) \\ &= \lambda\Big(1 - \frac{1}{(n)_2}\Big) + \sum\sum_{\alpha\ne\beta} c_\alpha c_\beta \mathrm{Cov}\,(I_\alpha, I_\beta).\end{aligned}$$

In order to compute the double sum, it is convenient to define, for $\alpha = \{(i,s),(j,t)\}$, the sets of indices

$$\begin{aligned}\Gamma_{\alpha 1}^+ &= \Big\{\{(k,u),(l,v)\} : \{(k,u),(l,v)\} \subset (H_{is} \cap H_{jt})\Big\}, \\ \Gamma_{\alpha 2}^+ &= \Big\{\{(i,s),(l,v)\} : (l,v) \in H_{jt}\Big\} \cup \Big\{\{(j,t),(l,v)\} : (l,v) \in H_{is}\Big\}\end{aligned}$$

and

$$\Gamma_\alpha^- = \Gamma \setminus (\Gamma_{\alpha 1}^+ \cup \Gamma_{\alpha 2}^+ \cup \{\alpha\}),$$

where the H_{uv} are as defined in the introduction to this chapter. Note that, for α, β such that $\beta \in \Gamma_{\alpha 1}^+$,

$$\mathrm{Cov}\,(I_\alpha, I_\beta) = \frac{1}{(n)_4} - \frac{1}{(n)_2^2} > 0,$$

and hence that

$$\begin{aligned}&\sum_\alpha \sum_{\beta\in\Gamma_{\alpha 1}^+} c_\alpha c_\beta \mathrm{Cov}\,(I_\alpha, I_\beta) \\ &= \Big(\frac{1}{(n)_4} - \frac{1}{(n)_2^2}\Big) \sum\sum_{i<j,k<l,} \ \sum\sum_{\{i,j\}\cap\{k,l\}=\emptyset} a_{ij}a_{kl} \sum\sum_{s\ne t,u\ne v,} \ \sum\sum_{\{s,t\}\cap\{u,v\}=\emptyset} b_{st}b_{uv} \\ &= \Big(\frac{1}{(n)_4} - \frac{1}{(n)_2^2}\Big) 4A_2B_2,\end{aligned}$$

because, by symmetry,

$$\sum\sum_{i<j,k<l,\ \{i,j\}\cap\{k,l\}=\emptyset}\sum\sum a_{ij}a_{kl} = \frac{1}{4}\sum\sum_{i\neq j,k\neq l,\ \{i,j\}\cap\{k,l\}=\emptyset}\sum\sum a_{ij}a_{kl}$$
$$= \frac{1}{4}\sum\sum_{i\neq j} a_{ij}(2a_{..} - 2a_{i.} - 2a_{.j} + 2a_{ij}) = a_{..}^2 - \sum_i a_{i.}^2 + a_{..} = A_2.$$

For a pair of indices α, β such that $\beta \in \Gamma_{\alpha 2}^+$,

$$\mathrm{Cov}\,(I_\alpha, I_\beta) = \Big(\frac{1}{(n)_3} - \frac{1}{(n)_2^2}\Big) > 0,$$

and summing over such pairs gives

$$\sum_\alpha \sum_{\beta\in\Gamma_{\alpha 2}^+} c_\alpha c_\beta = \sum\sum_{i<j} a_{ij} \sum\sum_{k\neq i,j} a_{ik} \sum\sum_{s\neq t} b_{st} \sum\sum_{u\neq s,t} b_{su} + \ldots$$
$$= \sum\sum_{i\neq j} a_{ij}(a_{i.} - a_{ij}) \sum\sum_{s\neq t} b_{st}(b_{s.} - b_{st}) = A_1 B_1.$$

All pairs of indices with $\beta \in \Gamma_\alpha^-$ have components in common in such a way that $I_\alpha I_\beta \equiv 0$, implying that the indicators are negatively correlated with

$$\mathrm{Cov}\,(I_\alpha, I_\beta) = -(n)_2^{-2}.$$

Furthermore,

$$\sum_\alpha \sum_{\beta\in\Gamma_\alpha^-} c_\alpha c_\beta = \sum_\alpha\sum_\beta c_\alpha c_\beta - \sum_\alpha c_\alpha^2 - \sum_\alpha \sum_{\beta\in\Gamma_{\alpha 1}^+} c_\alpha c_\beta - \sum_\alpha \sum_{\beta\in\Gamma_{\alpha 2}^+} c_\alpha c_\beta$$
$$= (2a_{..}b_{..})^2 - 2a_{..}b_{..} - 4A_2B_2 - A_1B_1.$$

A simple computation then shows that the sum of the covariances for these terms equals $-AB_-$ and one obtains the formula for $\mathrm{Var}\,W$ given in (b).

To prove (c), Theorem 2.C can be applied much as in the proof of Theorem 4.A, by introducing the following coupling. Whenever $I_\alpha = 1$, take $J_{\beta\alpha} \equiv I_\beta$, and otherwise make one or two transpositions in the random permutation $(\sigma_1, \ldots, \sigma_n)$, giving new indicators $J_{\beta\alpha}$. For any pair of indices α, β such that $\beta \in \Gamma_{\alpha 1}^+ \cup \Gamma_{\alpha 2}^+$, this construction gives $J_{\beta\alpha} \geq I_\beta$, and otherwise $J_{\beta\alpha} \equiv 0$. Thus Theorem 2.C can be applied with $\Gamma_\alpha^+ = \Gamma_{\alpha 1}^+ \cup \Gamma_{\alpha 2}^+$, Γ_α^- as above and $\Gamma_\alpha^0 = \emptyset$. The assertion (c) follows from the results used above in the derivation of the variance. □

Remark 4.4.1 The same estimate can also be established using Theorem 2.B and a coupling like that of the alternative proof of Theorem 4.A.

As an application of Theorem 4.B, consider the data of Knox (1964), in which the index i, $1 \leq i \leq n = 96$, indexes cases of childhood leukaemia, and $a_{ij} = 1$ if children i and j lived close to each other, $b_{ij} = 1$ if cases i and j presented at times close to one another. If $w = \sum_{i<j} a_{ij}b_{ij}$ is larger than $W = \sum_{i<j} a(i,j)b(\sigma_i, \sigma_j)$ is expected to be when $(\sigma_1, \ldots, \sigma_n)$ is drawn from the uniform distribution on the permutations of $1, \ldots, n$, there is evidence for an association between time and place of presentation, for instance as a result of infection.

In Knox's data, $a_{..} = 25$ and $b_{..} = 152$. Values of A_1 and B_1 are not given, but values of around 25 and 1000 respectively are not unreasonable. Theorem 4.B then gives the mean as 0.833 and the variance as 0.801, and an upper bound of 0.097 for the total variation distance from the Poisson distribution with the same mean. This compares favourably with the bound 0.71 given by Barbour and Eagleson (1983), who also used simulations to indicate that the true value of the variation distance is around 0.016. The value of w actually observed was 5, which is much larger than would be expected from $\mathrm{Po}(0.833)$.

It is unfortunately impossible to reconstruct the values of A_1 and B_1 from Knox's original data, since they were destroyed in the course of a burglary some years later.

5
Random graphs

In this chapter, we indicate how the general results of Chapter 2 can be used in problems involving random graphs. In Section 5.1 we consider the number of copies of a small graph G contained in $K_{n,p}$. Three cases are distinguished, counting all copies of G, induced copies of G and isolated copies of G, illustrating different ways of applying the basic Stein–Chen method. Counts of the numbers of vertices of specified degrees are considered in Section 5.2. It proves advantageous at times, when counting the vertices of degree m for m large, to introduce couplings which are rather more complicated than those so far employed, necessitating the use of extensions of Stein's method to multivariate distributional approximation along the lines described in Chapter 10. Then we study some graph colouring problems, including a general birthday problem, and questions arising therefrom concerning different versions of dissociation. Finally, we investigate two models where independent random variables are associated with the edges of a large complete graph, and random directed graphs are constructed as functions of these random variables. Related results and references on random graphs can be found in Barbour (1982), Bollobás (1985) and Janson (1986, 1987).

5.1 Subgraphs in $K_{n,p}$

Consider the complete graph K_n with n vertices and $\binom{n}{2}$ edges. Delete each edge with probability $1-p$ independently of the other edges. Thus we get the random graph $K_{n,p}$. Let $G \subset K_n$ be a given graph with $e(G)$ edges and $v(G)$ vertices. In order to avoid trivial complications, we assume that $e(G) > 1$ and that G has no isolated vertices. Consider the set Γ of all copies of G in K_n, i.e. all subgraphs of K_n isomorphic to G. For $\alpha \in \Gamma$ set

$$I_\alpha = I(\alpha \subset K_{n,p}).$$

Thus the number of subgraphs of $K_{n,p}$ isomorphic to G can be written

$$W = \sum_{\alpha \in \Gamma} I_\alpha.$$

For $\beta \in \Gamma$ set

$$J_{\beta\alpha} = I(\beta \subset K_{n,p} \cup \alpha) \geq I_\beta,$$

where the union of two graphs combines their vertex and edge sets. By the independence in $K_{n,p}$, we have

$$\mathcal{L}(J_{\beta\alpha}; \beta \in \Gamma) = \mathcal{L}(I_\beta; \beta \in \Gamma | I_\alpha = 1).$$

This coupling shows that the $(I_\alpha, \alpha \in \Gamma)$ are positively related, and so Corollary 2.C.4 is applicable. Clearly,

$$\pi_\alpha = \mathbb{E} I_\alpha = p^{e(G)}; \quad \lambda = \mathbb{E} W = \binom{n}{v(G)} \frac{(v(G))!}{a(G)} p^{e(G)},$$

where $a(G)$ is the number of elements in the automorphism group of G.

Theorem 5.A *W, the number of copies of a graph G in $K_{n,p}$, satisfies*

$$d_{TV}(\mathcal{L}(W), \mathrm{Po}(\lambda)) \leq (1 - e^{-\lambda})\Big(\frac{\mathrm{Var}\, W}{\lambda} - 1 + 2p^{e(G)}\Big). \qquad \square$$

In order to make use of Theorem 5.A, it is necessary to be able to find estimates of the variance of W. This is the substance of the following lemma.

Lemma 5.1.1

(a) *Let $\Gamma_\alpha^t \subset \Gamma$ be the set of subgraphs of K_n isomorphic to G with exactly t edges not in α. Then, for any $\alpha \in \Gamma$,*

$$\frac{\mathrm{Var}\, W}{\lambda} = 1 - p^{e(G)} + \sum_{t=1}^{e(G)-1} |\Gamma_\alpha^t|(p^t - p^{e(G)}).$$

(b) *If $c(G, H)$ is the number of copies of H in G,*

$$\frac{\mathrm{Var}\, W}{\lambda} - 1 + 2p^{e(G)} \leq \sum_H \frac{a(H)}{a(G)} c(G, H) n^{v(G)-v(H)} p^{e(G)-e(H)}$$

where we sum over all non-empty subgraphs $H \subsetneq G$ without isolated vertices.

Proof For part (a), direct computation yields

$$\begin{aligned}
\mathrm{Var}\, W &= \sum_\alpha \mathrm{Var}\, I_\alpha + \sum_\alpha \sum_{\beta \neq \alpha} \mathrm{Cov}\,(I_\alpha, I_\beta) \\
&= \lambda(1 - p^{e(G)}) + \sum_\alpha \sum_{t=1}^{e(G)} \sum_{\beta \in \Gamma_\alpha^t} \mathrm{Cov}\,(I_\alpha, I_\beta) \\
&= \lambda(1 - p^{e(G)}) + \sum_\alpha \pi_\alpha \sum_{t=1}^{e(G)} \sum_{\beta \in \Gamma_\alpha^t} \{\mathbb{E}(I_\beta | I_\alpha = 1) - p^{e(G)}\} \\
&= \lambda\Big(1 - p^{e(G)} + \sum_{t=1}^{e(G)} \sum_{\beta \in \Gamma_\alpha^t} (p^t - p^{e(G)})\Big),
\end{aligned}$$

as required. For part (b), let N_G and N_H be the number of copies of G and H, respectively, in K_n. Then there are $c(G,H)N_G$ pairs (G', H') with $H' \subset G' \subset K_n$ and $H' \cong H, G' \cong G$. Hence each copy of H is contained in $c(G,H)N_G/N_H$ copies of G, and it follows that, if $1 \leq t \leq e(G) - 1$,

$$|\Gamma_\alpha^t| < \sum_{e(H)=e(G)-t} c(G,H)N_G/N_H \leq \sum_{e(H)=e(G)-t} c(G,H)\frac{a(H)}{a(G)} n^{v(G)-v(H)},$$

where the sums may be restricted to $H \subset G$ without isolated vertices. □

Suppose that $v(H)/e(H) \leq v(G)/e(G)$ for some $H \subsetneq G$. Then, in the estimate of Lemma 5.1.1(b), there is a term of order at least $\lambda^{1-e(H)/e(G)}$ so that, assuming that λ is bounded away from zero, Lemma 5.1.1(b) does not imply a small error in Poisson approximation. This is not surprising, since it is shown in Ruciński and Vince (1985) that the distribution of W is then not close to Poisson. Thus, when deriving simpler estimates for the order of Poisson approximation, it is appropriate to restrict attention to graphs G for which no such H exists, which are said to be strictly balanced.

We introduce the following notation. Let

$$d(G) = \frac{e(G)}{v(G)}; \quad \alpha(G) = \min_H \frac{e(G) - e(H)}{v(G) - v(H)}, \tag{1.1}$$

$$\gamma(G) = \min_H\left(\frac{e(G)}{v(G)}v(H) - e(H)\right) = \min_H v(H) \cdot (d(G) - d(H)), \tag{1.2}$$

where the minima are taken over all subgraphs $H \subsetneq G$ with $e(H) > 0$ (it may also be assumed that H has no isolated vertices). The quantity $d(G)$ (or $2d(G)$, the usage varying from author to author) is called the degree of G; note that $2d(G)$ is the average of the degrees of the vertices of G. The graph G is then said to be *strictly balanced* if $d(H) < d(G)$ for all subgraphs $H \subsetneq G$, or, equivalently, if $\gamma(G) > 0$ or $\alpha(G) > d(G)$. The following theorem, partly given in Bollobás (1985) Chapter IV, provides simple bounds for the accuracy of Poisson approximation in terms of the quantities d, α and γ.

Theorem 5.B *Let G be a fixed strictly balanced graph. Then, with notation as above,*

(a) *as $n \to \infty$,*

$$\lambda = \mathbb{E}W \sim \frac{(np^{d(G)})^{v(G)}}{a(G)},$$

(b) *for some C depending on G alone, we have the estimate*

$$d_{TV}(\mathcal{L}(W), \mathrm{Po}(\lambda)) \leq C \begin{cases} \min\{\lambda^{1-2/v(G)}p^{\gamma(G)}, \lambda^{1-1/e(G)}n^{-\gamma(G)/d(G)}, np^{\alpha(G)} + p\} & \text{if } \lambda \geq 1, \\ \min\{\lambda p^{\gamma(G)}, \lambda n^{-\gamma(G)/d(G)}, np^{\alpha(G)} + p\} & \text{if } \lambda < 1. \end{cases}$$

Proof Part (a) is immediate from the formula for $\mathbb{E}W$. For part (b), Theorem 5.A and Lemma 5.1.1 yield

$$d_{TV}(\mathcal{L}(W), \mathrm{Po}(\lambda)) \le C(1 \wedge \lambda) \sum_H n^{v(G)-v(H)} p^{e(G)-e(H)}. \tag{1.3}$$

Hence, with different constants C,

$$\begin{aligned} d_{TV}(\mathcal{L}(W), \mathrm{Po}(\lambda)) &\le C(1 \wedge \lambda) \sum_H \lambda^{1-v(H)/v(G)} p^{e(G)v(H)/v(G)-e(H)} \\ &\le \begin{cases} C\lambda^{1-2/v(G)} p^{\gamma(G)} & \text{if } \lambda \ge 1; \\ C\lambda p^{\gamma(G)} & \text{if } \lambda < 1, \end{cases} \end{aligned}$$

giving the first part of the estimate, and the second follows similarly. The third is trivial when $np^{\alpha(G)} > 1$, whereas, if $np^{\alpha(G)} \le 1$, we obtain the inequality

$$\begin{aligned} & d_{TV}(\mathcal{L}(W), \mathrm{Po}(\lambda)) \\ & \le C\Big(\sum_{H:v(H)<v(G)} (np^{\alpha(G)})^{v(G)-v(H)} + \sum_{H:v(H)=v(G)} p^{e(G)-e(H)} \Big) \\ & \le C(np^{\alpha(G)} + p), \end{aligned}$$

from (1.3) and (1.1). □

Remark 5.1.2 For the rates obtained in Theorem 5.B, we observe that $v(G)\gamma(G)$ is an integer, and thus

$$\gamma(G) \ge 1/v(G); \quad \gamma(G)/d(G) \ge 1/e(G).$$

Example 5.1.3 Let G be a cycle with v vertices (and edges). Then

$$\lambda = \mathbb{E}W = \tfrac{1}{2}\binom{n}{v}(v-1)!p^v,$$

and $d(G) = 1$, $\alpha(G) = \frac{v-1}{v-2}$, $\gamma(G) = 1$. Thus it follows from Theorem 5.B(b) that

$$d_{TV}(\mathcal{L}(W), \mathrm{Po}(\lambda)) \le Cn^{v-2}p^{v-1}.$$

In particular, if $n \to \infty, p \to 0$ in such a way that $np = c$, $0 < c < \infty$, then $\lambda \to \lambda_\infty = \frac{1}{2v}c^v$ and $W \xrightarrow{\mathcal{D}} \mathrm{Po}(\lambda_\infty)$, at rate $O(p) = O(n^{-1})$.

Counting the number of copies of G in $K_{n,p}$ is very similar to counting the number of *induced* copies of G. α is said to index an induced copy of G in $K_{n,p}$ if it is a copy of G in K_n and if the edges of $K_{n,p}$ with both endpoints in the vertex set $V(\alpha)$ of α coincide with the edges of α: when this is so, we now set $I_\alpha = 1$. There are, however, important differences. First, for induced copies, both p small and p near 1 give non-trivial possibilities of Poisson approximation, but, since the case $p > 1/2$ can now be reduced to the case $p < 1/2$ by complementation, it is enough to consider small p. More importantly for us, the I_α are now no longer positively related. However, the indicator I_α is independent of the set of indicators $(I_\beta, |V(\alpha) \cap V(\beta)| \leq 1)$, since they are generated by disjoint sets of edges, making possible an application of Corollary 2.C.5. This leads to the following estimate.

Theorem 5.C *Let W denote the number of induced copies of G in $K_{n,p}$, set $\lambda = \mathbb{E}W$ and assume that $p \leq 1/2$. Then*

$$d_{TV}(\mathcal{L}(W), \mathrm{Po}(\lambda)) \leq (1 - e^{-\lambda})\Big(p^{e(G)} \binom{v(G)}{2} \frac{v(G)!}{a(G)} n^{v(G)-2} + \sum_{t=1}^{e(G)-1} |\Gamma_\alpha^t| p^t\Big). \quad (1.4)$$

Proof By direct application of Corollary 2.C.5, it follows that

$$d_{TV}(\mathcal{L}(W), \mathrm{Po}(\lambda)) \leq \lambda^{-1}(1 - e^{-\lambda})\Big(\sum_{\alpha\in\Gamma} \pi_\alpha^2 + \sum_{\alpha\in\Gamma} \sum_{\beta\in\Gamma_\alpha^0} (\mathbb{E}(I_\alpha I_\beta) + \pi_\alpha\pi_\beta)\Big),$$

where $\pi_\alpha \leq p^{e(G)}$ and $\Gamma_\alpha^0 = \{\beta \neq \alpha : |V(\alpha) \cap V(\beta)| \geq 2\}$. The factor $\binom{v(G)}{2} \frac{v(G)!}{a(G)} n^{v(g)-2}$ is an upper estimate for $|\Gamma_\alpha^0| + 1$, $\pi_\alpha^{-1}\mathbb{E}(I_\alpha I_\beta) \leq p^t$ if α and β have exactly t edges not in common, and $\mathbb{E}(I_\alpha I_\beta) = 0$ if β has edges with both endpoints in $V(\alpha)$ which are not edges of α. □

Note that, in view of Lemma 5.1.1(a), the estimate differs from that of Theorem 5.A only in the contribution from the $p^{e(G)}$ terms. Since (1.3) already contains a term of larger order, arising from the choice of a single edge for H in the sum, the conclusions of Theorem 5.B(b) also hold for W the number of induced copies of G.

Next, consider the number of isolated copies of G in $K_{n,p}$: that is, the random variable $W = \sum_{\alpha\in\Gamma} I_\alpha$, where

$$I_\alpha = I(\alpha \text{ is an isolated copy of } G \text{ in } K_{n,p}).$$

Introduce the edge indicators

$$\xi_e = I(e \in K_{n,p}) \text{ for } e \in K_n,$$

which are independent Be (p) random variables. Set

$$E_{1\alpha} = \{e; e \text{ joins two vertices in } \alpha\}, E_{3\alpha} = \{e; e \text{ has no endpoint in } \alpha\},$$
$$E_{2\alpha} = \{e; e \text{ has exactly one endpoint in } \alpha\}, G_\alpha = \{\beta \in \Gamma; \beta \cap \alpha \neq \emptyset\},$$

noting that $E_{1\alpha} \cup E_{2\alpha} \cup E_{3\alpha}$ is a partition of the edges of K_n. Obviously I_α is a function only of $(\xi_e, e \in E_{1\alpha} \cup E_{2\alpha})$ and decreasing as a function of $(\xi_e, e \in E_{2\alpha})$. For $\beta \notin G_\alpha$, the indicator I_β is a function only of $(\xi_e, e \in E_{2\alpha} \cup E_{3\alpha})$ and decreasing in $(\xi_e, e \in E_{2\alpha})$. Hence Corollary 2.F.1 can be applied. It is easily seen that

$$\lambda = \mathbb{E}W = |\Gamma|\pi_\alpha = \binom{n}{v(G)} \frac{v(G)!}{a(G)} \pi_\alpha$$

with

$$\pi_\alpha = \mathbb{E}I_\alpha = p^{e(G)}(1-p)^{(n-v(G))v(G)+v(G)(v(G)-1)/2-e(G)},$$

leading to the next result.

Theorem 5.D *Let W be the number of isolated copies of G in $K_{n,p}$. Then*

$$d_{TV}(\mathcal{L}(W), \text{Po}(\lambda)) \leq (1-e^{-\lambda})\Big(\frac{\text{Var}\, W}{\lambda} - 1 + 2\pi_\alpha|G_\alpha|\Big). \qquad \square$$

The rate estimates of this section have been framed in the context of keeping a fixed graph G while $K_{n,p}$ changes as $n \to \infty$. There is, however, no reason why Theorem 5.A and Lemma 5.1.1 should not be combined to give rate estimates when G is allowed to grow with n. We content ourselves with just one example: more general results can be found in Karoński (1984), Theorems 2.9 and 2.10.

Example 5.1.4 *The number of large complete subgraphs.* For $v \geq 3$, we approximate the distribution of the number W of copies of K_v in $K_{n,p}$ by Po(λ), with $\lambda = \mathbb{E}W = \binom{n}{v}p^{\binom{v}{2}}$, using Theorem 5.A and Lemma 5.1.1. We assume throughout that $p \geq n^{-2/(v-1)}$, since otherwise $\lambda < 1/v!$, which is typically too small to be of interest. If, in addition, $v^3 = O(n^{1-3\varepsilon})$ for some $\varepsilon > 0$, it then follows that

$$d_{TV}(\mathcal{L}(W), \text{Po}(\lambda)) = O\Big(v^{3/2}\Big\{\frac{ne}{v}p^{\frac{v+1}{2}}\Big\}^{v-2} + nvp^{v-1}\Big). \qquad (1.5)$$

In order to prove this, it is enough to consider the range $\frac{ne}{v}p^{\frac{v+1}{2}} \leq 1$, since otherwise the assertion is vacuous. From Theorem 5.A and Lemma 5.1.1,

$$\begin{aligned} d_{TV}(\mathcal{L}(W), \text{Po}(\lambda)) &\leq \Big(\frac{\text{Var}\, W}{\lambda} - 1 + 2p^{\binom{v}{2}}\Big) \\ &\leq p^{\binom{v}{2}} + \sum_{s=2}^{v-1}\binom{v}{s}\binom{n}{v-s}p^{\binom{v}{2}-\binom{s}{2}}, \end{aligned} \qquad (1.6)$$

and only the sum needs examination.

Consider first the range $2 \le s \le [v/2]$, in which the sum is less than

$$\sum_{s=2}^{[v/2]} \frac{v^s n^{v-s}}{s!(v-s)!} p^{\binom{v}{2}-\binom{s}{2}} \le \sum_{s=2}^{[v/2]} \frac{v^{2s} n^{v-s}}{s!v!} p^{\binom{v}{2}-\binom{s}{2}} = \sum_{s=2}^{[v/2]} a_s,$$

say. Direct calculation then shows that

$$\begin{aligned} \frac{a_s}{a_2} &= \frac{2}{s!} v^{2(s-2)} n^{2-s} p^{1-\binom{s}{2}} \\ &\le \frac{e^2}{s^2}\left(\frac{ev^2}{ns} p^{-\frac{s+1}{2}}\right)^{s-2} \le \frac{e^2}{s^2}\left(\frac{ev^2}{ns} n^{\frac{s+1}{v-1}}\right)^{s-2}, \end{aligned} \tag{1.7}$$

using $p \ge n^{-2/(v-1)}$. If $3 \le s \le \frac{1}{3}(v-1)-1$, then

$$\frac{ev^2}{ns} n^{\frac{s+1}{v-1}} \le v^2 n^{-2/3} \le 1 \qquad \text{for all} \quad n \ge n_0,$$

for some n_0, because $v^3 = O(n^{1-3\varepsilon})$. If $\frac{1}{3}(v-1)-1 < s \le [v/2]$ and $s \ge 3$, then necessarily $v \ge 6$, implying $\frac{v}{9} < s \le \frac{2}{3}(v-1)$ and hence

$$\frac{ev^2}{ns} n^{\frac{s+1}{v-1}} \le \frac{9ev}{n} \cdot n^{2/3} \le 1 \qquad \text{for all} \quad n \ge n_1,$$

say. Thus $a_s/a_2 \le e^2/s^2$ for all $n \ge n_0 \vee n_1$ and $s \le [v/2]$, which yields

$$\sum_{s=2}^{[v/2]} a_s \le \frac{\pi^2 e^2 v^4}{12 v!} n^{v-2} p^{\binom{v}{2}-1} \le \frac{\pi^2 e^4}{12\sqrt{2\pi}} v^{3/2} \left(\frac{ne}{v} p^{\frac{v+1}{2}}\right)^{v-2}, \tag{1.8}$$

in $n \ge n_0 \vee n_1$.

For the remaining range $[v/2] < s \le v-1$, write $k = v - s$, and estimate the corresponding sum as

$$\sum_{k=1}^{[\frac{v-1}{2}]} \binom{v}{k}\binom{n}{k} p^{\binom{v}{2}-\binom{v-k}{2}} \le \sum_{k=1}^{[\frac{v-1}{2}]} \frac{v^k n^k}{(k!)^2} p^{\binom{v}{2}-\binom{v-k}{2}} = \sum_{k=1}^{[\frac{v-1}{2}]} b_k,$$

say. Then it follows that

$$\frac{b_k}{b_1} = \frac{(vn)^{k-1}}{(k!)^2} p^{\binom{v-1}{2}-\binom{v-k}{2}} \le \frac{e^2}{2\pi k^3}\left(\frac{vne^2}{k^2} p^{v-\frac{k}{2}-1}\right)^{k-1}.$$

Take $\delta > 0$ such that $(3-2\delta)(\frac{1}{3}-\varepsilon) < 1-2\delta$. In the range $\frac{1}{2}k+2 \le \delta(v+1)$,

$$p^{v-\frac{k}{2}-1} \le p^{(v+1)(1-\delta)} \le \left(\frac{v}{ne}\right)^{2(1-\delta)},$$

since we have assumed that $\frac{ne}{v}p^{\frac{v+1}{2}} \leq 1$, and hence, for such k,

$$\frac{vne^2}{k^2}p^{v-\frac{k}{2}-1} \leq e^{2\delta}v^{3-2\delta}n^{-(1-2\delta)} \leq 1$$

for all $n \geq n_2$, say. In the remaining range of k, the inequality $k \leq [\frac{v-1}{2}]$ implies that $\frac{1}{2}k+2 \leq \frac{15}{32}(v+1)$ for all $v \geq 6$, and $\frac{1}{2}k+2 > \delta(v+1)$ implies that $k > \frac{1}{3}\delta(v+1)$. Hence, if $\frac{1}{2}k+2 > \delta(v+1)$ and $k \leq [\frac{v-1}{2}]$,

$$\begin{aligned}\frac{vne^2}{k^2}p^{v-\frac{k}{2}-1} &\leq \frac{9e^2n}{\delta^2 v}p^{\frac{17(v+1)}{32}} \\ &\leq \frac{9e^2n}{\delta^2 v}\cdot\left(\frac{v}{en}\right)^{17/16} \leq 1 \qquad \text{for all} \quad n \geq n_3,\end{aligned}$$

say, provided that $v \geq 6$, because we have taken $\frac{ne}{v}p^{\frac{v+1}{2}} \leq 1$. Hence, in $v \geq 6$ and for all $n \geq n_2 \vee n_3$, we have

$$\sum_{k=1}^{[\frac{v-1}{2}]} b_k \leq \frac{e^2}{2\pi}b_1 \sum_{k=1}^{[\frac{v-1}{2}]} k^{-3} \leq 2nvp^{v-1}. \tag{1.9}$$

This, together with the inequality (1.8), establishes (1.5) in $v \geq 6$ for all $n \geq \max(n_0, n_1, n_2, n_3)$. For smaller n, the order expression in (1.5) is bounded away from zero, because then v is bounded and $p \geq n^{-2/(v-1)}$, so that (1.5) therefore holds for all n.

If $3 \leq v \leq 5$, use (1.7) for $2 \leq s \leq v-2$, giving the estimate

$$b_1 + \sum_{s=2}^{v-2}\frac{a_2 e^2}{s^2}\left(\frac{ev^2}{s}\right)^{s-2},$$

which once again is enough for (1.5).

Note that if $v^{3/2}\left\{\frac{ne}{v}p^{\frac{v+1}{2}}\right\}^{v-2} \to 0$ and $v^3 = O(n^{1-3\varepsilon})$, then it follows that $nvp^{v-1} \to 0$ also, since $\theta = \frac{ne}{v}p^{\frac{v+1}{2}}$ implies that

$$nvp^{v-1} = nv\left(\frac{\theta v}{en}\right)^{2-\frac{4}{v+1}} \leq v^3 n^{-(1-\frac{4}{v+1})}\left(\frac{\theta}{e}\right)^{2-\frac{4}{v+1}};$$

if v remains bounded, $\theta \to 0$ and hence also nvp^{v-1}, whereas, if $v \to \infty$, $\theta < 1$ and $\frac{4}{v+1} < 2\varepsilon$ for all n sufficiently large, again entailing $nvp^{v-1} \to 0$. Thus, for $p \geq n^{-2/(v-1)}$, Poisson convergence takes place provided only that $v^3 = O(n^{1-3\varepsilon})$ and that $v^{3/2}\left\{\frac{ne}{v}p^{\frac{v+1}{2}}\right\}^{v-2} \to 0$. Sufficient conditions for this are that v remains bounded and $p = o(n^{-2/(v+1)})$, or that $1 \ll v^3 \ll n^{1-3\varepsilon}$ and $p \leq n^{-2/(v+1)}$. In the case of bounded v, the condition for Poisson convergence is the same as requiring $np^{\alpha(K_v)} \to 0$, which is the condition that follows from Theorem 5.B(b).

5.2 Vertex degrees

Another group of quantities to which the method can easily be applied is that of counts of vertices whose degree lies in a given range. The three counts we consider here are the numbers of vertices i whose degree $D(i)$ in $K_{n,p}$ is greater than, equal to or less than a prescribed value m. The property of having a prescribed vertex degree is a 'semi-induced' property, as defined in Karoński and Ruciński (1987): they consider vertex degrees in their Section 3.1. It is immediate that the indicators $(I_{i1}, 1 \leq i \leq n)$ defined by $I_{i1} = I[D(i) \geq m]$ are increasing functions of the independent edge indicators $\{E_{ij}, 1 \leq i < j \leq n\}$, whereas the $I_{i2} = I[D(i) \leq m]$ are decreasing functions of the E_{ij}, so that the collections $(I_{i1}, 1 \leq i \leq n)$ and $(I_{i2}, 1 \leq i \leq n)$ are both positively associated. Thus they are positively related, by Theorem 2.G, and hence the conditions of Corollary 2.C.4 are satisfied for $W_1 = \sum_{i=1}^n I_{i1}$ and for $W_2 = \sum_{i=1}^n I_{i2}$. As a result, determining the accuracy of Poisson approximation for either of these quantities involves only the computation of their mean and variance. This is relatively simply accomplished.

We introduce the notation

$$\pi_1 = \pi_1(m) = \mathbb{P}[\mathrm{Bi}\,(n-1,p) \geq m]; \quad \pi_2 = \pi_2(m) = \mathbb{P}[\mathrm{Bi}\,(n-1,p) \leq m];$$
$$\pi_3 = \pi_3(m) = \mathbb{P}[\mathrm{Bi}\,(n-1,p) = m], \tag{2.1}$$

and set $\lambda_j = n\pi_j$, $1 \leq j \leq 3$. We then have the following theorem.

Theorem 5.E *Let W_1 count the number of vertices in $K_{n,p}$ with degree greater than or equal to m, W_2 the number with degree less than or equal to m. Then*

$$\delta_1 = d_{TV}(\mathcal{L}(W_1), \mathrm{Po}\,(\lambda_1)) \leq \pi_1 + \frac{m^2(1-p)\pi_3^2}{(n-1)p\pi_1}$$

and

$$\delta_2 = d_{TV}(\mathcal{L}(W_2), \mathrm{Po}\,(\lambda_2)) \leq \pi_2 + \frac{(n-1-m)^2 p\pi_3^2}{(n-1)(1-p)\pi_2}.$$

Proof In each case, the quantity $2\pi_j + \{\mathrm{Var}\, W_j/\mathbb{E}W_j - 1\}$ gives an upper bound for the accuracy of approximation in total variation, by Corollary 2.C.4. The stated inequalities come from evaluating this expression in the two cases. For example, for δ_1, note that, if $i \neq j$,

$$\mathbb{E}(I_{i1}I_{j1}) = p\mathbb{P}[\min(B_1, B_2) \geq m-1] + (1-p)\mathbb{P}[\min(B_1, B_2) \geq m],$$

where B_1 and B_2 are independent $\mathrm{Bi}\,(n-2,p)$ random variables. Thus, writing p_{m-1} for $\mathrm{Bi}\,(n-2,p)\{m-1\}$ and q_m for $\mathrm{Bi}\,(n-2,p)\{[m,\infty)\}$, it follows that

$$\begin{aligned}\mathbb{E}(I_{i1}I_{j1}) - \pi_1^2 &= p[p_{m-1} + q_m]^2 + (1-p)q_m^2 - \pi_1^2 \\ &= q_m^2 - \pi_1^2 + p[2p_{m-1}q_m + p_{m-1}^2]\end{aligned}$$

whenever $i \neq j$. However, since $\pi_1 = q_m + pp_{m-1}$, this in turn implies that $\mathrm{Cov}\,(I_{i1}, I_{j1}) = p(1-p)p_{m-1}^2$ for $i \neq j$, and the estimate of δ_1 now follows easily. □

Corollary 5.E.1 *Let $p = p(n) \leq 1/2$ and $m = m(n)$, where $n \to \infty$.*

(i) *If $m \geq 2$ and $np \to 0$, then $\delta_1 \to 0$.*

(ii) *If np is bounded away from 0 and $(np)^{-1/2}(m - np) \to \infty$, then $\delta_1 \to 0$.*

(iii) *If $(np)^{-1/2}(np - m) \to \infty$, then $\delta_2 \to 0$.*

Remark 5.2.1 If $p > 1/2$, the complementary graph with edge probability $1 - p$ can be used to provide the corresponding results.

Remark 5.2.2 If $m = 1$ and $np \to 0$, Poisson approximation cannot be good, since the number of vertices of degree one is essentially twice the (almost Poisson distributed) number of edges and $\mathrm{Var}\, W_1 \sim 2\mathbb{E}W_1$. In the other cases excluded, where π_1 or π_2 is not small, the mean and variance are also typically no longer asymptotically equivalent, and normal or binomial approximation is more appropriate.

Proof If $np \to 0$, we have $\pi_1 \sim \pi_3 \leq (np)^m/m!$, giving

$$\delta_1 = O((np)^{m-1}) = o(1)$$

if $m \geq 2$. Otherwise, if $\Delta = (np)^{-1/2}(m - np) \to \infty$, use Propositions A.2.2 and A.2.5 to give the inequalities

$$m\pi_3 \leq 6\pi_1 \Delta (np)^{1/2}$$

and

$$\begin{aligned} \pi_1 &\leq \frac{(m+1)(1-p)}{(m - np + 1 - p)} \pi_3 \leq \frac{m+1}{m - np} \mathbb{P}[\mathrm{Bi}\,(n,p) = m] \\ &\leq \frac{1}{\sqrt{2\pi}} \left(\frac{m+1}{m-np} \right) \sqrt{\frac{n}{m(n-m)}} \exp\left\{ -\frac{(m-np)^2}{2(m+np)} \right\}. \end{aligned}$$

For $m \leq 3n/4$, defining $y = m^{-1/2}(m - np) \to \infty$, this shows that $\pi_1 = O(y^{-1}e^{-y^2/4})$, and hence that

$$\delta_1 = O\{\pi_1(1 + (np)^{-1}(m\pi_3/\pi_1)^2)\} = O\{y^{-1}e^{-y^2/4}(1 + \Delta^2)\} = o(1),$$

since, if np is bounded away from zero and $y \to \infty$, $\Delta = O(y^2)$. If instead $m > 3n/4$, it follows immediately that

$$\frac{(m-np)^2}{2(m+np)} \geq \frac{n}{48},$$

implying easily that $\delta_1 \to 0$. The argument for δ_2 is similar. □

The third quantity to be analysed is the number $W_3 = \sum_{i=1}^n I_{i3}$ of vertices of degree exactly m. Here, it is not possible to use Corollary 2.C.4, since the indicators $I_{i3} = I[D(i) = m]$, $1 \le i \le n$, are neither positively nor negatively related. Instead, we use Theorem 1.B with an explicit coupling. Let

$$R_1 = \left[\frac{(n-1-m)}{(n-1)(1-p)} + \frac{m}{(n-1)p}\right]\mathbb{E}(m - D(1))^+;$$

$$R_2 = \frac{(n-1-m)}{(n-1)(1-p)}\left[1 + \frac{(n-m-2)p}{(m+1)(1-p)}\right]\mathbb{E}(D(1) - m)^+,$$

and suppose henceforth that $p \le 1/2$.

Theorem 5.F *If W_3 denotes the number of vertices of degree m in $K_{n,p}$, then*

$$\delta_3 = d_{TV}(\mathcal{L}(W_3), \mathrm{Po}(\lambda_3)) \le \pi_3(1 + R_1 + R_2),$$

where π_3 is as defined in (2.1). In particular, $\delta_3 \to 0$ if either
(i) *$np \to 0$ and $m \ge 2$;*
(ii) *np is bounded away from 0 and $(np)^{-1/2}|m - np| \to \infty$.*

Proof Since permuting the labels of vertices leaves the distribution of $K_{n,p}$ unchanged, it suffices to compute $\mathbb{E}|U - V|$, where $\mathcal{L}(U) = \mathcal{L}(W_3)$ and $\mathcal{L}(V + 1) = \mathcal{L}(W_3 | I_{13} = 1)$. Let $G = \{E_{ij}, 1 \le i < j \le n\}$ be sampled as usual, and take $U = W_3$. To determine V, construct a new graph $G' = \{E'_{ij}, 1 \le i < j \le n\}$, where $E'_{ij} = E_{ij}$ if $i \ge 2$. For $i = 1$ there are three possibilities:
(i) $D(1) = m$, in which case $E'_{1j} = E_{1j}$, $2 \le j \le n$;
(ii) $D(1) < m$, in which case $m - D(1)$ of those vertices j for which $E_{1j} = 0$ are chosen at random, and E'_{1j} set equal to 1, leaving $E'_{1j} = E_{1j}$ for all other j;
(iii) $D(1) > m$, in which case $D(1) - m$ of those vertices j for which $E_{1j} = 1$ are chosen at random, and E'_{1j} set equal to 0, leaving $E'_{1j} = E_{1j}$ for all other j.

V is then set equal to $W'_3 - 1$, where W'_3 is the value of W_3 obtained from G'.

Let $D_{jr} = I[D^*(j) = r]$, $j \ge 2$, where $D^*(j) = D(j) - E_{1j}$ is independent of $\{E_{1k}, E'_{1k}; 2 \le k \le n\}$, and let

$$F_j = I[E'_{1j} > E_{1j}]; \qquad G_j = I[E'_{1j} < E_{1j}].$$

Then, clearly,

$$|U - V| \le I[D(1) = m] + \sum_{j \ge 2}\{F_j(D_{j,m-1} + D_{jm}) + G_j(D_{jm} + D_{j,m+1})\}.$$

Now

$$\mathbb{E}D_{jm} = \frac{(n-m-1)p}{m(1-p)}\mathbb{E}D_{j,m-1} = \frac{(m+1)(1-p)}{(n-m-2)p}\mathbb{E}D_{j,m+1}$$
$$= \binom{n-2}{m}p^m(1-p)^{n-2-m} = \frac{(n-1-m)\pi_3}{(n-1)(1-p)},$$

for each $j \geq 2$, and

$$\sum_{j\geq 2}\mathbb{E}F_j = \mathbb{E}(m - D(1))^+; \qquad \sum_{j\geq 2}\mathbb{E}G_j = \mathbb{E}(D(1) - m)^+.$$

Thus it follows that $\mathbb{E}|U - V| \leq \pi_3(1 + R_1 + R_2)$, as required.

For the asymptotic results, take first the case $np \to 0$. In this case, $\pi_3 \leq (np)^m/m!$, $\mathbb{E}(m - D(1))^+ \sim m$ and $\mathbb{E}(D(1) - m)^+ = o(1)$. Thus

$$\delta_3 \leq \pi_3(1 + R_1 + R_2) \sim \pi_3 m^2/np \to 0$$

if $m \geq 2$.

Next, note that, if $m \geq (n-1)p$,

$$\mathbb{E}(D(1) - m)^+ \leq \mathbb{E}(D(1) - (n-1)p)^+ \leq \mathbb{E}|D(1) - \mathbb{E}D(1)| \leq \sqrt{np}.$$

Hence, under these circumstances, $\mathbb{E}(m - D(1))^+ \leq m - (n-1)p + (np)^{1/2}$. Thus, if np is bounded away from 0 and $(np)^{-1/2}(m - np) \to \infty$, it follows that $R_1 = O\{m(m-np)/np\}$ and $R_2 = o(R_1)$. Hence

$$\delta_3 = O\{m\pi_3(m - np)/np\} \to 0,$$

from Proposition A.2.5(i).

Finally, if $(np)^{-1/2}(np - m) \to \infty$, $R_2 = O\{np(np-m)/m\}$ and $R_1 = o(R_2)$ by a similar argument, and hence, if $m \geq 1$,

$$\delta_3 = O\{np\pi_3(np - m)/m\} \to 0$$

by Proposition A.2.5(i). The case $m = 0$ is already covered by Corollary 5.E.1 above. □

The asymptotic conclusions of Theorem 5.F reveal a weakness in the estimates obtained, in that they do not imply accurate Poisson approximation in all cases where π_3 is small, as, for instance, when $np \to \infty$ and $|m - np| = O(\sqrt{np})$, although Poisson approximation should presumably be good here too. The problem arises in this case because the natural coupling of U with V achieved by modifying G to G' typically yields a difference of order 1 between U and V, so that $\mathbb{E}|U - V| \asymp 1$ also: the coupling is not tight enough, and no advantage is taken of the possibility that the differences $U - V$ might be positive or negative with roughly equal probability. To cope with such examples, a different argument is required.

One approach is to construct a coupling of the distributions $\mathcal{L}(W_3+1)$ and $\mathcal{L}(W_3|I_{13} = 1)$ conditional on the numbers L_1 and L_2 of vertices in the graph $G_1 = \{E_{ij}, 2 \le i < j \le n\}$ of degree $m - 1$ and m respectively. The fundamental equation (2.0.3) can then be exploited in the form

$$\begin{aligned}\mathbb{P}[W \in A] - \mathrm{Po}(\lambda)\{A\} &= \sum_{\alpha\in\Gamma} \pi_\alpha\{\mathbb{E}g(W+1) - \mathbb{E}[g(W)|I_\alpha = 1]\} \\ &= \mathbb{E}\{\mathbb{E}[f(W_3+1)|L_1, L_2] - \mathbb{E}[f(W_3)|L_1, L_2, I_{13} = 1]\},\end{aligned}$$

where $f = \lambda g$ satisfies

$$\Delta f = \sup_{j\ge 0} |f(j+1) - f(j)| \le 1.$$

In particular, it follows that

$$d_{TV}(\mathcal{L}(W_3), \mathrm{Po}(\lambda_3)) \le \mathbb{E}\delta(L_1, L_2), \tag{2.2}$$

where

$$\begin{aligned}&\delta(l_1, l_2) \\ &\quad = d_W\Big(\mathcal{L}(W_3|L_1 = l_1, L_2 = l_2), \mathcal{L}(W_3 - 1|L_1 = l_1, L_2 = l_2, I_{13} = 1)\Big)\end{aligned} \tag{2.3}$$

and d_W denotes the Wasserstein distance

$$d_W(\mathcal{L}(X), \mathcal{L}(Y)) = \sup_{f:\Delta f\le 1} |\mathbb{E}f(X) - \mathbb{E}f(Y)|.$$

Thus the problem is reduced to one of computing differences of expectations for a new pair of random variables, in which once again Stein's method can be used. Note that it actually suffices to restrict attention to those test functions f for which, in addition, $\|f\| \le c(1 \vee \lambda^{1/2})$, though we shall not use the fact. We indicate briefly how the argument proceeds, as an illustration of the range of application of Stein's method, without going into detail.

Given the values $L_1 = l_1$ and $L_2 = l_2$, the number of vertices other than vertex 1 which have degree m in $K_{n,p}$ is $Z_1 + l_2 - Z_2$, where Z_1 and Z_2 are independent, and $\mathcal{L}(Z_i) = \mathrm{Bi}(l_i, p)$. Hence it follows that

$$d_W\Big(\mathcal{L}(W_3|L_1 = l_1, L_2 = l_2), \mathcal{L}(l_2 - Z_2 + Z_1)\Big) \le \pi_3. \tag{2.4}$$

On the other hand,

$$\mathcal{L}(W_3 - 1|L_1 = l_1, L_2 = l_2, I_{13} = 1) = \mathcal{L}(l_2 - Y_2 + Y_1),$$

where (Y_1, Y_2) has a bivariate hypergeometric distribution: m indices are chosen uniformly without replacement from the set $\{1, 2, \ldots, n-1\}$, Y_1 denotes the number less than or equal to l_1 and $Y_1 + Y_2$ the number less than or equal to $l_1 + l_2$. Thus it is enough to estimate

$$\sup_{f:\Delta f \leq 1} |\mathbb{E}f(Y_1 - Y_2) - \mathbb{E}f(Z_1 - Z_2)| \tag{2.5}$$

for each possible choice of l_1 and l_2.

Stein's method for multivariate (Poisson) approximation forms the subject of Chapter 10, and it would now be possible to borrow techniques from there to show that the distributions of (Y_1, Y_2) and (Z_1, Z_2) are both close to the product distribution $\mathrm{Po}(l_1 p) \times \mathrm{Po}(l_2 p)$. However, even for the approximation of $\mathcal{L}((Z_1, Z_2))$, an error of order at least p would be incurred when $(l_1 + l_2)p \geq 1$, whereas it should not be necessary to assume p to be small. It is therefore better to combine the multivariate method with the binomial approximation of Section 9.2, and to compare the distributions of (Y_1, Y_2) and (Z_1, Z_2) directly.

Accordingly, for any function $h : \mathbb{Z}^+ \times \mathbb{Z}^+ \to \mathbb{R}$, define the Stein operator $\mathcal{A}$ by

$$(\mathcal{A}h)(r) = \sum_{j=1}^{2} [r_j(1-p)\{h(r - e_j) - h(r)\} + (l_j - r_j)p\{h(r + e_j) - h(r)\}],$$

where $r = (r_1, r_2)$ and e_1, e_2 denote the coordinate unit vectors. Then $\mathbb{E}\{(\mathcal{A}h)((Z_1, Z_2))\} = 0$ for all h, and, if f satisfies $\Delta f \leq 1$, there exists an h such that $(\mathcal{A}h)(r) = f(r) - \mathbb{E}f(Z_1, Z_2)$ for which the inequalities

$$\begin{aligned} \Delta h &= \sup_r \max_{j=1,2} |h(r + e_j) - h(r)| \leq 1; \\ \Delta_{jk} h &= \sup_r |h(r + e_j + e_k) - h(r + e_j) - h(r + e_k) + h(r)| \\ &\leq c\{1 \wedge [p(l_j \vee l_k)]^{-1/2}\} \end{aligned} \tag{2.6}$$

hold for some constant c, uniformly in $l_1, l_2 \in \mathbb{Z}^+$ and in $p \leq 1/2$. These inequalities are proved, using the representation (10.1.5) for h and coupling arguments, in a way reminiscient of the proofs of Lemmas 10.2.3 and 10.2.5, with the immigration–death processes replaced by Ehrenfest urn processes. The further inequality

$$\Delta' h = \sup_r |h(r + e_1 + e_2) - h(r)| \leq c(1 \wedge [p(l_1 \vee l_2)]^{-1/2}) \tag{2.7}$$

is special to functions of the form $f(r) = f(r_1 - r_2)$, and is proved in similar style: note that, if h inherited this form (it doesn't), $h(r + e_1 + e_2) - h(r)$

would be zero for all r. Note also that the class $\mathcal{F}_W$ of test functions f appropriate to d_W approximation is not the same as the class used for total variation approximation, as observed already in the Introduction. The estimates of Lemma 1.1.5 have an additional factor of $\lambda^{1/2}$ when compared to those of Lemma 1.1.1, and this is also why (2.6) is less good than might at first sight be expected from Lemmas 10.2.3 and 10.2.5.

In order to estimate $\mathbb{E}(\mathcal{A}h)((Y_1, Y_2))$, we use a technique which is in essence Stein's (1986) idea of the 'exchangeable pair'. Observe that the distribution of (Y_1, Y_2) can be realized, not only by a single sample of indices, but also as the equilibrium distribution of the process in which, after the initial selection, at times determined by a Poisson process of rate $m(n-m-1)/(n-1)$, a randomly chosen member of the current sample of indices is replaced by a new index, chosen at random from those no longer in the sample. The generator $\mathcal{A}'$ of this process is defined by

$$
\begin{aligned}
(\mathcal{A}'h)(r) = \sum_{j=1}^{2}\Big[&r_j\Big(1 - \frac{m + (l_1 - r_1) + (l_2 - r_2)}{n-1}\Big)\{h(r - e_j) - h(r)\} \\
&+ (l_j - r_j)\frac{m}{n-1}\Big(1 - \frac{r_1 + r_2}{m}\Big)\{h(r + e_j) - h(r)\}\Big] \\
+ (n-1)^{-1}\big[&r_1(l_2 - r_2)\{h(r + e_2 - e_1) - h(r)\} \\
&+ r_2(l_1 - r_1)\{h(r + e_1 - e_2) - h(r)\}\big],
\end{aligned}
$$

and

$$\mathbb{E}(\mathcal{A}'h)(Y_1, Y_2) = 0.$$

Hence

$$
\begin{aligned}
&\mathbb{E}(\mathcal{A}h)((Y_1, Y_2)) \\
&= \mathbb{E}\Big\{\sum_{j=1}^{2}\Big[Y_j\Big(\frac{m + (l_1 - Y_1) + (l_2 - Y_2)}{n-1} - p\Big)\{h(Y - e_j) - h(Y)\} \\
&\qquad + (l_j - Y_j)\Big(p - \frac{m}{n-1} + \frac{Y_1 + Y_2}{n-1}\Big)\{h(Y + e_j) - h(Y)\}\Big] \\
&\quad - \frac{1}{n-1}\Big[Y_1(l_2 - Y_2)\{h(Y + e_2 - e_1) - h(Y)\} \\
&\qquad + Y_2(l_1 - Y_1)\{h(Y + e_1 - e_2) - h(Y)\}\Big]\Big\},
\end{aligned}
$$

where Y denotes (Y_1, Y_2).

Thus, if $f(r) = f(r_1 - r_2)$ satisfies $\Delta f \leq 1$ and h is the solution of

$\mathcal{A}h = f - \mathbb{E}f(Z_1 - Z_2)$, it follows that

$$\begin{aligned}
|\mathbb{E}f(Y_1 - Y_2) - \mathbb{E}f(Z_1 - Z_2)| &= |\mathbb{E}(\mathcal{A}h)((Y_1, Y_2))| \\
&\le \sum_{j,k=1}^{2} \Delta_{jk}h \frac{l_j \mathbb{E}Y_k}{n-1} \\
&\quad + \left|p - \frac{m}{n-1}\right| \left\{ \sum_{j=1}^{2} \mathbb{E}Y_j \Delta_{jj}h + l_1(\Delta_{12}h + \Delta' h) + |l_1 - l_2|\Delta h \right\} \\
&\le c \sum_{j,k=1}^{2} \left[\frac{(l_j l_k)^{3/4} p^{1/2}}{n-1} \wedge \frac{l_j l_k p}{n-1} \right] \\
&\quad + c\left|p - \frac{m}{n-1}\right| \left[\sum_{j=1}^{2} ((l_j/p)^{1/2} \wedge l_j) + |l_1 - l_2| \right],
\end{aligned}$$

where (2.6) and (2.7) have been used in the last inequality. Recalling (2.2) – (2.5), it remains to add π_3, replace (l_1, l_2) by (L_1, L_2) and take expectations, giving an estimate of δ_3 which, for $p \le 1/2$ and $|m - (n-1)p| \le np/2$, can be expressed in the form

$$\delta_3 \le \pi_3 + c[(np\pi_3)^{1/2} \wedge np\pi_3]\{\pi_3 + (np)^{-1}|(n-1)p - m|\}. \tag{2.8}$$

If the quantity $np\pi_3$ is taken as an upper bound for the minimum, the estimate (2.8) is of the same order as that obtained in Theorem 5.F. However, when $np\pi_3$ is large compared to one, as is the case when $np \to \infty$ and $|m - np| = O((np)^{1/2})$ or when $|m - np|$ grows very slightly faster, (2.8) shows that the error in Poisson approximation is at most of order $(np)^{-1/4}$. This, for $|m - np| = O((np)^{1/2})$, is equivalent to an error of order $\pi_3^{1/2}$. We do not know if an estimate of order approaching π_3 can be obtained in this case.

5.3 Coloured graphs and birthday problems

Let Γ be a fixed graph having N edges and vertices $1, \ldots, n$ with degrees $d_1, \ldots, d_n$. For an edge $\alpha \in \Gamma$ joining vertices i and j, say, we define $d_\alpha = d_i \wedge d_j$ and $D_\alpha = d_i + d_j$. Colour the vertices of Γ at random, independently of each other and with the same probability distribution over the colours. Let X_i denote the colour of vertex i, set $p_r = \mathbb{P}[X_i = r]$, and take $I_\alpha = I[X_i = X_j]$. Then $W = \sum_{\alpha \in \Gamma} I_\alpha$ is the number of edges connecting two vertices of the same colour. This structure arises in the birthday problem, where the vertices of the graph Γ denote people, the colours X_i their birthdays, and the edges pairs of people who are acquainted: W then counts the number of pairs of people who are acquainted and have the same birthday. It is usually reasonable to assume that the X_i's are independent, and sometimes, as a first approximation, that birthdays are uniformly distributed over the year. Applying the results of Section 2.3, we obtain the following estimate.

Theorem 5.G *With the above assumptions and definitions,*

$$\text{(i)}\quad d_{TV}(\mathcal{L}(W), \mathrm{Po}(\lambda)) \le (1 - e^{-\lambda})\Big\{\pi + (\sigma_1 + \pi)N^{-1}\sum_{i=1}^{n} d_i(d_i - 1)\Big\},$$

where $\pi = \sum_r p_r^2$, $\lambda = N\pi$ *and* $\sigma_1 = \sum_r p_r^3 / \sum_r p_r^2$;

(ii) *if, in addition, the* p_r*'s are all equal to* $1/m$, *we have the stronger estimate*

$$d_{TV}(\mathcal{L}(W), \mathrm{Po}(\lambda)) \le 2m^{-1}(1 - e^{-\lambda})\Big(N^{-1}\sum_{\alpha\in\Gamma} d_\alpha - \frac{1}{2}\Big) \le \sqrt{8}\lambda N^{-1/2},$$

where now $\lambda = N/m$.

Proof The first estimate is Corollary 2.N.1. If the p_r's are all equal, the I_α are strongly dissociated, and the second estimate follows from Theorem 2.O. □

If the birthday distribution is uniform on m days and there are N acquaintances in all, but Γ is otherwise arbitrary, then it follows from Theorem 5.G that, provided $\sqrt{N}/m$ is small, W's distribution is approximately $\mathrm{Po}(N/m)$. However, if the distribution is only approximately uniform, the second part of the theorem is not applicable, although one might still expect better approximation than is given by the first part. This can be achieved using an explicit coupling, as follows.

The proofs of both parts of Theorem 5.G rely on Corollary 2.C.5, but, tracing the argument in Chapter 2 back, one sees that Theorem 2.A is implicitly used with two different couplings, which can be realized as follows. In order to obtain the variable V_α with the distribution $\mathcal{L}(W - 1 | I_\alpha = 1)$, we change the colours of the endpoints of α so that the new colours are the same and have the right distribution: that is, the conditional distribution of the original two colours given that they are equal. For the first part, we do this completely ignoring the original colours of the vertices, but for the second we let one endpoint (the one with highest degree) keep its colour, and only change the colour of the other. If the colour distribution is not uniform, we cannot do this, because the new colour distribution differs from the original one, but we can still keep the colour of one vertex as often as possible. This leads to the following result, which generalizes Theorem 5.G.

Assume that there are m colours $1, \dots, m$, with corresponding probabilities $\mathbb{P}[X_i = r] = p_r$.

Theorem 5.H *With notation as above, and with* $\pi = \sum_{r=1}^{m} p_r^2$,

$$d_{TV}(\mathcal{L}(W), \mathrm{Po}(\lambda)) \le (1 - e^{-\lambda}) \left(\frac{\sum_{i=1}^{n} d_i(d_i - 1)}{N} \sum_{r=1}^{m} p_r \left| \frac{p_r^2}{\pi} - p_r \right| + \frac{\sum_{\alpha \in \Gamma} (2d_\alpha - 1)}{N} \pi \right).$$

Proof We construct a coupling for each edge α as follows. Let α have endpoints i and j, where $d_i \ge d_j$. Let (Ξ_k) be random variables such that:

(i) $\Xi_k = X_k$ when $k \neq i, j$;
(ii) the pair (Ξ_i, Ξ_j) has the distribution $\mathcal{L}((X_i, X_j) | X_i = X_j)$: that is, $\Xi_i = \Xi_j$ and $\mathbb{P}(\Xi_i = r) = \mathbb{P}(X_i = r | X_i = X_j) = p_r^2/\pi$;
(iii) (Ξ_i, Ξ_j) is independent of $(X_k)_{k \neq i,j}$;
(iv) Ξ_i and X_i are maximally coupled; that is,

$$\mathbb{P}(\Xi_i = X_i = r) = \mathbb{P}(\Xi_i = r) \wedge \mathbb{P}(X_i = r) \text{ for every } r.$$

Finally, let $V_\alpha = \sum_{(k,l) \in \Gamma \setminus \{\alpha\}} I(\Xi_k = \Xi_l)$. Then $\mathcal{L}(V_\alpha + 1) = \mathcal{L}(W | I_\alpha = 1)$, and, with $U_\alpha = W = \sum_{(k,l) \in \Gamma} I(X_k = X_l)$, $E_i = \{k \neq j : k \text{ is adjacent to } i\}$ and $E_j = \{k \neq i : k \text{ is adjacent to } j\}$,

$$\begin{aligned}
\mathbb{E}|U_\alpha - V_\alpha| &\le \mathbb{E} I_\alpha + \sum_{E_i} \mathbb{E}|I(X_k = X_i) - I(X_k = \Xi_i)| \\
&\quad + \sum_{E_j} \mathbb{E}|I(X_k = X_j) - I(X_k = \Xi_j)| \\
&\le \pi + (d_i - 1) \sum_{r=1}^{m} p_r \left| p_r - \frac{p_r^2}{\pi} \right| + (d_j - 1) \sum_{r=1}^{m} p_r \left(p_r + \frac{p_r^2}{\pi} \right) \\
&\le \pi + (d_i + d_j - 2) \sum_{r=1}^{m} p_r \left| \frac{p_r^2}{\pi} - p_r \right| + 2(d_j - 1) \sum_{r=1}^{m} p_r^2 \\
&= (D_\alpha - 2) \sum_{r=1}^{m} p_r \left| \frac{p_r^2}{\pi} - p_r \right| + (2d_\alpha - 1)\pi
\end{aligned}$$

The result now follows from Theorem 2.A. □

Remark 5.3.1 The birthday problem was analysed by Diaconis and Stein (personal communication, 1986), who obtained the same result as ours for the uniform birth distribution. The theorems of Chapter 2 could similarly be applied to related problems such as finding the probability that the time of birth differs by at most 24 hours for some pair of persons who know each other.

Note that in Theorem 5.G(ii), because of pairwise independence, we have $\mathrm{Var}\, W = \mathbb{E}W - Np^2$, and so $1 - \frac{\mathrm{Var}\, W}{\lambda} = p = \frac{\lambda}{N}$, while the theorem only gives the estimate $\lambda\sqrt{8/N}$ for the error. However, $O(\frac{1}{\sqrt{N}})$ is the correct order for a general graph, as the following example with the standard birthday problem shows.

Example 5.3.2 Let Γ be the complete graph with n vertices and with $N = \binom{n}{2} \sim n^2/2$ edges. Colour the vertices independently with $m = n^2$ colours, all colours having the same probability $p = 1/n^2$. Let I_α, as before, take the value 1 when the colours at both of α's vertices are the same. It is easily seen that this family of edge indicators is strongly dissociated with $\mathbb{E}I_\alpha = p$ and $\mathbb{E}W = \lambda \sim 1/2$. Now

$$\begin{aligned}\mathbb{P}(W=0) &= \mathbb{P}(\text{all colours different}) = (1-\tfrac{1}{m})(1-\tfrac{2}{m})\cdots(1-\tfrac{n-1}{m}) \\ &= \exp\Big(\sum_{k=1}^{n-1}\log(1-\tfrac{k}{m})\Big) = \exp(-\lambda - \tfrac{1}{6n} + O(n^{-2})).\end{aligned}$$

Hence, when $n \sim \sqrt{2N} \to \infty$,

$$\begin{aligned}d_{TV}(\mathcal{L}(W), \mathrm{Po}(\lambda)) &\geq |\mathbb{P}(W=0) - e^{-\lambda}| \\ &= \frac{e^{-\lambda}}{6n} + O(n^{-2}) = \frac{e^{-\lambda}}{6\sqrt{2}} \times \frac{1}{\sqrt{N}} + O(N^{-1}).\end{aligned}$$

A more detailed analysis of this example can be found as Example 9.3.5.

We end this section with some comments on the property 'strongly dissociated'. The definition is asymmetric, in that we demand that one variable I_α be independent of a family (I_β). It might seem more natural to require either that

(i) the variables I_α and I_β are independent for any edges α and β with at most one common vertex, which is just pairwise independence,

or

(ii) the families $(I_\alpha : \alpha \in H_1)$ and $(I_\beta : \beta \in H_2)$ are independent for any subgraphs H_1 and H_2 with at most one common vertex: we call variables with this property 2-dissociated.

In the particular case where $I_\alpha = \phi(X_i, X_j)$ for a given symmetric function ϕ, it turns out that strong dissociation and pairwise independence are equivalent.

Proposition 5.3.3 *Let $X_1, \ldots, X_n$ be independent and identically distributed random variables, and define, for an edge $\alpha \in \Gamma$ with endpoints i and j, $I_\alpha = \phi(X_i, X_j)$, where ϕ is a given symmetric function. Then the following statements are equivalent:*

(i) *the (I_α) are strongly dissociated;*

(ii) *the (I_α) are pairwise independent;*

(iii) *if i is a vertex with $d_i \geq 2$, then $\phi(X_i, X_j)$ and X_i are (stochastically) independent when $j \neq i$.*

Proof The implications (iii) $\Longrightarrow$ (i) and (i) $\Longrightarrow$ (ii) are immediate. Assuming (ii), let i be a vertex connected to at least two other vertices, say to j and k. Then

$$\mathbb{E}(\phi(X_i, X_j)\phi(X_i, X_k)) = \mathbb{E}(\mathbb{E}(\phi(X_i, X_j)|X_i)^2)$$

and

$$\mathbb{E}\phi(X_i, X_j)\mathbb{E}\phi(X_i, X_k) = (\mathbb{E}(\mathbb{E}\phi(X_i, X_j)|X_i))^2.$$

Since these are equal by (ii), $\mathbb{E}(\phi(X_i, X_j)|X_i)$ is almost surely constant, which implies (iii). □

However, even in this case, 2-dissociation is a distinct property.

Example 5.3.4 Let Γ be a complete graph with $n \geq 5$ vertices, let the X_i be independent and uniformly distributed on $\{1, 2, 3\}$, and let $\phi(X_i, X_j) = I(X_i + X_j = 4)$. Then $\{\phi(X_i, X_j)\}$ is strongly dissociated by Proposition 5.3.3, but not 2-dissociated, because the events

$$\{\phi(X_1, X_2) = \phi(X_1, X_3) = \phi(X_2, X_3) = 1\} = \{X_1 = X_2 = X_3 = 2\}$$

and

$$\{\phi(X_3, X_4) = \phi(X_3, X_5) = \phi(X_4, X_5) = 1\} = \{X_3 = X_4 = X_5 = 2\}$$

are not independent.

Thus, although 2-dissociation implies strong dissociation, it is an unnecessarily stringent requirement for Theorem 2.O.

On the other hand, strong dissociation and pairwise independence are not in general equivalent if the I_α's are not as in Proposition 5.3.3, and pairwise independence need not be sufficient to obtain the conclusion of Theorem 2.O, as is shown by the following example.

Example 5.3.5 Let Γ be the N-star defined in Example 2.3.4. Let X_1 and X_2 be two independent variables, uniformly distributed on $\{1, \ldots, N\}$, and define

$$I_\alpha = I(\{X_1 = i \text{ and } X_2 > i\} \quad \text{or} \quad \{X_1 \leq i \text{ and } X_2 = i\})$$

if α is the edge connecting i and $N + 1$. It is easily checked that the (I_α) are dissociated and pairwise independent, with $\mathbb{P}(I_\alpha = 1) = \frac{1}{N}$. However, they are not strongly dissociated, and furthermore $W = \sum I_\alpha$ converges in distribution as $N \to \infty$ to a two-point distribution on $\{0, 2\}$, and not to any Poisson distribution.

5.4 Complete graphs with repulsions

In this section, we consider a further random graph model. Take the complete graph K_{n+1} on the vertices $0, 1, \ldots, n$, and to each edge $\{i, j\}$ attach a random variable X_{ij} ($= X_{ji}$), the *repulsion* between the vertices i and j. The X's are assumed to be independent and to have continuous but not necessarily identical distributions. An *outgoing arrow* is drawn from vertex i to that vertex j for which $X_{ij} = \min_{k \neq i} X_{ik}$: that is, to the vertex with smallest repulsion, which is unique with probability one. Let W_l be the number of *ingoing arrows* at vertex l. Clearly $0 \leq W_l \leq n$ and $\sum_{l=0}^{n} W_l = n + 1$. What can be said about $\mathcal{L}(W_l)$?

By symmetry it is sufficient to study W_0. Set $I_i = 1$ if there is an arrow from i to 0 and $I_i = 0$ otherwise: that is,

$$I_i = I(X_{0i} < \min_{j \neq 0} X_{ij}), \qquad i = 1, 2, \ldots, n.$$

As both I_i and I_k are functions of X_{ik}, the I's are in general dependent random variables. However, since the X's are independent and all X_{ij} and $\min_{j \neq 0} X_{ij}$ are increasing functions of $\{X_{ij}; j \neq 0\}$, it follows that the random variables $\min_{j \neq 0} X_{ij}, \{X_{ij}; j \neq 0\}$ are associated and independent of X_{0i} and $(X_{jk}; j, k \neq i)$. By Theorem 2.H we can thus assume that for each fixed $i = 1, 2, \ldots, n$ there exist random variables $\{X'_{ij}; j \neq 0\}$ with

$$\mathcal{L}(X'_{ij}; j \neq 0) = \mathcal{L}(X_{ij}; j \neq 0 | I_i = 1) \quad \text{and} \quad X'_{ij} \geq X_{ij} \text{ for } j \neq 0,$$

independent of $(X_{jk}; j, k \neq i)$. In turn, this implies that, for each $k \neq i$,

$$J_{ki} = I(X_{0k} < X'_{ik} \wedge \min_{j \neq 0, i} X_{kj}) \geq I(X_{0k} < \min_{j \neq 0} X_{kj}) = I_k, \qquad (4.1),$$

with

$$\mathcal{L}(J_{ki}; k \neq i) = \mathcal{L}(I_k; k \neq i | I_i = 1),$$

that is, the I's are positively related. Thus Corollary 2.C.4 gives the following result.

Theorem 5.I *If W_0 denotes the number of vertices with smallest repulsion from vertex 0,*

$$d_{TV}(\mathcal{L}(W_0), \text{Po}(\lambda)) \leq \frac{1 - e^{-\lambda}}{\lambda} \Big(\sum_{i=1}^{n} \pi_i^2 + \sum\sum_{1 \leq i \neq k \leq n} (\pi_{ik} - \pi_i \pi_k) \Big),$$

where

$$\lambda = \mathbb{E} W_0 = \sum_{i=1}^{n} \pi_i = \sum_{i=1}^{n} \mathbb{P}[X_{0i} < \min_{j \neq 0} X_{ij}],$$

$$\pi_{ik} = \mathbb{P}[X_{0i} < \min_{j \neq 0} X_{ij}, \ X_{0k} < \min_{j \neq 0} X_{kj}]. \qquad \square$$

Newman, Rinott and Tversky (1983) considered the case of independent and identically distributed X's, and proved that $W_0 \xrightarrow{\mathcal{D}} \text{Po}(1)$ as $n \to \infty$. This is a consequence of the following result, which also gives a rate of convergence.

Corollary 5.I.1 *For identically distributed repulsions,*

$$\mathbb{E}W_0 = 1, \quad \operatorname{Var} W_0 = \frac{2n-2}{2n-1}, \quad d_{TV}(W_0, \operatorname{Po}(1)) \le \frac{3(1-e^{-1})}{2n} < \frac{1}{n}.$$

Proof Since the X's are identically distributed, it follows that $\pi_i = 1/n$ and $\pi_{ik} = 2/n(2n-1)$; the assertion now follows from the theorem above. □

Note that the I's are positively related and that for identically distributed repulsions $\operatorname{Var} W_0 < \mathbb{E}W_0$. For this symmetric case $\mathcal{L}(W_0)$ can be given in closed form. Direct computations show that $\operatorname{Bi}(2n-1, 1/(2n-1))$ gives an accurate approximation of $\mathcal{L}(W_0)$, much better than $\operatorname{Po}(1)$; see Holst (1989), where an explicit construction of the coupling random variables $\{X'_{ij}\}$ is also given.

Next suppose that for each vertex i there are outgoing arrows to the two vertices with the smallest and next smallest repulsions from i. Let Γ be the set of all sets of three vertices. For $\alpha \in \Gamma$, let $I_\alpha = 1$ if, for each ordered pair of vertices (i,j) in α, there is an arrow from i to j, and let $I_\alpha = 0$ otherwise. Hence, $I_\alpha = 1$ if and only if the three vertices in α are connected by a triangle of in- and outgoing arrows. Let $W_\Delta = \sum_{\alpha\in\Gamma} I_\alpha$ be the number of such triangles.

Theorem 5.J *If W_Δ denotes the number of triangles consisting of in- and outgoing arrows,*

$$d_{TV}(W_\Delta, \operatorname{Po}(\lambda)) \le \frac{1-e^{-\lambda}}{\lambda}\Big(\sum_{\alpha\in\Gamma} \pi_\alpha^2 + \sum\sum_{\alpha\cap\beta=\emptyset}(\pi_{\alpha\beta} - \pi_\alpha\pi_\beta) + \sum\sum_{1\le|\alpha\cap\beta|\le 2} \pi_\alpha\pi_\beta\Big).$$

Proof From the definitions, $I_\beta = 0$ for $\beta \in \{\gamma; 1 \le |\gamma\cap\alpha| \le 2\}$ whenever $I_\alpha = 1$: we set $J_{\beta\alpha} = 0$ for such β. As in the proof of Theorem 5.I, now using Remark 2.2.3, there exist random variables X'_{ij} for all $\{i,j\} \not\subset \alpha$ which satisfy

$$\mathcal{L}(X'_{ij}; \{i,j\} \not\subset \alpha) = \mathcal{L}(X_{ij}; \{i,j\} \not\subset \alpha | I_\alpha = 1)$$

and

$$X'_{ij} \ge X_{ij} \text{ for } \{i,j\} \not\subset \alpha; \quad X'_{ij} = X_{ij} \text{ when } \{i,j\} \cap \alpha = \emptyset.$$

Using these X'_{ij}, we define $J_{\beta\alpha} \ge I_\beta$ for $\beta \in \{\gamma; \gamma\cap\alpha = \emptyset\}$ in essence as in (4.1). Hence there exists a coupling with

$$\begin{aligned} J_{\beta\alpha} &\ge I_\beta \quad \text{for} \quad \beta \in \Gamma_\alpha^+ = \{\gamma; \gamma\cap\alpha = \emptyset\}; \\ J_{\beta\alpha} &\le I_\beta \quad \text{for} \quad \beta \in \Gamma_\alpha^- = \{\gamma; 1 \le |\gamma\cap\alpha| \le 2\}, \end{aligned}$$

and $\Gamma_\alpha^0 = \emptyset$. Thus the assertion follows from Theorem 2.C. □

Corollary 5.J.1 *For identically distributed repulsions,*

$$\lambda = \mathbb{E}W_\Delta = \frac{n(n+1)}{18(n-\frac{4}{3})(n-\frac{3}{2})}; \quad \mathrm{Var}\, W_\Delta = \frac{1}{18} + O(n^{-1});$$

$$d_{TV}(W_\Delta, \mathrm{Po}(\lambda)) = O(n^{-1}).$$

Proof To calculate π_α, note that the smallest and next smallest of the repulsions attached to all edges connected with vertices of α are the repulsions attached to two of the three edges connecting the vertices of α, and that the remaining repulsions are still exchangeably distributed. Therefore

$$\pi_\alpha = \left\{\frac{6}{(3(n-2)+3)(3(n-2)+2)}\right\} \cdot \left\{\frac{1}{2(n-2)+1}\right\}$$

and

$$\mathbb{E}W_\Delta = \binom{n+1}{3}\pi_\alpha = \frac{n(n+1)}{18(n-\frac{4}{3})(n-\frac{3}{2})}.$$

After some calculation, one finds that, for $\alpha \cap \beta = \emptyset$,

$$\pi_{\alpha\beta} = \pi_\alpha^2(1 + O(\frac{1}{n})).$$

These results and Theorem 5.J prove the assertion. □

The methods can also be applied to patterns other than triangles.

6
Occupancy and urn models

In this chapter, the general results on Poisson approximation are applied to problems connected with urn models; in particular occupancy problems are studied. Urn models have a long history going back to the early development of probability theory. General references on the subject are Johnson and Kotz (1977), Kolchin, Sevast'yanov and Chistyakov (1978) and the classic Feller (1968).

In Section 6.1, the Poisson approximation of the hypergeometric distribution is considered. The number of boxes with given contents in the urn model with a multinomial allocation rule is studied in detail in Section 6.2. In Section 6.3, urn models, negative association and general occupancy problems are discussed. Finally, a matrix occupancy problem is investigated in Section 6.4.

6.1 Drawing without replacement

Arrange m ones and $N-m$ zeros at random to form an N-vector, so that each of the different outcomes has the same probability $\binom{N}{m}^{-1}$. Set $I_i = 1$ if there is a one at the ith position, $I_i = 0$ otherwise. The total number of ones at positions $1, \ldots, n$ can be expressed as $W = I_1 + \cdots + I_n$, whose distribution is given by

$$\mathbb{P}[W=j] = \binom{m}{j}\binom{N-m}{n-j}\Big/\binom{N}{n} = \binom{n}{j}\binom{N-n}{m-j}\Big/\binom{N}{m};$$

that is, W is hypergeometrically distributed. For small m/N and n/N, this distribution is often approximated by a Poisson distribution. What is the error in the approximation?

Consider the following coupling, expressed, here and subsequently, using the notation of Chapter 2. If $I_k = 1$, set $J_{ik} \equiv I_i$. If $I_k = 0$, change a randomly chosen one to zero and then, for $k \neq i$, set $J_{ik} = 1$ if there is a one at position i, $J_{ik} = 0$ else. By construction, this coupling satisfies the hypothesis of Corollary 2.C.2. For the hypergeometric distribution we have

$$\lambda = \mathbb{E}W = \frac{nm}{N}, \quad \operatorname{Var} W = \frac{N-n}{N-1}\cdot\frac{nm}{N}\left(1-\frac{m}{N}\right).$$

This proves the following theorem.

Theorem 6.A *If W has the hypergeometric distribution given above,*

$$d_{TV}(\mathcal{L}(W), \operatorname{Po}(\lambda)) \le (1-e^{-\lambda})\frac{N}{N-1}\left(\frac{n}{N}+\frac{m}{N}-\frac{n}{N}\frac{m}{N}-\frac{1}{N}\right). \qquad \square$$

By construction, the I's are negatively related, and Theorem 3.D gives the lower bound

$$d_{TV}(\mathcal{L}(W), \mathrm{Po}(\lambda)) \geq \frac{\varepsilon_-}{11 + 3\max(0, \frac{\gamma_-}{\lambda\varepsilon_-})}.$$

Using the fourth cumulant of W, see for example Johnson and Kotz (1969) page 144, we find

$$\gamma_- = \varepsilon_- + \frac{\operatorname{Var} W}{\lambda} \frac{6N^2}{(N-2)(N-3)} \left(\frac{m}{N}\left(1 - \frac{m}{N}\right) + \frac{n}{N}\left(1 - \frac{n}{N}\right) - \frac{1}{N} + \frac{1}{N^2} - \frac{\lambda}{N}\frac{5N-6}{N}\frac{\operatorname{Var} W}{\lambda} \right).$$

When n/N and m/N are both small, it follows that $\varepsilon_- \approx (n+m)/N$ and $\gamma_- \approx 7\varepsilon_-$, implying that a lower bound is roughly $\min(\lambda, 1)\varepsilon_-/32$.

We remark that there exist independent, albeit not identically distributed, indicators whose sum is distributed as W. This is a special case of Vatutin and Mikhailov (1982) page 737. Hence, upper and lower bounds for the Poisson approximation follow from the results for sums of independent indicators. Corollary 3.D.1 gives the above bounds. Improved approximations can be obtained by the results of Chapter 9. In particular, approximation of the hypergeometric distribution by a binomial distribution can be investigated using Theorem 9.E.

Note that the vector $(I_1, \ldots, I_N)$ has a permutation distribution. Thus the indicators $I_1, \ldots, I_n$ are negatively associated by Proposition 2.2.12. By Theorem 2.I, this proves the existence of a coupling satisfying Corollary 2.C.2. However, the simple explicit construction given above is more illuminating.

Another way of proving that the I's are negatively associated is to use Proposition 2.2.9, because $\mathcal{L}(I_1, \ldots, I_N) = \mathcal{L}(X_1, \ldots, X_N | \sum_1^N X_i = m)$, where the X's are independent $\mathrm{Be}(p)$ random variables for any $0 < p < 1$. Using this approach one can generalize the results above by letting the X's be independent indicators which are not necessarily identically distributed, say $X_k \sim \mathrm{Be}(p_k)$. This corresponds to a certain kind of drawing without replacement with unequal probabilities. The mean of the corresponding W is

$$\mathbb{E}W = \sum_{k=1}^{n} \frac{p_k \mathbb{P}[\sum_{i \neq k} X_i = m-1]}{\mathbb{P}[\sum_{i=1}^{N} X_i = m]},$$

and a similar expression can be obtained for $\operatorname{Var} W$, giving an estimate of the error in the approximation of $\mathcal{L}(W)$ by $\mathrm{Po}(\mathbb{E}W)$.

6.2 Multinomial allocations

Let r balls be thrown independently of each other into n boxes, with probability p_k of hitting the kth box. Set X_k equal to the number of balls hitting box k. For the X's, the multinomial distribution holds:

$$\mathbb{P}[X_1 = j_1, \ldots, X_n = j_n] = \frac{r!}{j_1! \ldots j_n!} p_1^{j_1} \ldots p_n^{j_n}, \quad j_1 + \cdots + j_n = r.$$

We assess the accuracy of a Poisson approximation to the distribution of the number of boxes with given content in various cases, by giving bounds on variation distances. Kolchin, Sevast'yanov and Chistyakov (1978) Chapter III Sections 1–3 proved Poisson convergence in general situations, without giving any rates, by using the method of moments.

For equal p's, Arratia, Goldstein and Gordon (1989) Example 2 gives a bound on the variation distance between the distribution of the number of boxes with at least $m \geq 2$ balls (essentially) and an appropriate Poisson distribution. Poisson convergence for general symmetric situations was probably first obtained by von Mises (1939).

Following Kolchin, Sevast'yanov and Chistyakov (1978), two separate cases are distinguished: the left hand domain, where r/n is small, and the right hand domain, where r/n is large.

Left hand domain

We begin by setting $rp^* = r \max_{1 \leq k \leq n} p_k = \max_{1 \leq k \leq n} \mathbb{E}X_k$. For small rp^*, it seems reasonable to try to approximate the distribution of the number of boxes with at least $m\,(\geq 2)$ balls,

$$W^* = \sum_{k=1}^{n} I_k = \sum_{k=1}^{n} I[X_k \geq m],$$

using a Poisson distribution with the same mean, that is

$$\mathbb{E}W^* = \lambda^* = \sum_{k=1}^{n} \pi_k,$$

where

$$\pi_k = \mathbb{E}I_k = \mathbb{P}[X_k \geq m] = \sum_{j=m}^{r} \binom{r}{j} p_k^j (1 - p_k)^{r-j}.$$

The variance of W^* can also be expressed explicitly, albeit in a rather complicated form, as

$$\operatorname{Var} W^* = \sum\nolimits_k \pi_k(1 - \pi_k) + \sum\sum\nolimits_{i \neq k} (\pi_{ik} - \pi_i \pi_k),$$

where

$$\pi_{ik} = \mathbb{P}[X_i \geq m, X_k \geq m]$$
$$= \sum_{x,y \geq m}\sum \frac{r!}{x!y!(r-x-y)!} p_i^x p_k^y (1 - p_i - p_k)^{r-x-y}.$$

Theorem 6.B *If W^* denotes the number of boxes with at least m (≥ 2) balls, then*

$$d_{TV}(\mathcal{L}(W^*), \mathrm{Po}(\lambda^*)) \leq (1 - e^{-\lambda^*})(1 - \mathrm{Var}\, W^*/\lambda^*).$$

When $rp^* \leq m-1$, this yields

$$d_{TV}(\mathcal{L}(W^*), \mathrm{Po}(\lambda^*))$$
$$\leq (1 - e^{-\lambda^*})\left(1 - \frac{r+1}{m}p^*\right)^{-2}$$
$$\times \{m^2 + rp^* + (rp^*)^2\}\frac{1}{m!}(1-p^*)^{r-m}(rp^*)^{m-1}.$$

Proof Consider the following coupling:
(a) If $I_k = 1$, that is $X_k \geq m$, set $J_{ik} = I_i$, for all i.
(b) If $I_k = 0$, that is $X_k < m$, let $\tilde{X}_k$ be a random variable having the distribution $\mathcal{L}(X_k | X_k \geq m)$ and independent of the X's. Draw $\tilde{X}_k - X_k$ balls by simple random sampling without replacement from the boxes different from k, and put these balls into the kth box. If after this the ith box contains at least m balls, set $J_{ik} = 1$, and otherwise set $J_{ik} = 0$.

A little thought shows that this coupling satisfies

$$\mathcal{L}(J_{ik}; 1 \leq i \leq n) = \mathcal{L}(I_i, 1 \leq i \leq n | I_k = 1); \qquad J_{ik} \leq I_i, \text{ for all } i,$$

and hence the first inequality follows from Corollary 2.C.2.

The remainder of the proof consists of justifying the simplified estimate of $(1 - \mathrm{Var}\, W^*/\lambda^*)$. First, note that

$$\mathbb{E}W^* - \mathrm{Var}\, W^* = \sum_k \pi_k^2 + \sum_{i \neq k}\sum (\pi_i\pi_k - \pi_{ik})$$
$$\leq \max_k \pi_k \, \mathbb{E}W^* + \sum_{i \neq k}\sum (\pi_i\pi_k - \pi_{ik}). \qquad (2.1)$$

For $x, y \leq r$, we introduce the notation

$$\hat{\pi}_{kx} = \mathbb{P}[X_k = x] = \binom{r}{x} p_k^x (1 - p_k)^{r-x},$$

$$\hat{\pi}_{ikxy} = \mathbb{P}[X_i = x, X_k = y] = \frac{r!}{x!y!(r-x-y)!} p_i^x p_k^y (1 - p_i - p_k)^{r-x-y}.$$

Then direct calculation yields the inequalities

$$\begin{aligned}
\frac{\hat{\pi}_{ikxy}}{\hat{\pi}_{ix}\hat{\pi}_{ky}} &= \frac{(r-x)!(r-y)!}{r!(r-x-y)!} \frac{(1-p_i-p_k)^{r-x-y}}{(1-p_i)^{r-x}(1-p_k)^{r-y}} \\
&\geq \frac{(r-x)\cdots(r-x-y+1)}{r\cdots(r-y+1)} \frac{(1-p_i-p_k)^r}{(1-p_i)^r(1-p_k)^r} \\
&= \left(1 - \frac{x}{r}\right)\cdots\left(1 - \frac{x}{r-y+1}\right)\left(1 - \frac{p_i p_k}{(1-p_i)(1-p_k)}\right)^r \\
&\geq 1 - y\frac{x}{r-y+1} - r\frac{p_i p_k}{(1-p_i)(1-p_k)}.
\end{aligned}$$

Thus it follows that

$$\begin{aligned}
&\sum\sum_{i \neq k} (\pi_i \pi_k - \pi_{ik}) = \sum\sum_{i \neq k} \sum_{x \geq m} \sum_{y \geq m} (\hat{\pi}_{ix}\hat{\pi}_{ky} - \hat{\pi}_{ikxy}) \\
&\leq \sum\sum\sum\sum \hat{\pi}_{ix}\hat{\pi}_{ky} \left(\frac{xy}{r-y+1} + r\frac{p_i p_k}{(1-p_i)(1-p_k)}\right) \\
&\leq (r-m+1)\left(\sum_i \sum_{x \geq m} \frac{x}{r-x+1}\hat{\pi}_{ix}\right)^2 + r\left(\frac{p^*}{1-p^*}\right)^2 \left(\sum_i \sum_{x \geq m} \hat{\pi}_{ix}\right)^2.
\end{aligned} \tag{2.2}$$

We now observe that, for $m \leq x \leq r$,

$$\frac{\hat{\pi}_{k,x+1}}{\hat{\pi}_{kx}} = \frac{r-x}{x+1} \cdot \frac{p_k}{1-p_k} \leq \frac{r-m}{m+1} \cdot \frac{p^*}{1-p^*}, \tag{2.3}$$

and thus that

$$\begin{aligned}
0 \leq \pi_k - \hat{\pi}_{km} = \sum_{x=m+1}^{r} \hat{\pi}_{kx} &\leq \hat{\pi}_{km} \frac{(r-m)p^*}{(m+1)(1-p^*) - (r-m)p^*} \\
&\leq \hat{\pi}_{km} \frac{rp^*}{m+1-(r+1)p^*}.
\end{aligned}$$

Consequently, if $\lambda = \sum_k \hat{\pi}_{km}$, it follows that

$$\lambda \leq \mathbb{E}W^* \leq \left(1 + \frac{rp^*}{m+1-(r+1)p^*}\right)\lambda \leq \frac{m+1}{m+1-(r+1)p^*}\lambda, \tag{2.4}$$

and

$$\begin{aligned}
\pi^* &\leq \frac{m+1}{m+1-(r+1)p^*} \sup_k \hat{\pi}_{km} \\
&\leq \frac{m+1}{m+1-(r+1)p^*} \binom{r}{m} (p^*)^m (1-p^*)^{r-m}.
\end{aligned}$$

A similar comparison with a geometric series yields

$$\sum_{x\geq m}\frac{x}{r-x+1}\hat{\pi}_{ix}\leq\frac{m}{m-(r+1)p^*}\cdot\frac{m}{r-m+1}\pi_{im}. \tag{2.5}$$

We also need the estimate

$$\frac{1}{r-rp^*}\lambda\leq\frac{1}{r-m+1}\lambda\leq\frac{1}{r-m+1}\sum_k\binom{r}{m}p_k(p^*)^{m-1}(1-p^*)^{r-m}$$
$$\leq\frac{(rp^*)^{m-1}}{m!}(1-p^*)^{r-m}.$$

Combining the estimates above, we obtain

$$\begin{aligned}\mathbb{E}W^*-\operatorname{Var}W^* &\leq\frac{m+1}{m+1-(r+1)p^*}\binom{r}{m}(p^*)^m(1-p^*)^{r-m}\mathbb{E}W^*\\ &\quad+\left(\frac{m}{m-(r+1)p^*}\right)^2\frac{m^2}{r-m+1}\lambda^2+\frac{r(p^*)^2}{(1-p^*)^2}(\mathbb{E}W^*)^2\\ &\leq\mathbb{E}W^*\left(1-\frac{r+1}{m}p^*\right)^{-2}\left(\frac{1}{m!}(rp^*)^m(1-p^*)^{r-m}\right.\\ &\qquad\left.+m^2\frac{1}{r-m+1}\lambda+\frac{(rp^*)^2}{r-rp^*}\lambda\right)\\ &\leq\mathbb{E}W^*\left(1-\frac{r+1}{m}p^*\right)^{-2}(1-p^*)^{r-m}\left(\frac{1}{m!}(rp^*)^m\right.\\ &\qquad\left.+\frac{m^2}{m!}(rp^*)^{m-1}+\frac{1}{m!}(rp^*)^{m+1}\right)\\ &=\mathbb{E}W^*\left(1-\frac{r+1}{m}p^*\right)^{-2}\{m^2+rp^*+(rp^*)^2\}\frac{1}{m!}(1-p^*)^{r-m}(rp^*)^{m-1}.\end{aligned}$$

□

The last inequality in Theorem 6.B is sharp to within a constant if m is fixed, $rp^*\to 0$ and $p^*=O(\Sigma p_k^2)$, and asymptotic equality holds if m is fixed, $rp^*\to 0$ and all the p_k are equal. On the other hand, it is not sharp in cases where $rp^*\sim m\to\infty$. For such cases we can use other estimates of $\operatorname{Var}W^*$ and obtain for example the following corollary, which implies Poisson convergence for W^* when $\lambda^*\to\lambda_\infty$, $\max\pi_k\to 0$ and $m=o(\sqrt{r})$.

Corollary 6.B.1 *If $r\geq 2m$ and $rp^*\leq m$ then*

$$d_{TV}(\mathcal{L}(W^*),\operatorname{Po}(\lambda^*))\leq(1-e^{-\lambda^*})\Big\{\big((1-m^{-1/3})^{-6}+2\big)\lambda^*\frac{m^2}{r-m}+\max\pi_k\Big\}.$$

Proof Let $n=[m+m^{2/3}+1]\leq r$. We obtain, using (2.5) and the fact that $\hat{\pi}_{ix}$ is decreasing for $x\geq m\geq rp^*$, as is implied by (2.3),

$$\begin{aligned}\sum_{x\geq n}\frac{x}{r-x+1}\hat{\pi}_{ix}&\leq\frac{n}{n-(r+1)p^*}\frac{n}{r-n+1}\hat{\pi}_{in}\\ &\leq\frac{n}{n-(r+1)p^*}\frac{n}{r-n+1}\frac{1}{n-m+1}\pi_i.\end{aligned}$$

Hence it follows that

$$\sum_{x \geq m} \frac{x}{r-x+1} \hat{\pi}_{ix} \leq \frac{n}{r-n+1} \pi_i + \frac{n}{n-m-p^*} \frac{n}{r-n+1} \frac{1}{n-m+1} \pi_i$$

$$\leq \frac{n}{r-n+1}\left(1 + \frac{n}{(n-m)^2}\right) \pi_i \leq \frac{m}{r-m+1}(1-m^{-1/3})^{-3} \pi_i.$$

Since $rp^* \leq m$, this, (2.1) and (2.2) yield

$$1 - \frac{\operatorname{Var} W^*}{\lambda^*} \leq \max \pi_k + \frac{m^2}{r-m}(1-m^{-1/3})^{-6}\lambda^* + r\frac{m^2}{(r-m)^2}\lambda^*.$$

The result now follows from Theorem 6.B. □

Example 6.2.1 The binary digital tree (Devroye 1984) can be used to model a computer storage system. Given r numbers $\xi_1, \ldots, \xi_r$ from $[0,1)$, h is chosen as small as possible, subject to the requirement that the binary expansions of the ξ_j's to length h are all different. Each ξ_j is associated with the leaf in the binary tree of height h corresponding to its length h binary expansion. The tree is then pruned as much as it can be, consistent with the requirement that the subtree of the original tree rooted at any leaf in the reduced tree contains exactly one of the ξ_j's: thus the leaves of the reduced tree can be used as the storage locations of the ξ_j's. The retrieval time of a ξ_j is then the length of the path from the root to the corresponding leaf in the reduced tree, and the maximal retrieval time is just h.

Suppose now that $\xi_1, \ldots, \xi_r$ are chosen independently at random from a distribution F over $[0,1)$: what is the distribution of H_r, the height of the tree? It is immediate that the event $\{H_r \leq l\}$ is the same as the event that no dyadic interval of the form $K_j = [j.2^{-l}, (j+1)2^{-l})$, $0 \leq j < 2^l$, contains two or more of the ξ_j's. Thus $\mathbb{P}[H_r \leq l] = \mathbb{P}[W^* = 0]$, where $W^* = W^*(l)$ is as above with $m = 2$ and $n = 2^l$, and the p_j's are given by

$$p_{j+1} = p_{j+1}(l) = F\{K_j\}, \qquad 0 \leq j < 2^l.$$

Hence, from Theorem 6.B,

$$|\mathbb{P}[H_r \leq l] - e^{-\lambda^*(l)}| \leq (1 - e^{-\lambda^*(l)})(1 - \operatorname{Var} W^*(l)/\lambda^*(l)),$$

where $\lambda^*(l) = \mathbb{E}W^*(l)$.

If, for instance, F is uniform over $[0,1)$,

$$\lambda^*(l) = 2^l\Big(1 - \Big\{1 - r2^{-l} + \binom{r}{2}2^{-2l} + O(r^3 2^{-3l})\Big\}$$
$$- r2^{-l}\{1 - (r-1)2^{-l} + O(r^2 2^{-2l})\}\Big)$$
$$= \binom{r}{2}2^{-l} + O(r^3 2^{-2l}) = r^2 2^{-l-1} + O(r2^{-l} + r^3 2^{-2l}),$$

and Corollary 6.B.1 yields, for $4 \le r \le 2^{l+1}$,

$$|\mathbb{P}[H_r \le l] - e^{-\lambda^*(l)}| \le Cr^{-1}\lambda^*(l) = O(r2^{-l}).$$

Furthermore, if $4 \le r \le \frac{1}{3}2^{-l}$, it follows that

$$\lambda^*(l) \ge r^2 2^{-l-1} - r^2 2^{-l-2} = r^2 2^{-l-2},$$

and hence that

$$\begin{aligned}\left|e^{-\lambda^*(l)} - e^{-r^2 2^{-l-1}}\right| &\le |\lambda^*(l) - r^2 2^{-l-1}| e^{-r^2 2^{-l-2}} \\ &= O\left(r2^{-l}(1 + r^2 2^{-l}) e^{-r^2 2^{-l-2}}\right) = O(r2^{-l}).\end{aligned}$$

Since this estimate is trivial for $r > \frac{1}{3}2^{-l}$, we obtain

$$\left|\mathbb{P}[H_r \le l] - e^{-r^2 2^{-l-1}}\right| = O(r2^{-l}),$$

uniformly in $r \ge 1$, $l \ge 1$. In particular, if $r = 2^k$ and $l = 2k + s$, where $s \in \mathbb{Z}$ is fixed,

$$\mathbb{P}[H_r \le 2k + s] = e^{-2^{-s-1}} + O(2^{-s-k}).$$

The above estimate also allows allows one to approximate the distribution of H in total variation, for F the uniform distribution. Let X have the extreme value distribution

$$\mathbb{P}[X \le x] = e^{-e^{-x}}, \qquad -\infty < x < \infty,$$

and define

$$Z_r = \left\lceil \frac{X + 2\log r}{\log 2} \right\rceil = \left\lceil \frac{X}{\log 2} + 2\log_2 r \right\rceil;$$

thus Z_r is integer valued, and

$$\mathbb{P}[Z_r \le l] = \mathbb{P}[X < (l+1)\log 2 - 2\log r] = e^{-r^2 2^{-l-1}}, \quad l \in \mathbb{Z}.$$

Let l_0 be the integer satisfying $2r^{-2}\log r \le 2^{-l_0} < 4r^{-2}\log r$. Then $l_0 \ge 1$, and we obtain

$$\begin{aligned}&d_{TV}(\mathcal{L}(H_r), \mathcal{L}(Z_r)) \\ &\le \frac{1}{2}\sum_{l=l_0}^{\infty}\Big|\{\mathbb{P}[H_r \le l+1] - \mathbb{P}[H_r \le l]\} - \{\mathbb{P}[Z_r \le l+1] - \mathbb{P}[Z_r \le l]\}\Big| \\ &\qquad\qquad + \frac{1}{2}\{\mathbb{P}[H_r \le l_0] + \mathbb{P}[Z_r \le l_0]\} \\ &\le \sum_{l=l_0}^{\infty} |\mathbb{P}[H_r \le l] - \mathbb{P}[Z_r \le l]| + \mathbb{P}[Z_r \le l_0] \\ &= O\left(r2^{-l_0} + e^{-r^2 2^{-l_0-1}}\right) = O\left(\frac{\log r}{r}\right).\end{aligned}$$

Similar estimates can also be proved when the distribution of F is not uniform. In particular, if F has a density f and $\int_0^1 f^2 < \infty$, then it follows that $d_{TV}(\mathcal{L}(H_r), \mathcal{L}(Z_r)) \to 0$, where now we define

$$Z_r = \left[\frac{X}{\log 2} + 2\log_2 r + \log_2 \int_0^1 f^2\right],$$

and if further f is Lipschitz continuous of order $\frac{1}{4}$, one obtains the same rate $O(r^{-1}\log r)$ as above.

Returning to the general, Theorem 6.B can also be used to derive quick error estimates for the Poisson approximation of exact box counts, by means of the following simple lemma.

Lemma 6.2.2 *Let U_1, V_1, U_2, V_2 be integer valued random variables. Then*

$$|d_{TV}(\mathcal{L}(U_1+V_1), \mathcal{L}(U_2+V_2)) - d_{TV}(\mathcal{L}(U_1), \mathcal{L}(U_2))| \le \mathbb{E}|V_1| + \mathbb{E}|V_2|.$$

Proof By the triangle inequality,

$$d_{TV}(\mathcal{L}(U_1), \mathcal{L}(U_2+V_2)) \le d_{TV}(\mathcal{L}(U_1), \mathcal{L}(U_2)) + d_{TV}(\mathcal{L}(U_2), \mathcal{L}(U_2+V_2)),$$

and by the elementary coupling inequality,

$$d_{TV}(\mathcal{L}(U_2), \mathcal{L}(U_2+V_2)) \le \mathbb{P}[U_2 \ne U_2+V_2] = \mathbb{P}[V_2 \ne 0] \le \mathbb{E}|V_2|.$$

Hence

$$d_{TV}(\mathcal{L}(U_1), \mathcal{L}(U_2+V_2)) \le d_{TV}(\mathcal{L}(U_1), \mathcal{L}(U_2)) + \mathbb{E}|V_2|.$$

Thus by symmetry

$$\begin{aligned} d_{TV}(\mathcal{L}(U_1+V_1), \mathcal{L}(U_2+V_2)) &\le d_{TV}(\mathcal{L}(U_1), \mathcal{L}(U_2+V_2)) + \mathbb{E}|V_1| \\ &\le d_{TV}(\mathcal{L}(U_1), \mathcal{L}(U_2)) + \mathbb{E}|V_1| + \mathbb{E}|V_2|, \end{aligned}$$

and

$$d_{TV}(\mathcal{L}(U_1), \mathcal{L}(U_2)) \le d_{TV}(\mathcal{L}(U_1+V_1), \mathcal{L}(U_2+V_2)) + \mathbb{E}|V_1| + \mathbb{E}|V_2|,$$

proving the assertion. □

The number of boxes with exactly m balls can be written as

$$W = \sum_{k=1}^n I[X_k = m] = \sum_{k=1}^n I[X_k \ge m] - \sum_{k=1}^n I[X_k > m],$$

with

$$\mathbb{E}W = \lambda = \sum_{k=1}^n \binom{r}{m} p_k^m (1-p_k)^{r-m}.$$

By Theorem 6.B, (2.4) and Lemma 6.2.2 we now have

Theorem 6.C *For W the number of boxes with exactly m balls,*

$$\begin{aligned} d_{TV}(\mathcal{L}(W), \mathrm{Po}(\lambda)) &\le d_{TV}(\mathcal{L}(W^*), \mathrm{Po}(\lambda^*)) + 2(\mathbb{E}W^* - \mathbb{E}W) \\ &\le (1-e^{-\lambda^*})(1 - \frac{r+1}{m}p^*)^{-2}(m^2 + rp^* + (rp^*)^2) \\ &\quad \times \frac{1}{m!}(1-p^*)^{r-m}(rp^*)^{m-1} + 2\frac{rp^*}{(m+1)-(r+1)p^*}\lambda, \end{aligned}$$

when $rp^* \le m-1$. □

The following result extends that of Kolchin, Sevast'yanov and Chistyakov (1978) Chapter III Section 3 Theorem 1 by giving a rate of convergence. Their Theorem 2 can be extended in a similar fashion.

Corollary 6.C.1 *If $m \ge 2$ is fixed, $n \to \infty$, $rp^* \to 0$ and $\sum_{k=1}^{n} \frac{(rp_k)^m}{m!} \to \lambda_\infty \ge 0$, then*

$$d_{TV}(\mathcal{L}(W^*), \mathrm{Po}(\lambda^*)) = O((rp^*)^{m-1}), \quad d_{TV}(\mathcal{L}(W), \mathrm{Po}(\lambda)) = O(rp^*),$$
$$\mathcal{L}(W), \mathcal{L}(W^*) \xrightarrow{\mathcal{D}} \mathrm{Po}(\lambda_\infty), \quad n \to \infty. \qquad □$$

Normal convergence for W^* is obtained if $\mathbb{E}W^* \to \infty$ and $rp^* \to 0$, and for W if $\lambda = \mathbb{E}W \to \infty$ and $rp^* = o(\lambda^{-1})$. Results could also be obtained for $\sum_k I[X_k \ge m_k]$, where the m's are not all equal.

For the symmetric case where $p_k = 1/n$ for all k, $r/n \to 0$ and $m \ge 2$ is fixed, the convergence of λ to $\lambda_\infty > 0$ is equivalent to

$$r = (\lambda_\infty m!)^{\frac{1}{m}}\, n^{1-\frac{1}{m}}(1 + o(1)).$$

The bound given by Corollary 6.C.1 for $d_{TV}(\mathcal{L}(W), \mathrm{Po}(\lambda))$ can then be evaluated as $O(r^{-\frac{1}{m-1}})$, the same order of magnitude as obtained in Arratia, Goldstein and Gordon (1989) Example 2, for a slightly different random variable. For $d_{TV}(\mathcal{L}(W^*), \mathrm{Po}(\lambda^*))$ we have the better bound $O(\frac{1}{r})$. Furthermore, the lower bound in Theorem 3.D shows that $\frac{1}{r}$ is the correct order of magnitude for $d_{TV}(\mathcal{L}(W^*), \mathrm{Po}(\lambda^*))$ in the symmetric case, because straightforward calculations using Proposition 3.2.12 yield

$$\delta_k = -\binom{k}{2}\frac{m^2}{r} + o\left(\frac{1}{r}\right)$$

and

$$\varepsilon_- = \lambda^* \frac{m^2}{r}(1 + o(1)) \quad \text{and} \quad \gamma_- = 7\lambda^* \frac{m^2}{r}(1 + o(1)).$$

This suggests that the true order of magnitude of $d_{TV}(\mathcal{L}(W), \mathrm{Po}(\lambda))$ may also be $\frac{1}{r}$, even though the present approach only yields a much cruder estimate. That this is indeed the case is shown in Remark 6.2.4 below. Note that we cannot apply Theorem 2.C to bound $d_{TV}(\mathcal{L}(W), \mathrm{Po}(\lambda))$ directly, because we do not have the monotone coupling structure for W.

We expect that the bounds in Theorem 6.B and Corollary 6.B.1 are of the right order of magnitude in asymmetric cases, under weak restrictions, but we have not attempted to compute the γ_- required for an application of Theorem 3.D. Note that Theorem 3.A immediately shows that the bounds in many situations are sharp to within logarithmic factors.

Right hand domain

Consider the number of boxes with at most m (≥ 0) balls. This random variable can be written as

$$W_* = \sum_{k=1}^{n} I_k = \sum_{k=1}^{n} I[X_k \leq m],$$

with

$$\mathbb{E}W_* = \lambda_* = \sum_{k=1}^{n} \pi_k; \quad \mathrm{Var}\, W_* = \sum_k \pi_k(1-\pi_k) + \sum\sum_{i \neq k} (\pi_{ik} - \pi_i \pi_k),$$

where now

$$\pi_k = \mathbb{P}[X_k \leq m] = \sum_{j=0}^{m} \binom{r}{j} p_k^j (1-p_k)^{r-j},$$

and

$$\begin{aligned} \pi_{ik} &= \mathbb{P}[X_i \leq m,\ X_k \leq m] \\ &= \sum_{x,y;\ x,y \leq m} \sum \frac{r!}{x!y!(r-x-y)!} p_i^x p_k^y (1-p_i-p_k)^{r-x-y}. \end{aligned}$$

Set $rp_* = r\ \min_{1 \leq k \leq n} p_k = \min_{1 \leq k \leq n} \mathbb{E}X_k$. For large rp_*, it seems reasonable to approximate the distribution of W_* by using a Poisson random variable U_* with $\mathbb{E}U_* = \lambda_*$.

Theorem 6.D *For W_* the number of boxes with at most m balls,*

$$\begin{aligned} d_{TV}(\mathcal{L}(W_*), \mathrm{Po}(\lambda_*)) &\leq (1-e^{-\lambda_*})(1 - \mathrm{Var}\, W_*/\lambda_*) \\ &\leq (1-e^{-\lambda_*}) \Big\{ \max_k \pi_k + \frac{r}{\lambda_*} \Big(\frac{\log r + m \log\log r + 5m}{r - \log r - m \log\log r - 4m} \lambda_* + \frac{4}{r} \Big)^2 \Big\}, \end{aligned}$$

when $r > \log r + m \log\log r + 4m$.

Proof The proof of Theorem 6.B can be used with only slight modification. If $X_k > m$, take a random variable $\tilde{X}_k$ which is independent of the X's and has distribution $\mathcal{L}(X_k | X_k \leq m)$. Take $X_k - \tilde{X}_k$ balls from box k and re-allocate these balls among the other boxes using probabilities

$p_i/(1-p_k)$ for $i \neq k$. If box i contains at most m balls after this, set $J_{ik} = 1$, and otherwise set $J_{ik} = 0$. This coupling satisfies the conditions of Corollary 2.C.2.

We obtain as in (2.1) and (2.2), with $\hat{\pi}_{kx}$ as in the proof of Theorem 6.B, that

$$\begin{aligned} &\mathbb{E}W_* - \mathrm{Var}W_* \\ &\quad \leq \lambda_* \max \pi_k + \sum_i \sum_k \sum\sum_{x,y \leq m} \frac{xy}{r-y+1} \hat{\pi}_{ix} \hat{\pi}_{ky} + r\Big(\sum_k \frac{p_k}{1-p_k}\pi_k\Big)^2 \\ &\quad \leq \lambda_* \max \pi_k + \frac{m^2}{r-m}\lambda_*^2 + r\Big(\sum_k \frac{p_k}{1-p_k}\pi_k\Big)^2. \end{aligned} \tag{2.6}$$

We estimate the last term as follows. Let $p = \frac{1}{r}(\log r + m \log\log r + 4m)$. Clearly,

$$\sum_{k: p_k \leq p} \frac{p_k}{1-p_k}\pi_k \leq \frac{p}{1-p}\lambda_*. \tag{2.7}$$

The remaining terms are small because the corresponding π_k are small, as is seen by the following argument.

Assume that $m \geq 1$; the case $m = 0$ is similar but simpler. Note that $r \geq 4m$. Suppose that $p_k > p \geq 4m/r$. If $x \leq m$ then, as for (2.3),

$$\frac{\hat{\pi}_{k,x-1}}{\hat{\pi}_{kx}} \leq \frac{m}{r-m+1}\frac{1-p_k}{p_k} \leq \frac{m}{r-m}\frac{r-4m}{4m} < \frac{1}{4},$$

and thus

$$\pi_k = \sum_{x=0}^{m} \hat{\pi}_{kx} \leq \frac{4}{3}\pi_{km}.$$

Since $p^m(1-p)^{r-1-m}$ is decreasing for $\frac{m}{r-1} \leq p \leq 1$, this gives

$$\begin{aligned} \frac{\pi_k}{1-p_k} &\leq \frac{4}{3}\binom{r}{m} p_k^m (1-p_k)^{r-1-m} \leq \frac{4}{3}\binom{r}{m} p^m (1-p)^{r-1-m} \\ &\leq \frac{4}{3}\frac{1}{m!}(rp)^m e^{-(r-1-m)p} \leq \frac{4}{3}\frac{e^m}{m^m}(rp)^m e^{m+1} e^{-rp} \\ &< 4\Big(\frac{rpe^2}{m}\Big)^m e^{-rp} = 4\Big(\frac{rpe^2}{m}\Big)^m \frac{1}{r}(e^4 \log r)^{-m} < \frac{4}{r}, \end{aligned}$$

because $rpe^2/(me^4 \log r) < e^{-2}(1 + \frac{\log\log r + 4}{\log r}) < 1$ in $r \geq 4m > e$. Consequently

$$\sum_{k: p_k > p} \frac{p_k}{1-p_k}\pi_k < \sum_{k: p_k > p} \frac{4}{r} p_k \leq \frac{4}{r}, \tag{2.8}$$

and the result now follows from (2.6), (2.7) and (2.8). □

The number of boxes with exactly m balls can be written as

$$W = \sum_{k=1}^{n} I[X_k \le m] - \sum_{k=1}^{n} I[X_k < m].$$

Using Lemma 6.2.2, we immediately have

Theorem 6.E *For W the number of boxes with exactly m balls,*

$$\begin{aligned} d_{TV}(\mathcal{L}(W), \mathrm{Po}(\lambda)) &\le (1 - e^{-\lambda_*})\Big(1 - \frac{\mathrm{Var}\, W_*}{\lambda_*}\Big) + 2\mathbb{E}(W_* - W) \\ &\le (1 - e^{-\lambda_*})\Big(\max_k \pi_k + r\Big(\frac{\log r + m \log\log r + 5m}{r - \log r - m \log\log r - 4m}\lambda_* + \frac{4}{r}\Big)^2 / \lambda_*\Big) \\ &\quad + \frac{2m}{rp_* - m}\lambda, \end{aligned}$$

when $r > \log r + m \log\log r + 4m$ and $rp_* > m$. □

From the results above we get the following corollary, which extends Kolchin, Sevast'yanov and Chistyakov (1978) Chapter III Section 3 Theorem 3.

Corollary 6.E.1 *If, for some fixed m, $n \to \infty$, $rp_* \to +\infty$ and $\sum_{k=1}^{n} \frac{(rp_k)^m}{m!}(1 - p_k)^{r-m} \to \lambda_\infty > 0$, then*

$$d_{TV}(\mathcal{L}(W_*), \mathrm{Po}(\lambda_*)) = O\Big(\frac{(\log r)^2}{r} + \max \pi_k\Big);$$

$$d_{TV}(\mathcal{L}(W), \mathrm{Po}(\lambda)) = O\Big(\frac{(\log r)^2}{r} + \frac{1}{rp_*}\Big); \quad \mathcal{L}(W),\ \mathcal{L}(W_*) \xrightarrow{\mathcal{D}} \mathrm{Po}(\lambda_\infty).$$

□

As in the left hand domain, we could obtain convergence to the normal distribution. We could also consider different m_k's, and let $m \to \infty$.

For the symmetric case $p_k = \frac{1}{n}$ and $\frac{r}{n} \to \infty$ (m fixed), the convergence $\mathbb{E}W_* \to \lambda_\infty > 0$ is equivalent to

$$r = n(\log n + m \log\log n - \log \lambda_\infty - \log m! + o(1)).$$

Theorems 6.D and 6.E then yield

$$\begin{aligned} d_{TV}(\mathcal{L}(W_*), \mathrm{Po}(\lambda_*)) &\le (1 - e^{-\lambda_\infty})\lambda_\infty \frac{(\log r)^2}{r}(1 + o(1)) \\ &= (1 - e^{-\lambda_\infty})\lambda_\infty \frac{\log n}{n}(1 + o(1)), \end{aligned}$$

and

$$d_{TV}(\mathcal{L}(W), \mathrm{Po}(\lambda)) \leq \frac{2m\lambda_\infty + o(1)}{\log r} = \frac{2m\lambda_\infty + o(1)}{\log n}.$$

After some calculation using Proposition 3.2.12, one finds for W_* that

$$\varepsilon_- = \lambda_\infty \frac{\log n}{n}(1 + o(1)), \quad \gamma_- = 7\lambda_\infty \frac{\log n}{n}(1 + o(1)).$$

Hence the true order of magnitude for $d_{TV}(\mathcal{L}(W_*), \mathrm{Po}(\lambda_*))$ is $n^{-1}\log n$, in view of Theorem 3.D.

For $d_{TV}(\mathcal{L}(W), \mathrm{Po}(\lambda))$, Corollary 6.E.1 only yields the crude estimate $O(\frac{1}{\log n})$. The estimate is improved to $O(n^{-1}\log n)$ in Remark 6.2.4 below.

Exact box counts

As observed above, the bounds obtained in Theorems 6.C and 6.E for exact box counts do not appear to be sharp. In this section, we give sharper bounds, though the argument required is substantially more complicated.

Let $I_k = I[X_k = m]$, so that $\pi_k = \mathbb{E}I_k = \binom{r}{m} p_k^m (1-p_k)^{r-m}$, and define $W = \sum_{k=1}^n I_k$ and $\lambda = \mathbb{E}W = \sum_{k=1}^n \pi_k$: W now counts the number of boxes containing exactly m balls. Using Theorem 2.A, $d_{TV}(\mathcal{L}(W), \mathrm{Po}(\lambda))$ can be estimated by

$$d_{TV}(\mathcal{L}(W), \mathrm{Po}(\lambda)) \leq (1 \wedge \lambda^{-1}) \sum_{k=1}^n \pi_k \mathbb{E}|W - V_k|, \tag{2.9}$$

where V_k is any random variable with the distribution $\mathcal{L}(W - 1 | I_k = 1)$, chosen in practice to make the estimate in (2.9) as small as possible. In this case, we have no natural monotone coupling of W and V_k, so we apply (2.9) directly to obtain the following theorem.

Theorem 6.F *Suppose that*

$$\min_{1 \leq k \leq n} (1 - p_k)^m \geq 3/4 \tag{2.10}$$

and that $m \leq \frac{r}{2}$. Then, for W the number of boxes with exactly m balls,

$$d_{TV}(\mathcal{L}(W), \mathrm{Po}(\lambda)) \leq (1 \wedge \lambda^{-1}) \Big[6r(\sum_{k=1}^n p_k \pi_k)^2 + \sum_{k=1}^n \pi_k^2 + 6\lambda^2 m^2 / r \Big]. \tag{2.11}$$

Remark 6.2.3 The assumption (2.10) is used only to keep the argument simple.

Proof For any given realization $X_1, \ldots, X_n$, construct V_k as follows:

(i) if $X_k = m + t, t \geq 1$, t balls from box k are re-distributed among the other boxes independently with probabilities $p_j/(1 - p_k)$, $j \neq k$;

(ii) if $X_k = m - t, 1 \leq t \leq m$, t balls are removed by simple random sampling without replacement from the boxes different from k, and are put into box k;

(iii) if $X_k = m$ no re-distribution is made.

Let X_{jk} denote the count in box j after re-distribution, and set $V_k = \sum_{j \neq k} I[X_{jk} = m]$. Thus $V_k = W - A_k + B_k$, where

(i) if $X_k = m + t$, $t \geq 1$,

$$A_k = \sum_{l=1, l \neq k}^{n} I[X_l = m] I[X_{lk} > m],$$

and

$$B_k = \sum_{l=1, l \neq k}^{n} I[X_l < m] I[X_{lk} = m];$$

(ii) if $X_k = m - t$, $1 \leq t \leq m$,

$$A_k = \sum_{l=1, l \neq k}^{n} I[X_l = m] I[X_{lk} < m],$$

and

$$B_k = \sum_{l=1, l \neq k}^{n} I[X_l > m] I[X_{lk} = m],$$

(iii) if $X_k = m$, $A_k = 1$ and $B_k = 0$.

Since also

$$\mathbb{E}|W - V_k| \leq \mathbb{E}A_k + \mathbb{E}B_k = 2\mathbb{E}A_k + \mathbb{E}(V_k - W), \tag{2.12}$$

it suffices to bound $\mathbb{E}A_k$ and $\mathbb{E}(V_k - W)$.

First, consider $\mathbb{E}(A_k I[X_k = m + t])$ for any $t \geq 1$: we have

$$\begin{aligned} &\mathbb{E}(I[X_k = m + t] I[X_l = m] I[X_{lk} > m]) \\ &= \pi_l \mathbb{P}[X_k = m + t | X_l = m] \mathbb{P}[\mathrm{Bi}\,(t, p_l/(1 - p_k)) \geq 1] \\ &\leq \pi_l \mathbb{P}[X_k = m + t | X_l = m] t p_l/(1 - p_k). \end{aligned} \tag{2.13}$$

Hence, adding over t and l,

$$\mathbb{E}(A_k I[X_k > m]) \leq \frac{4}{3} \sum_{l \neq k} \pi_l p_l \mathbb{E}\{(B_{lk} - m)_+\},$$

where $B_{lk} \sim \mathrm{Bi}\,(r-m, p_k/(1-p_l))$; this gives the crude estimate

$$\mathbb{E}(A_k I[X_k > m]) \le \frac{16}{9} r p_k \sum_l \pi_l p_l. \tag{2.14}$$

Next, take $\mathbb{E}(A_k I[X_k = m-t])$ for $1 \le t \le m$, and observe that

$$\begin{aligned}
&\mathbb{E}(I[X_k = m-t] I[X_l = m] I[X_{lk} < m]) \\
&\quad \le \pi_l \mathbb{P}[X_k = m-t | X_l = m] \mathbb{E}(m - X_{lk} | X_k = m-t, X_l = m) \\
&\quad = \pi_l \mathbb{P}[X_k = m-t | X_l = m] mt/(r-m+t) \\
&\quad \le (2mt\pi_l/r) \mathbb{P}[X_k = m-t | X_l = m];
\end{aligned}$$

adding over t and l gives the crude estimate

$$\mathbb{E}(A_k I[X_k < m]) \le 2m^2\lambda/r. \tag{2.15}$$

Finally, $\mathbb{E}(A_k I[X_k = m]) = \pi_k$. Thus, collecting the estimates,

$$\mathbb{E}A_k \le \pi_k + 2rp_k \sum_l \pi_l p_l + 2m^2\lambda/r,$$

and so

$$2\sum_{k=1}^n \pi_k \mathbb{E}A_k \le 2\sum_{k=1}^n \pi_k^2 + 4r\Big(\sum_{k=1}^n \pi_k p_k\Big)^2 + 4m^2\lambda^2/r. \tag{2.16}$$

The contribution from the remaining term in (2.12) is estimated by

$$\begin{aligned}
\sum_{k=1}^n \pi_k \mathbb{E}(V_k - W) &= \mathrm{Var} W - \mathbb{E}W \\
&\le \sum_{k=1}^n \sum_{l \ne k} \pi_k \pi_l [\{(1-p_k)(1-p_l)\}^{-m} - 1] - \sum_{k=1}^n \pi_k^2 \\
&\le \frac{16}{9} \sum_{k=1}^n \sum_{l \ne k} \pi_k \pi_l m (p_k + p_l) - \sum_{k=1}^n \pi_k^2.
\end{aligned} \tag{2.17}$$

Hence, from (2.12), (2.16) and (2.17),

$$\sum_{k=1}^n \pi_k \mathbb{E}|W - V_k| \le \sum_{k=1}^n \pi_k^2 + 4r(\sum_{k=1}^n \pi_k p_k)^2 + 4m^2\lambda^2/r + 4m\lambda \sum_{k=1}^n \pi_k p_k,$$

which, with the Cauchy–Schwarz inequality, proves the theorem. □

Remark 6.2.4 When all the p_k are equal, estimate (2.11) simplifies to

$$d_{TV}(\mathcal{L}(W), \mathrm{Po}(\lambda)) \leq \frac{\lambda \wedge \lambda^2}{n}[6rn^{-1} + 1 + 6nm^2 r^{-1}], \tag{2.18}$$

valid for $m \leq (\log \frac{4}{3})((n-1) \wedge r)$. Thus, if $r/n \to 0$ and $\lambda \to \lambda_\infty$, then $d_{TV}(\mathcal{L}(W), \mathrm{Po}(\lambda)) = O(n^{-1} \vee m^2 r^{-1})$, and if $r/n \to \infty$ and $\lambda \to \lambda_\infty$, $d_{TV}(\mathcal{L}(W), \mathrm{Po}(\lambda)) = O(rn^{-2} \vee m^2 r^{-1})$. These estimates give the same rates r^{-1} and $n^{-1}\log n$ in the cases where $\lambda \to \lambda_\infty$ for fixed m as have already been obtained for $\mathcal{L}(W^*)$ and $\mathcal{L}(W_*)$ respectively, improving on the rates given in Corollaries 6.C.1 and 6.E.1.

The estimate (2.11) can be improved, without using (2.10), at the cost of greater complexity. By estimating

$$\mathbb{E}\{(B_{lk} - m)_+\} \leq (\mathbb{E}B_{lk} - m)_+ + \sqrt{\mathrm{Var} B_{lk}}$$

following (2.13), and by making a similar estimate in place of (2.15), one arrives at

$$\begin{aligned} &d_{TV}(\mathcal{L}(W), \mathrm{Po}(\lambda)) \\ &\leq (1 \wedge \lambda^{-1})\Big\{2\sum_{k=1}^{n} \pi_k^2 + \mathrm{Var}\, W - \mathbb{E}W \\ &\quad + \sum_{k,l} \pi_k \pi_l \Big\{\frac{2p_l}{1-p_k}\Big[\Big(\frac{rp_k}{1-p_l} - m\Big)_+ + \sqrt{\frac{rp_k}{1-p_l}}\Big] \\ &\quad + \frac{4m}{r}\Big[\Big(m - \frac{(r-m)p_k}{1-p_l}\Big)_+ + \sqrt{\frac{rp_k}{1-p_l}}\Big]\Big\}\Big\} \end{aligned} \tag{2.19}$$

for all $m \leq r/2$. For example, in the case where all the p_k are equal, if $m \to \infty$ in such a way that $\liminf \lambda > 0$, it follows from this estimate that

$$d_{TV}(\mathcal{L}(W), \mathrm{Po}(\lambda)) = O(\lambda m[\psi + 1][\psi + m^{1/2}]/r),$$

where $\psi = m^{-1/2}|rn^{-1} - m|$. Thus, in this case, if $m \to \infty$ fast and $\lambda \to \lambda_\infty$, as is possible if m is near r/n and $r = O(n^3)$, (2.19) gives a better rate than (2.18): indeed, (2.18) is insufficient to establish Poisson convergence if $n^2 = O(r)$. However, computation of $\lambda^{-1}(\mathrm{Var}\, W - \lambda)$ suggests that the true rate may be near $\lambda m\{n^{-3} + r^{-1}[1 + \psi^2]\}$, which is significantly smaller when $\psi \ll m^{1/2}$. Note that the method used here to couple W and V_k typically involves the redistribution of roughly $\psi m^{1/2}$ balls into or from $n-1$ boxes, roughly λ of which are sensitive, in their contribution to W, to the arrival or departure of one or two balls, so that an error estimate no better than order $\lambda n^{-1}\psi m^{1/2} \asymp \lambda\psi m^{3/2} r^{-1}$ is to be expected. More precise rate estimates would therefore require the use of a better method of coupling, perhaps in the spirit of the argument used to derive (5.2.8), where a similar problem arises in the context of vertex degrees.

6.3 Urn models and general occupancy

An urn initially contains N balls of n different colours in proportions $\pi_1, \ldots, \pi_n$. Balls are drawn at random from the urn in such a way that at each draw the probability of obtaining a specific ball is the same for all balls. Suppose that at each draw we replace the drawn ball together with c of the same colour. This is repeated r times, independently. Let X_i be the number of times a ball of colour i is drawn. The cases $c = -1, 0, d \geq 1$ correspond to drawing without replacement, with replacement and Pólya sampling. Let $Y_1, \ldots, Y_n$ be independent with $Y_i \sim \mathrm{Bi}\,(N\pi_i, \theta)$, $\mathrm{Po}(\theta\pi_i)$, $\mathrm{NBi}\,(N\pi_i/d, \theta)$ respectively, where θ is arbitrary. We then have

$$\mathcal{L}(X_1, \ldots, X_n) = \mathcal{L}\Big(Y_1, \ldots, Y_n \Big| \sum_{i=1}^{n} Y_i = r\Big)$$

for the three different urn models. As the probability functions of the first two of the above families of distributions are *log concave,* and of the third also provided that $d \leq \min_i N\pi_i$, it follows by Proposition 2.2.9 that the X's are negatively associated. Using Theorem 2.I and Corollary 2.C.2, Poisson approximation of sums of indicators of the form $I[X_i \geq m_i]$ can be studied. Alternatively, as in Barbour and Holst (1989), one can construct explicit couplings, similar to those in Sections 6.1 and 6.2, for all these urn models, now without any restriction on d in the third case, to show that such indicators are negatively related. From a somewhat different point of view, Joag-Dev and Proschan (1983) Section 3.1 discuss the above distributions, that is, multivariate hypergeometric, multinomial and Dirichlet compound multinomial.

As an illustration, consider Pólya sampling with $d = 1$; that is, each ball drawn is replaced together with a new ball of the same colour. Recall that drawing with and without replacement was studied in the previous sections. With $I_i = I[X_i = 0]$, the random variable $W = I_1 + \cdots + I_n$ is the number of colours which have not appeared in the first r drawings. It is easily shown that

$$p_i = \mathbb{E}I_i = \binom{-N(1-\pi_i)}{r}\Big/\binom{-N}{r};$$

$$p_{ik} = \mathbb{E}I_i I_k = \binom{-N(1-\pi_i-\pi_k)}{r}\Big/\binom{-N}{r}, \; i \neq k,$$

and

$$\mathbb{E}W = \sum_{i=1}^{n} p_i, \; \mathrm{Var}\, W = \sum_{i=1}^{n} p_i(1-p_i) + \sum\sum_{i\neq k}(p_{ik} - p_i p_k). \tag{3.1}$$

As $I_1, \ldots, I_n$ are negatively related indicators, the following result follows from Corollary 2.C.2.

Theorem 6.G *For W the number of colours not appearing in the first r drawings by Pólya sampling, and for $\mathbb{E}W$,* $\operatorname{Var} W$ *as in (3.1),*

$$d_{TV}(\mathcal{L}(W), \operatorname{Po}(\mathbb{E}W)) \leq (1 - \exp(-\mathbb{E}W))(1 - \operatorname{Var} W/\mathbb{E}W). \qquad \square$$

Consider a vector of *zeros* of length n. Choose a position at random, with probability π_i of getting the ith, where $\sum_{i=1}^{n} \pi_i = 1$, and change the *zero* to a *one* in the position selected. It is easily seen from the definition of negative association that the vector so obtained is negatively associated. Adding r independent such vectors we get a new vector $(X_1, \ldots, X_n)$ which is negatively associated by Proposition 2.2.7; see also Joag-Dev and Proschan (1983) Section 3.1. This gives another proof that the multinomial distribution is negatively associated. The situation can be considerably generalized. We can, for example, change the *zero* to a *one* at position i only with probability ρ_i (and keep the *zero* with probability $1 - \rho_i$) if position i is selected. This yields a vector which is also negatively associated, by Proposition 2.2.6. The probabilities can also be different for the different independent vectors which are added. In general, the sum is no longer multinomially distributed, but it is still negatively associated. Generalizing further, we could just assume that the vectors which are added are independent and negatively associated. Then their sum $(X_1, \ldots, X_n)$ is negatively associated. Thus, for example, the indicators $I[X_1 \leq m_1], \ldots, I[X_n \leq m_n]$ are negatively related, and bounds on the variation distance between the distribution of the sum W of these indicators and $\operatorname{Po}(\mathbb{E}W)$ could be given.

6.4 A matrix occupancy problem

A special case of the general occupancy situation in the previous section is the following. In an $r \times n$ matrix, put *ones* in row i at s_i positions chosen by simple random sampling without replacement, and let each *one* be changed to a *zero* with probability $1 - p_i$, independently of the others; *zeros* are put at the other positions of the row. This is done independently for each row in the matrix. Let X_k be the number of *ones* in column k and consider $W_m = \sum_{k=1}^{n} I(X_k \leq m)$, that is the number of columns with at most m *ones*. To obtain a useful expression for the exact distribution of W_m seems impossible in general, and so approximations are important. For cases with expectations small relative to n, one could contemplate using Poisson approximations.

There are many other ways of formulating the situation described above. Such models have been called matrix occupancy, committee, random allocations, allocation by complexes, capture–recapture; see the papers by Holst (1980b), Johnson and Kotz (1977) and Kolchin, Sevast'yanov and Chistyakov (1978).

The case $m = 0$, corresponding to the number of empty columns, has received most attention in the literature, mainly with $p_i = 1$. For this case, Barbour and Holst (1989) and Vatutin and Mikhailov (1982) give bounds on the variation distance when the *ones* are not changed. Except for the results in these papers, we are not aware of any other rate of convergence results. However, proofs of Poisson convergence are given for example in Harris (1989), Holst (1980b) and Kolchin, Sevast'yanov and Chistyakov (1978) Chapter VII.

Theorem 6.H *For W_m the number of columns with at most m ones,*

$$d_{TV}(\mathcal{L}(W_m), \mathrm{Po}(\mathbb{E}W_m)) \leq (1 - \exp(-\mathbb{E}W_m))(1 - \mathrm{Var}\, W_m/\mathbb{E}W_m).$$

Proof As the allocation of the s_i *ones* in the ith row is given by a permutation distribution, it follows by Proposition 2.2.12 that the entries of the ith row are negatively associated. By Proposition 2.2.6 this also holds after randomly changing some of the *ones* to *zeros*. As the rows are independent, Proposition 2.2.7 shows that the sum of the rows, $(X_1, \ldots, X_n)$, is negatively associated. Hence it follows from Theorem 2.K that the indicators $I[X_1 \leq m], \ldots, I[X_n \leq m]$ are negatively related. Finally, the assertion follows by Corollary 2.C.2. □

The explicit expressions for the means and variances of W_m are in general complicated. However, in the important case $m = 0$ the formulae simplify.

Corollary 6.H.1 *For W_0 the number of columns with only zeros,*

$$d_{TV}(\mathcal{L}(W_0), \mathrm{Po}(\mathbb{E}W_0)) \leq (1 - \exp(-\mathbb{E}W_0))(1 - \mathrm{Var}\, W_0/\mathbb{E}W_0),$$

where $\mathbb{E}W_0 = n\prod_{i=1}^r (1 - p_i s_i/n)$ *and*

$$\mathrm{Var}\ W_0 = n\prod_{i=1}^{r}(1 - p_i s_i/n)\Big(1 - \prod_{i=1}^{r}(1 - p_i s_i/n)\Big)$$
$$-n(n-1)\Big\{\prod_{i=1}^{r}(1 - p_i s_i/n)^2 - \prod_{i=1}^{r}\Big(1 - 2p_i s_i/n + p_i^2 s_i(s_i - 1)/n(n-1)\Big)\Big\}.$$

□

The following special case improves a result of Harris (1989), by giving an estimate of the rate of convergence.

Corollary 6.H.2 *If $s_1 = \cdots = s_r = s$ and $p_1 = \cdots = p_r = p$, with W_0 as above,*

$$d_{TV}(\mathcal{L}(W_0), \mathrm{Po}(\mathbb{E}W_0)) \leq \mathbb{E}W_0(1 + p\log n)/(n - sp),$$

where $\mathbb{E}W_0 = n\left(1 - \frac{sp}{n}\right)^r$ *and*

$$\operatorname{Var} W_0 = n\left(\left(1 - \frac{sp}{n}\right)^r - \left(1 - \frac{sp}{n}\right)^{2r}\right) - n(n-1)\left(\left(1 - \frac{sp}{n}\right)^{2r} - \left(1 - 2\frac{sp}{n} + \frac{s(s-1)p^2}{n(n-1)}\right)^r\right).$$

If $r > 1$, $n \to \infty$ *and* $\mathbb{E}W_0 \to \lambda_\infty$, *then* $\mathcal{L}(W_0) \xrightarrow{\mathcal{D}} \operatorname{Po}(\lambda_\infty)$.

Proof The explicit formulae for $\mathbb{E}W_0$ and $\operatorname{Var}W_0$ follow from Corollary 6.H.1. From these expressions we get

$$\mathbb{E}W_0 = n\left(1 - \frac{sp}{n}\right)^r < n \exp\left(-\frac{rsp}{n}\right),$$

which implies $rsp/n < \log n - \log \mathbb{E}W_0$ and thus

$$\begin{aligned}
\mathbb{E}W_0 - \operatorname{Var} W_0 &= \frac{(\mathbb{E}W_0)^2}{n} \\
&\quad + n(n-1)\left(\left(1 - 2\frac{sp}{n} + \frac{s^2p^2}{n^2}\right)^r - \left(1 - 2\frac{sp}{n} + \frac{s(s-1)}{n(n-1)}p^2\right)^r\right) \\
&\le \frac{(\mathbb{E}W_0)^2}{n} + n(n-1)r\left(1 - \frac{2sp}{n} + \frac{s^2p^2}{n^2}\right)^{r-1}\left(\frac{s^2p^2}{n^2} - \frac{s(s-1)}{n(n-1)}p^2\right) \\
&= \frac{(\mathbb{E}W_0)^2}{n} + (\mathbb{E}W_0)^2 \frac{r}{(n-sp)^2}\frac{s(n-s)}{n}p^2 \\
&\le \frac{(\mathbb{E}W_0)^2}{n} + (\mathbb{E}W_0)^2 \frac{1}{n-sp}\frac{rsp}{n}p \\
&\le \frac{(\mathbb{E}W_0)^2}{n} + \frac{(\mathbb{E}W_0)^2}{n-sp}p \log n + \frac{(\mathbb{E}W_0)^2 \log(1/\mathbb{E}W_0)}{n-sp}p.
\end{aligned}$$

If $\mathbb{E}W_0 \ge 1$, we thus get

$$d_{TV}(\mathcal{L}(W_0), \operatorname{Po}(\mathbb{E}W_0)) \le \mathbb{E}W_0 \frac{1 + p \log n}{n - sp},$$

whereas, if $\mathbb{E}W_0 \le 1$, we get

$$\begin{aligned}
&d_{TV}(\mathcal{L}(W_0), \operatorname{Po}(\mathbb{E}W_0)) \\
&\quad \le \frac{1}{n-sp}\left\{(\mathbb{E}W_0)^2\left(1 + \log\frac{1}{\mathbb{E}W_0}\right) + (\mathbb{E}W_0)^2 p \log n\right\} \\
&\quad \le \mathbb{E}W_0 \frac{1 + p\log n}{n - sp}.
\end{aligned}$$

Using the formula for $\mathbb{E}W_0$ in the estimate of the distance, we see that $\mathbb{E}W_0 \to \lambda_\infty$ implies that $d_{TV}(\mathcal{L}(W_0), \operatorname{Po}(\mathbb{E}W_0)) \to 0$, and hence also that $\mathcal{L}(W_0) \xrightarrow{\mathcal{D}} \operatorname{Po}(\lambda_\infty)$. □

Remark 6.4.1 When $p_i = 1$ for all i, Vatutin and Mikhailov (1982) proved that W_0 has the same distribution as a sum of independent Bernoulli random variables. This implies that, up to a positive constant, the bound given by $\varepsilon = (1 \wedge \mathbb{E}W_0)(1 - \operatorname{Var} W_0/\mathbb{E}W_0)$ is the best possible. For $\mathbb{E}W_0 \to \lambda_\infty > 0$ and $s_i = s \leq (1-\eta)n$, with $\eta > 0$ fixed, the bound is of order $n^{-1}\log n$.

7
Spacings

Take n points at random from a uniform distribution on the circumference of a circle of radius $\frac{1}{2\pi}$. Let $S_1, \ldots, S_n$ be the successive arc-length distances between these points, that is the spacings. The m-spacings are then defined by

$$S_{km} = S_k + \cdots + S_{k+m-1}, \qquad k = 1, \ldots, n,$$

for any $1 \leq m \leq n/2$. Note that the spacings are correlated even if their indices are far apart.

Different aspects of spacings have been considered previously; see Holst and Hüsler (1984) and the references therein. In Barbour and Holst (1989), Poisson approximation with a rate of convergence was obtained for the number of big spacings; some weaker bounds are given in Holst, Kennedy and Quine (1988). Below, we obtain such results for big and small m-spacings. First, suitable couplings are constructed in Section 7.1, and then the cases with $m = 1$, that is ordinary spacings, and $m > 1$ are considered in Sections 7.2 and 7.3. We use the notation

$$W_{Bm} = \sum_{k=1}^{n} I(S_{km} > a), \quad W_{Sm} = \sum_{k=1}^{n} I(S_{km} < a), \quad 0 < a < 1, \qquad (0.1)$$

for the numbers of big and small m-spacings respectively; for $m = 1$ we just write W_B or W_S.

Spacings in higher dimensions can be introduced in different ways. Some results for such generalizations are obtained in Section 7.4. The chapter concludes with an application to DNA breakage.

7.1 Couplings

The spacings can be considered as being generated by a sample of size $n-1$ from a uniform distribution on the unit interval. Let the order statistic of the sample be $0 = U_0 < U_1 < \cdots < U_{n-1} < U_n = 1$. Thus $S_k = U_k - U_{k-1}$ and $S_{km} = U_{k+m-1} - U_{k-1}$, with $U_{-1} \equiv U_{n-1} - 1$, $U_{n+1} \equiv U_1 + 1$ and so on. Since, by symmetry, all the S_{km} with the same m have the same distribution, we have

$$\mathbb{E}W_{Bm} = n\mathbb{P}(S_{1m} > a) = n\mathbb{P}(U_m > a) = n \sum_{j=0}^{m-1} \binom{n-1}{j} a^j (1-a)^{n-1-j}$$

$$= n\int_a^1 \frac{(n-1)!}{(m-1)!(n-1-m)!}u^{m-1}(1-u)^{n-1-m}du; \tag{1.1}$$

$$\mathbb{E}W_{Sm} = n\mathbb{P}(S_{1m} < a) = n\mathbb{P}(U_m < a) = n\sum_{j=m}^{n-1}\binom{n-1}{j}a^j(1-a)^{n-1-j}$$

$$= n\int_0^a \frac{(n-1)!}{(m-1)!(n-1-m)!}u^{m-1}(1-u)^{n-1-m}du. \tag{1.2}$$

For the conditional distributions needed in applying Theorem 2.C, it suffices by symmetry to consider $\mathcal{L}(S_1,\ldots,S_n \mid U_m > a)$ for W_{Bm} and $\mathcal{L}(S_1,\ldots,S_n \mid U_m < a)$ for W_{Sm}. Introduce the distribution functions

$$F(x) = \mathbb{P}[U_m \le x], \quad G_B(x) = \mathbb{P}[U_m \le x \mid U_m > a], \quad H_B(x) = G_B^{-1}(F(x)).$$

Conditional on U_m, the random variables

$$\frac{S_{m+1}}{1-U_m},\ldots,\frac{S_n}{1-U_m}$$

are spacings from a sample of size $n-m$ on the unit interval, and are thus independent of U_m. Hence

$$\mathcal{L}(S_{m+1},\ldots,S_n \mid U_m > a) = \mathcal{L}\Big(\frac{1-H_B(U_m)}{1-U_m}(S_{m+1},\ldots,S_n)\Big).$$

For $0 < x < 1$ we have $H_B(x) > x$, that is, $(1-H_B(x))/(1-x) < 1$. Thus, for $m+1 \le i \le n-m+1$,

$$J_i = I\Big(\frac{1-H_B(U_m)}{1-U_m}S_{im} > a\Big) \le I(S_{im} > a) = I_i. \tag{1.3}$$

Similarly, for small m-spacings, taking $G_S(x) = \mathbb{P}(U_m \le x \mid U_m < a)$ and $H_S(x) = G_S^{-1}(F(x)) \le x$, we find for $m+1 \le i \le n-m+1$ that

$$I\Big(\frac{1-H_S(U_m)}{1-U_m}S_{im} < a\Big) \le I(S_{im} < a), \tag{1.4}$$

the indicators having the appropriate distributions. Hence we can use

$$\Gamma_1^+ = \emptyset,\ \Gamma_1^- = \{m+1,\ldots,n-m+1\},\ \Gamma_1^0 = \{n-m+2,\ldots,n,2,\ldots,m\}, \tag{1.5}$$

in Corollary 2.C.1, and so on.

Remark 7.1.1 We have here constructed explicit couplings. Alternatively, note that $\mathcal{L}(S_1,\ldots,S_n) = \mathcal{L}(X_1,\ldots,X_n \mid \sum_{i=1}^n X_i = 1)$, where the X_i are independent random variables with a common exponential distribution, and thus Proposition 2.2.10 implies that the S_i are negatively associated. The existence of couplings with the above properties then follows easily from Theorem 2.J: for $m = 1$, we can use Theorem 2.I.

7.2 Ordinary spacings

In this section we study the case $m = 1$. The number of big spacings is

$$W_B = \sum_{k=1}^{n} I_k = \sum_{k=1}^{n} I(S_k > a)$$

with $\mathbb{E}W_B = n(1-a)^{n-1}$ and

$$\operatorname{Var} W_B = n(1-a)^{n-1} - n(1-a)^{2n-2} + n(n-1)((1-2a)_+^{n-1} - (1-a)^{2n-2}). \tag{2.1}$$

Using the coupling (1.3), we see that the I's are negatively related. Thus the assumptions of Corollary 2.C.2 are satisfied, and we have:

Theorem 7.A *If W_B denotes the number of big spacings,*

$$d_{TV}(\mathcal{L}(W_B), \mathrm{Po}(\mathbb{E}W_B)) \leq (1 - \exp(-\mathbb{E}W_B))\Big(1 - \frac{\operatorname{Var} W_B}{\mathbb{E}W_B}\Big). \qquad \square$$

The same result is obtained in Barbour and Holst (1989) Theorem 6.1, using a slightly different coupling.

For $a \leq \frac{1}{2}$ we have

$$1 - \frac{\operatorname{Var} W_B}{\mathbb{E}W_B} \leq e^{-(n-1)a}\Big\{\Big(1 + \Big[\frac{(n-1)a}{1-a}\Big]^2\Big) \wedge n\Big\}. \tag{2.2}$$

It is easily seen that, if $n \to \infty$ and $a \to 0$, then $\lambda = \mathbb{E}W_B \to \lambda_\infty > 0$ if and only if $na = \log n - \log \lambda_\infty + o(1)$, in which case, for each j,

$$\mathbb{E}\{I_1 \dots I_j\} = (1 - ja)_+^{n-1} = (\mathbb{E}I_1)^j \big(1 - \tbinom{j}{2} na^2 + O(n^2 a^4)\big).$$

Hence, using Proposition 3.2.12, we obtain

$$\varepsilon_- \sim \lambda n a^2, \quad \gamma_- \sim 7\varepsilon_-,$$

and thus, by Theorem 3.D, $na^2 \sim (\log n)^2/n$ is the correct order of Poisson convergence. This can also be seen from the explicit calculation

$$\begin{aligned} d_{TV}(\mathcal{L}(W_B), \mathrm{Po}(\mathbb{E}W_B)) &\geq |\mathbb{P}[W_B = 0] - \exp(-\mathbb{E}W_B)| \\ &= \Big|\sum_{j=0}^{\infty}(-1)^j \binom{n}{j}(1 - ja)_+^{n-1} - \sum_{j=0}^{\infty}(-1)^j \frac{n^j}{j!}(1-a)_+^{j(n-1)}\Big| \\ &\sim cn^{-1}(\log n)^2, \end{aligned}$$

for some $c > 0$.

If $na \to \infty$, (2.2) and Theorem 7.A imply that

$$\delta_B = d_{TV}(\mathcal{L}(W_B), \mathrm{Po}(\mathbb{E}W_B)) \to 0.$$

Computing the variance from (2.1), we find that $\liminf na < \infty$ entails $\limsup \operatorname{Var} W_B/\mathbb{E}W_B < 1$. If furthermore $\liminf \mathbb{E}W_B > 0$, then $\delta_B \not\to 0$, by Theorem 3.A. (If $\mathbb{E}W_B \to 0$, then trivially $\delta_B \to 0$.) These results are summarized as follows.

Corollary 7.A.1 *For big spacings,*
(a) *if* $\liminf \mathbb{E}W_B > 0$, *then*

$$d_{TV}(\mathcal{L}(W_B), \mathrm{Po}(\mathbb{E}W_B)) \to 0 \iff \frac{\mathrm{Var}\, W_B}{\mathbb{E}W_B} \to 1 \iff na \to \infty;$$

(b) *if* $\mathbb{E}W_B \to \lambda_\infty > 0$, *that is* $na = \log n - \log \lambda_\infty + \eta$, *where* $\eta = o(1)$ *as* $n \to \infty$, *then* $W_B \xrightarrow{\mathcal{D}} \mathrm{Po}(\lambda_\infty)$ *at rate* $O(|\eta| + (\log n)^2/n)$;
(c) *if* $na \to \infty$ *and* $\mathbb{E}W_B \to \infty$, *that is* $na - \log n \to -\infty$, *then*

$$\frac{W_B - \mathbb{E}W_B}{\sqrt{\mathrm{Var}\, W_B}} \xrightarrow{\mathcal{D}} N(0,1). \qquad \square$$

For the number of small spacings

$$W_S = \sum_{k=1}^{n} I_k = \sum_{k=1}^{n} I(S_k < a)$$

we have

$$\mathbb{E}W_S = \mathbb{E}(n - W_B) = n(1-(1-a)^{n-1}), \ \mathrm{Var}\, W_S = \mathrm{Var}\,(n - W_B) = \mathrm{Var}\, W_B.$$

Using the coupling (1.4) and Corollary 2.C.2, we obtain the following theorem.

Theorem 7.B *If* W_S *denotes the number of small spacings,*

$$d_{TV}(\mathcal{L}(W_S), \mathrm{Po}(\mathbb{E}W_S)) \le (1 - \exp(-\mathbb{E}W_S))\left(1 - \frac{\mathrm{Var}\, W_S}{\mathbb{E}W_S}\right). \qquad \square$$

In circumstances where $n \to \infty$ and $a \to 0$, $\mathbb{E}W_S \to \lambda_\infty > 0$ if and only if $na = n^{-1}(\lambda_\infty + o(1))$, in which case $1 - \mathrm{Var}\, W_S/\mathbb{E}W_S \sim 2\lambda_\infty/n$. Furthermore, if $\mathbb{E}W_S \to \lambda_\infty$, then

$$\begin{aligned}\mathbb{P}[W_S = 0] &= \mathbb{P}[S_k > a,\ k = 1, \ldots, n] \\ &= (1 - na)_+^{n-1} = \exp(-\mathbb{E}W_S)\left(1 - \frac{\lambda_\infty^2 + o(1)}{n}\right).\end{aligned}$$

Thus

$$d_{TV}(\mathcal{L}(W_S), \mathrm{Po}(\mathbb{E}W_S)) \ge |\mathbb{P}[W_S = 0] - \exp(-\mathbb{E}W_S)| \sim \frac{\lambda_\infty^2 e^{-\lambda_\infty}}{n}.$$

Therefore n^{-1} is the correct order of convergence for W_S to $\mathrm{Po}(\lambda_\infty)$. This can also be proved using Theorem 3.D and Proposition 3.2.12.

Elementary calculations using the mean value theorem show that

$$1 - (1-a)^{n-1} \le 1 - \mathrm{Var}\, W_S/\mathbb{E}W_S \le 2na, \qquad (2.3)$$

and thus $\mathrm{Var}\, W_S/\mathbb{E}W_S \to 1 \iff na \to 0$. This, together with Theorem 3.A, enables us to prove an analogue of Corollary 7.A.1.

Corollary 7.B.1 *For small spacings,*

(a) *if* $\liminf \mathbb{E}W_S > 0$, *then* $d_{TV}(\mathcal{L}(W_S), \mathrm{Po}(\mathbb{E}W_S)) \to 0$ *if and only if* $na \to 0$;

(b) *if* $\mathbb{E}W_S \to \lambda_\infty \geq 0$, *that is* $na = (\lambda_\infty + \eta)/n$ *with* $\eta = o(1)$, *then* $W_S \xrightarrow{\mathcal{D}} \mathrm{Po}(\lambda_\infty)$ *at rate* $O(|\eta| + \frac{1}{n})$;

(c) *if* $na \to 0$ *and* $\mathbb{E}W_S \to \infty$, *that is* $n^2 a \to \infty$, *then*

$$\frac{W_S - \mathbb{E}W_S}{\sqrt{\mathrm{Var}\, W_S}} \xrightarrow{\mathcal{D}} N(0,1). \qquad \square$$

Remark 7.2.1 Using the estimates of the variation distances, Corollaries 7.A.1(c) and 7.B.1(c) could be complemented with rates of convergence to the normal distribution.

Remark 7.2.2 The distributional convergence of W_B and W_S has been studied before, see Holst and Hüsler (1984), but the rates of convergence seem to be new.

Remark 7.2.3 In the remaining case, where $n \to \infty$ and $a \to 0$ in such a way that $na \to \alpha$, $0 < \alpha < \infty$, there is normal convergence; see Holst and Hüsler (1984). Note that then

$$\frac{\mathbb{E}W_B}{n} = 1 - \frac{\mathbb{E}W_S}{n} = (1-a)^{n-1} \to e^{-\alpha};$$
$$\frac{\mathrm{Var}\, W_B}{n} = \frac{\mathrm{Var}\, W_S}{n} \to e^{-\alpha}(1-e^{-\alpha}) - \alpha^2 e^{-2\alpha},$$

so that $\mathbb{E}W_B \not\sim \mathrm{Var}\, W_B$, and, as remarked above, Poisson approximation cannot hold because of Theorem 3.A.

7.3 m-Spacings

In this section we only consider the spacings $\{S_{km}\}$ with $m > 1$ fixed; W_{Bm} (W_{Sm}) is then the number of them that are bigger (smaller) than a, as in (0.1) above. The order terms of this section could depend on m, but not on n or a. We assume that $n \geq 2m$.

Applying Corollary 2.C.1 as described in (1.5), we immediately obtain the following theorem.

Theorem 7.C *For big m-spacings,*

$$d_{TV}(\mathcal{L}(W_{Bm}), \mathrm{Po}(\mathbb{E}W_{Bm}))$$
$$\leq (1 - \exp(-\mathbb{E}W_{Bm}))\Big(1 - \frac{\mathrm{Var}\, W_{Bm}}{\mathbb{E}W_{Bm}} + 4\sum_{j=2}^{m} \mathbb{P}[S_{jm} > a \mid S_{1m} > a]\Big).$$

For small m-spacings,

$$d_{TV}(\mathcal{L}(W_{Sm}), \mathrm{Po}(\mathbb{E}W_{Sm}))$$
$$\leq (1-\exp(-\mathbb{E}W_{Sm}))\Big(1 - \frac{\mathrm{Var}\, W_{Sm}}{\mathbb{E}W_{Sm}} + 4\sum_{j=2}^{m} \mathbb{P}[S_{jm} < a \mid S_{1m} < a]\Big). \quad \square$$

With $I_k = I(S_{km} > a)$, $\lambda = \mathbb{E}W_{Bm}$, we see that the theorem implies that

$$d_{TV}(\mathcal{L}(W_{Bm}), \mathrm{Po}(\lambda)) \leq \frac{1-e^{-\lambda}}{\lambda}\Big(\frac{2m\lambda^2}{n} - n^2 \mathrm{Cov}(I_1, I_{m+1}) + 2n\sum_{j=2}^{m} \mathbb{E}I_1 I_j\Big). \tag{3.1}$$

The same holds if W_{Bm} is replaced by W_{Sm} and if $I_k = I(S_{km} < a)$. In the following lemmas, the different quantities appearing in (3.1) are estimated.

Lemma 7.3.1 *If $n \to \infty$ and $a \to 0$ in such a way that $na \to +\infty$, then*

$$\mathbb{E}W_{Bm} = n\frac{(na)^{m-1}}{(m-1)!}(1-a)^n(1+O((na)^{-1})+O(a)),$$

and $\mathbb{E}W_{Bm} \to \lambda_\infty$, $0 \leq \lambda_\infty \leq \infty$, if and only if

$$na - \log n - (m-1)\log\log n + \log(m-1)! \to -\log\lambda_\infty.$$

Proof From formula (1.1) and the hypothesis,

$$\begin{aligned}
\mathbb{E}W_{Bm} &= n\sum_{j=0}^{m-1}\binom{n-1}{j}a^j(1-a)^{n-1-j} \\
&= n\sum_{j=0}^{m-1}\frac{(na)^j}{j!}(1-a)^n(1+O(n^{-1})+O(a)) \\
&= n\frac{(na)^{m-1}}{(m-1)!}(1-a)^n(1+O((na)^{-1})+O(a)),
\end{aligned}$$

which implies that

$$\begin{aligned}
na - \log n - (m-1)\log na + \log(m-1)! - n(\log(1-a)+a) \\
= -\log\mathbb{E}W_{Bm} + o(1).
\end{aligned}$$

If $\mathbb{E}W_{Bm} \to \lambda_\infty > 0$, then it is clear from this formula that $na \sim \log n$ and hence that

$$na - \log n - (m-1)\log\log n + \log(m-1)! = -\log\lambda_\infty + o(1),$$

from which the other two cases follow easily. $\square$

Lemma 7.3.2 *If $n \to \infty, a \to 0$ and $na \to \infty$, then, for $I_j = I(S_{jm} > a)$,*

$$\mathbb{E}I_1 I_{j+1} = \frac{(na)^{m-j-1}(1-a)^n}{(m-j-1)!}\binom{2j}{j}(1+O((na)^{-1})+O(a)), \quad 1 \le j < m,$$

and

$$\mathrm{Cov}\,(I_1, I_{m+1}) = \mathbb{P}[S_{1m} > a]^2\Big(-1 + (1-(\tfrac{a}{1-a})^2)^n + O(n^{-1}) + O(a)\Big).$$

Proof We have

$$\mathbb{E}I_1 I_{j+1} = \mathbb{P}[S_{1m} > a \cap S_{j+1,m} > a] = \mathbb{P}[S_{j+1,m-j} > a]$$
$$+\mathbb{P}\big[\{S_{j+1,m-j} < a\} \cap \{S_{1j} + S_{j+1,m-j} > a\} \cap \{S_{j+1,m-j} + S_{m+1,j} > a\}\big],$$

and, as in Lemma 7.3.1,

$$\mathbb{P}[S_{j+1,m-j} > a] = \frac{(na)^{m-j-1}(1-a)^n}{(m-j-1)!}(1+O((na)^{-1})+O(a)).$$

Assuming that $a \le 1/2$, direct computation now shows that

$$\mathbb{P}\big[\{S_{j+1,m-j} < a\} \cap \{S_{1j} + S_{j+1,m-j} > a\} \cap \{S_{j+1,m-j} + S_{m+1,j} > a\}\big]$$
$$= \int_0^a f_{S_{j+1,m-j}}(y)\mathbb{P}\big[\{S_{1j} > a-y\} \cap \{S_{m+1,j} > a-y\} \mid S_{j+1,m-j} = y\big]\,dy$$
$$= \int_0^a \frac{(n-1)!y^{m-j-1}(1-y)^{n-1-m+j}}{(m-j-1)!(n-1-m+j)!}$$
$$\times \sum_{k=0}^{j-1}\sum_{l=0}^{j-1} \frac{(n-1-m+j)!}{k!l!(n-1-m+j-k-l)!}$$
$$\times \left(1 - \frac{2(a-y)}{1-y}\right)^{n-1-m+j-k-l}\left(\frac{a-y}{1-y}\right)^{k+l} dy$$
$$= \sum_{k=0}^{j-1}\sum_{l=0}^{j-1} \frac{n^{k+l+m-j}(1-a)^n}{k!l!(m-j-1)!}(1+O(n^{-1})+O(a))$$
$$\times \int_0^a y^{m-j-1}(a-y)^{k+l}\left(1-\frac{a-y}{1-a}\right)^{n-1-m+j-k-l} dy$$
$$= \sum_0^{j-1}\sum_0^{j-1}\binom{k+l}{l}\frac{(na)^{k+l+m-j}}{(k+l)!(m-j-1)!}$$
$$\times \int_0^1 (1-t)^{m-j-1}t^{k+l}\left(1-\frac{at}{1-a}\right)^n dt \cdot (1-a)^n(1+O(n^{-1})+O(a))$$

$$= \sum_0^{j-1}\sum_0^{j-1}\binom{k+l}{l}\frac{(na)^{k+l+m-j}}{(m-j-1)!}\sum_{s=0}^{m-j-1}\binom{m-j-1}{s}(-1)^s(na)^{-(k+l+s+1)}$$

$$\times\int_0^{\frac{na}{1-a}}\frac{u^{k+l+s}}{(k+l)!}\Big(1-\frac{u}{n}\Big)^n du\cdot(1-a)^n(1+O(n^{-1})+O(a))$$

$$= \frac{(na)^{m-j-1}(1-a)^n}{(m-j-1)!}\{1+O(a)+O((na)^{-1})\}$$

$$\times\sum_0^{j-1}\sum_0^{j-1}\binom{k+l}{l}\int_0^{\frac{na}{1-a}}\frac{u^{k+l}}{(k+l)!}\Big(1-\frac{u}{n}\Big)^n du$$

$$= \frac{(na)^{m-j-1}(1-a)^n}{(m-j-1)!}\sum_0^{j-1}\sum_0^{j-1}\binom{k+l}{l}\{1+O(a)+O((na)^{-1})\}.$$

Thus, using the identity

$$1+\sum_0^{j-1}\sum_0^{j-1}\binom{k+l}{l}=\binom{2j}{j},$$

the first statement of the lemma follows. For proving the second, note that

$$\begin{aligned}\mathbb{E}I_1I_{m+1} &= \mathbb{P}[S_{1m}>a\cap S_{n-m+1,m}>a]\\ &= \mathbb{P}[\text{at most } m-1 \text{ points in each of } (0,a) \text{ and } (1-a,1)].\end{aligned}$$

Hence

$$\begin{aligned}\operatorname{Cov}(I_1,I_{m+1}) &= \sum_{s=0}^{m-1}\sum_{t=0}^{m-1}\frac{(n-1)!}{s!t!(n-1-s-t)!}a^{s+t}(1-2a)^{n-1-s-t}\\ &\qquad -\Big(\sum_{u=0}^{m-1}\frac{(n-1)!}{u!(n-1-u)!}a^u(1-a)^{n-1-u}\Big)^2\\ &= \sum_{s=0}^{m-1}\sum_{t=0}^{m-1}\frac{(na)^{s+t}}{s!\,t!}\Big\{(1+O(n^{-1})+O(a))(1-2a)^n\\ &\qquad -(1+O(n^{-1})+O(a))(1-a)^{2n}\Big\}\\ &= (\mathbb{P}[S_{1m}>a])^2\Big(-1+(1-(\frac{a}{1-a})^2)^n+O(n^{-1})+O(a)\Big),\end{aligned}$$

as required. □

Using the estimates of Lemmas 7.3.1 and 7.3.2 in (3.1) gives the next result.

Corollary 7.C.1 *For big m-spacings with $m > 1$ fixed, if $n \to \infty$ and $a \to 0$ in such a way that $na \to \infty$, then*

(a) $d_{TV}(\mathcal{L}(W_{Bm}), \mathrm{Po}(\mathbb{E}W_{Bm}))$

$$\le (1 - \exp(-\mathbb{E}W_{Bm}))(\frac{4(m-1)}{na} + O((\frac{1}{na})^2) + O(\frac{1}{n}));$$

(b) *if $\mathbb{E}W_{Bm} \to \lambda_\infty \ge 0$, then $W_{Bm} \xrightarrow{\mathcal{D}} \mathrm{Po}(\lambda_\infty)$ at rate*

$$O(|\mathbb{E}W_{Bm} - \lambda_\infty| + \frac{1}{\log n});$$

(c) *if $\mathbb{E}W_{Bm} \to \infty$, then*

$$\frac{W_{Bm} - \mathbb{E}W_{Bm}}{\sqrt{\mathrm{Var}\, W_{Bm}}} \xrightarrow{\mathcal{D}} N(0,1). \qquad \square$$

Small spacings are now investigated in a similar way.

Lemma 7.3.3 *Suppose $n \to \infty$ and $a \to 0$ in such a way that $na \to 0$. Then*

$$\mathbb{E}W_{Sm} = n\frac{(na)^m}{m!}(1 + O(na) + O(n^{-1})),$$

and, for every $0 \le \lambda_\infty \le \infty$,

$$\mathbb{E}W_{Sm} \to \lambda_\infty \iff n^{1+\frac{1}{m}}a \to (m!\lambda_\infty)^{\frac{1}{m}}.$$

Proof We immediately have

$$\begin{aligned}\mathbb{E}W_{Sm} = n\mathbb{P}[U_m < a] &= \frac{n!}{(m-1)!(n-m-1)!}\int_0^a u^{m-1}(1-u)^{n-m-1}du \\ &= n^{m+1}(1 + O(n^{-1}))\int_0^a (1 + O(na))\frac{u^{m-1}}{(m-1)!}du \\ &= n\frac{(na)^m}{m!}(1 + O(na) + O(n^{-1})),\end{aligned}$$

proving the first statement, and the second then follows directly. $\square$

Lemma 7.3.4 *Suppose $n \to \infty$, $a \to 0$ and $na \to 0$. Then, for $1 \le j < m$ and with $I_j = I(S_{jm} < a)$,*

$$\mathbb{E}I_1 I_{j+1} = \mathbb{P}[S_{1m} < a](na)^j \frac{m!}{(m+j)!}\binom{2j}{j}(1 + O(na) + O(n^{-1})),$$

and

$$\mathrm{Cov}\,(I_1, I_{m+1}) = -m^2 n^{-1}(\mathbb{P}[S_{1m} < a])^2(1 + O(na) + O(n^{-1})).$$

Proof With $A = \{(x, y, z) : x + y < a, y + z < a, x, y, z \geq 0\}$ we compute, assuming $a \leq 1/2$,

$$\begin{aligned}
\mathbb{E} I_1 I_{j+1} &= \mathbb{P}[S_{1m} < a \cap S_{j+1,m} < a] \\
&= \frac{(n-1)!}{((j-1)!)^2 (m-j-1)!(n-m-j-1)!} \\
&\quad \times \iiint_A x^{j-1} y^{m-j-1} z^{j-1} (1-x-y-z)^{n-m-j-1} dx\, dy\, dz \\
&= \frac{n^{m+j}(1+O(\frac{1}{n}))}{((j-1)!)^2 (m-j-1)!} \\
&\quad \times \int_0^a y^{m-j-1} \iint_{0<x,z<a-y} x^{j-1} z^{j-1} (1-x-y-z)^{n-m-j-1} dx\, dz\, dy \\
&= \frac{n^{m+j}}{(j!)^2 (m-j-1)!} \int_0^a y^{m-j-1} (a-y)^{2j} (1 + O(na) + O(n^{-1}))\, dy \\
&= \frac{(na)^{m+j}}{(m+j)!} \binom{2j}{j} (1 + O(na) + O(n^{-1})).
\end{aligned}$$

Thus the first statement follows from the previous lemma. In a similar way, we get

$$\begin{aligned}
&\operatorname{Cov}(I_1, I_{m+1}) \\
&= \frac{(n-1)!}{((m-1)!)^2 (n-1-2m)!} \int_0^a \int_0^a x^{m-1} y^{m-1} (1-x-y)^{n-1-2m} dx\, dy \\
&\quad - \Big(\frac{(n-1)!}{(m-1)!(n-m-1)!}\Big)^2 \Big(\int_0^a x^{m-1}(1-x)^{n-1-m} dx\Big)^2 \\
&= \frac{n^{2m}(1+O(\frac{1}{n}))}{((m-1)!)^2} \Big[\int_0^a \int_0^a x^{m-1} y^{m-1} (1-x-y)^{n-1-2m} dx\, dy \\
&\quad - \Big(1 + \frac{m^2}{n} + O\Big(\frac{1}{n^2}\Big)\Big) \int_0^a \int_0^a x^{m-1} y^{m-1} (1-x-y+xy)^{n-1-m} dx\, dy\Big] \\
&= \frac{n^{2m}(1+O(\frac{1}{n}))}{((m-1)!)^2} \\
&\quad \times \int_0^a \int_0^a x^{m-1} y^{m-1} ((1-x-y)^{n-1-2m} - (1-x-y+xy)^{n-1-m})\, dx\, dy \\
&\quad - \Big(\frac{m^2}{n} + O\Big(\frac{1}{n^2}\Big)\Big) (\mathbb{P}[S_{1m} < a])^2 \\
&= (\mathbb{P}[S_{1m} < a])^2 (O(a) + O(na^2) - m^2 n^{-1} + O(n^{-2})). \qquad \square
\end{aligned}$$

The next corollary is obtained by combining these estimates with (3.1).

Corollary 7.C.2 *For small m-spacings with $m > 1$ fixed, if $n \to \infty$ and $a \to 0$ in such a way that $na \to 0$, then*

(a) $d_{TV}(\mathcal{L}(W_{Sm}), \mathrm{Po}(\mathbb{E}W_{Sm})) \leq (1 - \exp(-\mathbb{E}W_{Sm}))(\frac{4na}{m+1} + O((na)^2))$;

(b) *if* $\mathbb{E}W_{Sm} \to \lambda_\infty \geq 0$, *then* $W_{Sm} \xrightarrow{\mathcal{D}} \mathrm{Po}(\lambda_\infty)$ *at rate*
$O(|\mathbb{E}W_{Sm} - \lambda_\infty| + n^{-\frac{1}{m}})$;

(c) *if* $\mathbb{E}W_{Sm} \to \infty$, *then*

$$\frac{W_{Sm} - \mathbb{E}W_{Sm}}{\sqrt{\mathrm{Var}\, W_{Sm}}} \xrightarrow{\mathcal{D}} N(0,1). \qquad \square$$

Remark 7.3.5 In the remaining case where $n \to \infty$ and $a \to 0$ in such a way that $na \to \alpha$, $0 < \alpha < \infty$, normal convergence holds; see Holst (1979).

Remark 7.3.6 The Poisson convergence of W_{Bm} and W_{Sm} is proved in Holst (1980a) and Cressie (1977) respectively. No rates of convergence are given in these papers.

Remark 7.3.7 It follows from the results obtained in Section 3.1 that the rates in Corollaries 7.C.1 and 7.C.2 are almost sharp. Note in this connection that the proof of Proposition 3.1.6 can be modified, using the couplings of Section 7.1, to bound an exponential moment of W_{Bm} or W_{Sm}.

Remark 7.3.8 There is a great difference in the rates of Poisson convergence for W_{Bm} between the cases $m = 1$ and $m > 1$; the rate is $(\log n)^2/n$ for $m = 1$ as compared to $1/\log n$ for $m > 1$. This is because of the relatively high tendency for clusters of large m-spacings to occur when $m > 1$, and very much improved approximation could be expected using a compound Poisson distribution.

7.4 Higher dimensions

It is possible to pose analogous problems in higher dimensions, though there are many different ways in which to do so. The circle is most naturally replaced by a d-dimensional compact homogeneous Riemannian manifold T, such as a sphere or a torus, upon which n points $Y_1, \ldots, Y_n$ are placed independently and at random, being sampled from the uniform distribution over T. The spacing associated with the ith point can then be defined to be the distance to the nearest neighbour:

$$S_i = \min_{j \neq i} d(Y_j, Y_i), \tag{4.1}$$

for a suitable choice of metric d. This formula does not reduce in one dimension to the previous definition, which also makes use of the circular ordering; in particular, using (4.1) as definition, the two smallest spacings are necessarily equal, so that a Poisson distribution is not a good model for the number of small spacings. The simplest modification in this case is to consider instead the number of small interpoint distances, which, if small is defined in such a way that there are relatively few of them, comes to almost the same as half the number of small spacings. Since small interpoint distances are considered in Section 2.3 and, in greater detail, in Section 9.3, we concentrate here upon the number of large spacings.

Fix $a > 0$, and define $I_i = I[S_i > a]$ and $W = \sum_{i=1}^n I_i$. Let $\pi_1 = \mathbb{E}I_1$ and $\lambda = \mathbb{E}W = n\pi_1$ as usual, and let s denote the fraction of the area (or volume) $|T|$ taken up by each set of the form $B_a(y) = \{y' : d(y', y) \le a\}$: thus $\pi_1 = (1-s)^{n-1}$. In the argument that follows, it is assumed that a is so small that the following two conditions are satisfied:

$$\text{(i)} \qquad |B_{2a}(y)| \le (2^d + 1)|B_a(y)|;$$

$$\text{(ii)} \qquad |B_a(y) \cap B_a(y')| \le \frac{1}{2}|B_a(y)| \quad \text{if} \quad a \le d(y, y') \le 2a.$$

This requires that $a \le a_0$, for some $a_0 = a_0(T)$ depending on the injectivity radius of T and on a bound for its sectional curvature, and we also assume that a_0 is such that $|B_{a_0}(y)| < |T|/2$: for spheres, for instance, this last condition is already enough. Then we have the following theorem.

Theorem 7.D *With W the number of large spacings as defined above,*

$$d_{TV}(\mathcal{L}(W), \mathrm{Po}(\lambda)) \le (1 - e^{-\lambda}) \times \left[\pi_1\left\{1 + \frac{ns}{1-s} + \frac{2n^2s^2}{(1-s)^2}\right\} + \pi_1^{1/2} 2^d \left\{\frac{nse}{1-s}\right\}\right],$$

provided that $a \le a_0(T)$ and $n\lambda \ge 2^{2d}$.

Proof The idea is to apply Theorem 2.A, using an explicit coupling to realize random variables U and V on the same probability space in such a way that $\mathcal{L}(U) = \mathcal{L}(W)$ and $\mathcal{L}(V+1) = \mathcal{L}(W \mid I_1 = 1)$ and that $\mathbb{E}|U - V|$ is small: by exchangeability, $(1 - e^{-\lambda})\mathbb{E}|U - V|$ is then an upper estimate for $d_{TV}(\mathcal{L}(W), \mathrm{Po}(\lambda))$.

To accomplish this, realize the Y_i's as usual, and use them to construct the I_i's and W. Then construct a new set of points $(Y_i', 1 \le i \le n)$ by defining $Y_1' = Y_1$ and

$$Y_i' = \begin{cases} Y_i & \text{if } d(Y_i, Y_1) > a; \\ Z_i & \text{otherwise,} \end{cases} \qquad 2 \le i \le n, \tag{4.2}$$

where $(Z_i, 2 \le i \le n)$ are independent of each other and of the Y_i's, and are distributed uniformly over $T \setminus \{y : d(y, Y_1) \le a\}$. The Y_i''s are now used to construct random variables I_i' and W' in the same way as the I_i's and W were constructed from the Y_i's, U is set equal to W and $V = W' - 1$.

It is clear from the construction that U and V have the right distributions. Furthermore, most of the Y_i''s are the same as the corresponding Y_i's, so that U and V differ only on account of differences between indicators I_i and I_i' occasioned by the re-definition of relatively few of the Y_i's. Considering the various ways in which moving points can change W, it follows that

$$\begin{aligned}|U-V| \le I_1 &+ \sum_{i=2}^{n} I[Y_i' \neq Y_i]\Big\{\sum_{j\neq 1,i}^{n} I_j I[d(Y_i',Y_j)\le a] + I_i'\Big\} \\ &+ \sum_{i\neq 1} I[a < d(Y_1,Y_i) \le 2a] \prod_{j\neq i,1} I\big[\{d(Y_j,Y_i) > a\}\cup\{d(Y_j,Y_1)\le a\}\big]. \end{aligned} \tag{4.3}$$

Now, in order to evaluate $\mathbb{E}|U-V|$, observe that

$$\mathbb{E}\{I[Y_i' \neq Y_i] I_j I[d(Y_i',Y_j)\le a]\} \le \pi_1 s^2/(1-s)^2,$$

by conditioning first on $I_j = 1$, and that

$$\mathbb{E}\{I[Y_i'\neq Y_i]I_i'\} \le s\Big[\big(1-\frac{s}{1-s}\big)^{n-2} + \frac{2^d s}{1-s}\big(1-\frac{s}{2(1-s)}\big)^{n-2}\Big],$$

where the second term arises because, if $d(Y_i',Y_1) < 2a$, each of the Y_j''s has probability bigger than $(1-s/(1-s))$ of not being close to Y_i'. Finally, for any y such that $a < d(Y_1,y)\le 2a$,

$$\mathbb{E}\Big(\prod_{j\neq i,1} I\big[\{d(Y_j,Y_i)>a\}\cup\{d(Y_j,Y_1)\le a\}\big]\,\Big|\, Y_i = y\Big) \le (1-\tfrac{1}{2}s)^{n-2},$$

so that

$$\begin{aligned}&\mathbb{E}\Big(\sum_{i\neq 1} I[a<d(Y_1,Y_i)\le 2a]\prod_{j\neq i,1} I\big[\{d(Y_j,Y_i)>a\}\cup\{d(Y_j,Y_1)\le a\}\big]\Big)\\ &\le n2^d s(1-\tfrac{1}{2}s)^{n-2} \le \pi_1^{1/2}2^d\Big\{\frac{ns}{1-s}\Big\}\exp\{\tfrac{1}{4}(n-1)s^2\}.\end{aligned}$$

Putting these estimates into (4.3) gives the inequality

$$\begin{aligned}\mathbb{E}|U-V| &\le \pi_1\{1+n^2s^2/(1-s)^2 + ns/(1-s)\}\\ &+ \pi_1^{1/2}2^d\Big\{\frac{ns}{1-s}\Big\}\Big(\frac{s}{1-s} + \exp\{\tfrac{1}{4}(n-1)s^2\}\Big).\end{aligned} \tag{4.4}$$

The theorem now follows because of the restriction on the size of $n\lambda$: note, in particular, that as a consequence $(n-1)s^2/4 \le 1$. □

Remark 7.4.1 When $n \to \infty$ and $\pi_1 \to 0$, the leading term in Theorem 7.D is that of order $ns\pi_1^{1/2}$. This, when compared with (2.2) for the circle, gives only the square root of the precision obtained in approximating the distribution of the number of ordinary large spacings. The difference reflects the change in the meaning of a spacing. Using the definition given in (4.1) on the circle, requiring that a point i have $S_i > a$ just means that its two immediate neighbours must be at least a distance a away from it. The conditional probability that either of these neighbours is similarly isolated is then almost exactly $\pi_1^{1/2}$, since the event only forces away one further immediate neighbour. Indeed, modifying the argument of Theorem 7.D in the case of the circle, one obtains a leading term which is asymptotically $2\pi_1^{1/2}$. For spheres of dimension $d \geq 2$, the leading term can immediately be improved to order $ns\pi^{\gamma(d)}$ for $\frac{1}{2} < \gamma(d) < 1$, where $\gamma(d)$ is given by $\gamma(d) = \lim_{a\to 0} \gamma(d, a)$ and

$$\gamma(d, a) = 1 - |B_a(y) \cap B_a(y')|/|B_a(y)|,$$

for any pair y, y' such that $d(y, y') = a$: for example,

$$\gamma(2) = \tfrac{1}{3} + \tfrac{\sqrt{3}}{2\pi}.$$

However, if more accurate approximation is needed, it would be better to compare the distribution to a suitably chosen compound Poisson distribution.

Remark 7.4.2 If one takes $s = n^{-1}\alpha \log n$, it follows that $\pi_1 \sim n^{-\alpha}$ and that $\lambda \sim n^{1-\alpha}$. Hence, for $0 < \alpha \leq 1$, the error estimate is of order $n^{-\alpha/2} \log n$.

The coupling used here is quite different in spirit from that used for Theorem 7.A. For Theorem 7.A, the coupling was based on two configurations of points, the second obtained from the first by shifting all but one of the points a very short distance. Here, only a few points are shifted, but once a point is shifted there is no restriction on how far it may be moved. It is therefore interesting to ask whether a coupling of the kind used for Theorem 7.A could have been used here too. It is relatively easy, for instance if T is a sphere and d the Euclidean metric, to construct an analogous explicit transformation which moves points short distances (in this case, radially as seen from Y_1) in such a way as to transform a typical configuration with $S_1 = a_1$ into a typical configuration with $S_1' = a_2$. This transformation could be combined with Theorem 2.A to yield an error estimate, but would require a rather more complicated argument than the one above. The main advantage obtained from the coupling used in proving Theorem 7.A was, however, that it was monotone, and hence that Corollary 2.C.2 was directly applicable. Here, by considering areas, if $a_2 > a_1$, any transformation which preserves the right distributions must bring many pairs of points closer together — this is for instance the case for pairs of points on a common radius from Y_1 in the transformation alluded to above — but points at distances between a_1 and a_2 from Y_1 must be moved further away from Y_1. Hence a monotone coupling cannot in general be achieved, and the proof as given is preferred to one based on moving almost all the points a very short distance, because it avoids the geometrical difficulties involved with transformations of the latter kind.

7.5 An application to DNA breakage

Cowan, Collis and Grigg (1987) have studied an interesting stochastic process which models a common type of enzymatic or radiation damage to certain kinds of DNA in bacteria, where the double-stranded DNA is topologically a circle. The agent of damage creates single-stranded breaks, known as 'nicks'. These nicks have independent positions equally likely to occur on each of the two strands and uniformly distributed around a strand. If two nicks on different strands occur within a critical distance d of each other, the circle of DNA breaks and becomes topologically linear. Let the random variable M be defined as the number of nicks needed to break the circle. Cowan, Collis and Grigg, using results for spacings, found that

$$\mathbb{P}(M > n) = \sum_{k=0}^{\infty} \binom{n}{2k} 2^{-n+1}(1 - kb)_+^{n-1},$$

where b is the ratio of $2d$ to the circumference of the circle.

As b gets smaller, the computations with the explicit expression for $\mathbb{P}(M > n)$ become too lengthy for DNA applications. Asymptotic results concerning M as $b \to 0$ are therefore of interest as approximations. Such approximations are given by Cowan, Culpin and Gates (1990) using the above explicit formula for $\mathbb{P}(M > n)$ and the saddle-point method. Holst (1990) simplified the derivation of the formula above and obtained an approximation using the result of Barbour and Eagleson (1984) on dissociated statistics; that is, Theorem 2.N. We now give some further results.

Consider a circle of circumference 2 on which points $Y_1, Y_2, \ldots,$ are taken independently of each other and with a uniform distribution. Here one half of the circle corresponds to one strand and the other half to the other strand. Set

$$I_{ij} = I\left(1 - \frac{b}{2} < |Y_i - Y_j| < 1 + \frac{b}{2}\right),$$

which indicates a break by the points Y_i, Y_j.

The number of breaks by n points can be written as

$$W_n = \sum\sum_{1 \le i < j \le n} I_{ij}.$$

Clearly the I's are *dissociated*, with $\mathbb{E}W_n = \binom{n}{2}b/2$. Furthermore, for any given i_0, the distribution of $\{|Y_i - Y_j|\}_{1\le i<j\le n}$ given the value of Y_{i_0} is the same as its unconditional distribution, and hence the I's are also *strongly dissociated*. Hence Theorem 2.O can be used, giving the estimate

$$d_{TV}\left(\mathcal{L}(W_n), \mathrm{Po}\left(\tfrac{1}{4}n(n-1)b\right)\right) \le \tfrac{1}{2}(1 - e^{-n(n-1)b/4})(2n-3)b,$$

which is roughly half the bound which would follow using dissociation alone, in conjunction with Theorem 2.N. Thus, in particular,

$$\mathbb{P}(M > n) = \mathbb{P}(W_n = 0) \approx e^{-n(n-1)b/4},$$

an approximation which is of interest when $n^2 b \asymp 1$.

It might be more natural to let the number of points, N, be random with a Poisson distribution with mean μ, say. The number of breaks is then W_N, with mean

$$\mathbb{E}W_N = \mathbb{E}\binom{N}{2}\frac{b}{2} = \frac{\mu^2 b}{4}.$$

It now follows from Corollary 2.P.1 that

$$d_{TV}(\mathcal{L}(W_N), \mathrm{Po}(\tfrac{1}{4}\mu^2 b)) \le (1 - e^{-\mu^2 b/4})\mu b.$$

It may also be of interest to consider more general models, for which explicit closed formulas are not available. The critical distances could, for example, be random or the positions be generated by a non-uniform distribution. The approach to Poisson approximation used above can easily be extended to such cases, the important assumption in the model being the independence between the 'nicks'. In these more general situations, the indicators are dissociated but no longer strongly dissociated, and the bounds obtained are correspondingly less tight.

8
Exceedances and extremes

Consider a collection of random variables $\{X_\alpha, \alpha \in \Gamma\}$, and let W count for how many α's X_α exceeds z_α, where the levels $\{z_\alpha, \alpha \in \Gamma\}$ are any real numbers. If the probabilities of exceedance are small and there is no clustering among the X's, a Poisson approximation for $\mathcal{L}(W)$ is likely to be appropriate. This can for example be used to derive asymptotic distributions for the extreme values of the X's. A comprehensive treatment of extreme value theory for dependent random variables is to be found in Leadbetter, Lindgren and Rootzén (1983). The Stein–Chen method was first used to simplify and sharpen some of the classical results by Smith (1988), who applied the local approach to the Stein–Chen method in conjunction with blocking arguments, as mentioned in Chapter 1. A slightly different approach, exploiting the sort of conditions advocated by Berman (1987), is discussed in the context of random fields by Barbour and Greenwood (1991).

In this chapter, we study the Poisson approximation of $\mathcal{L}(W)$ using the Stein–Chen method and coupling. In Section 8.1, the X's are assumed to have a jointly Gaussian distribution. The upper bound for the variation distance between $\mathcal{L}(W)$ and $\mathrm{Po}(\mathbb{E}W)$ given in Theorem 2.L is further developed, and applied in the context of stationary Gaussian sequences. In Section 8.2 we study the scan statistic, that is, the number of clusters of close points, for the Poisson process. The couplings needed are explicitly constructed in a way similar to that used for spacings in Chapter 7. Associated random variables, in particular moving averages of independent random variables with non-negative weights, are studied in Section 8.3. The existence of good couplings now follows directly from the FKG inequality (2.2.1). Although the scan statistic for the Poisson process is a special case of such moving averages, it is illuminating to consider it first, before discussing the general case, because the couplings can be made explicit. In Section 8.4, counts of runs and other patterns in sequences of independent random letters are examined, and the last section is concerned with the number of visits of an ergodic Markov chain to a rare set.

8.1 The Gaussian case

Let $\{X_\alpha : \alpha \in \Gamma\}$ be jointly normally distributed random variables. Let $\{t_\alpha^+, t_\alpha^- : \alpha \in \Gamma\}$ be positive numbers, and set

$$I_\alpha^+ = I[X_\alpha > t_\alpha^+], \qquad I_\alpha^- = I[X_\alpha < -t_\alpha^-], \qquad I_\alpha = I_\alpha^+ + I_\alpha^-;$$
$$\pi_\alpha^+ = \mathbb{P}[X_\alpha > t_\alpha^+], \qquad \pi_\alpha^- = \mathbb{P}[X_\alpha < -t_\alpha^-], \qquad \pi_\alpha = \pi_\alpha^+ + \pi_\alpha^-;$$

$$W = \sum_{\alpha\in\Gamma} I_\alpha, \qquad \lambda = \mathbb{E}W = \sum_{\alpha\in\Gamma} \pi_\alpha.$$

Define W_+, W_-, λ_+ and λ_- analogously, and set

$$S = \sum\sum_{\alpha,\beta\in\Gamma,\alpha\neq\beta} \{|\mathrm{Cov}(I_\alpha^+, I_\beta^+)| + |\mathrm{Cov}(I_\alpha^+, I_\beta^-)| + |\mathrm{Cov}(I_\alpha^-, I_\beta^+)| + |\mathrm{Cov}(I_\alpha^-, I_\beta^-)|\}. \tag{1.1}$$

Then we have the following result.

Theorem 8.A *For W defined as above,*

$$d_{TV}(\mathcal{L}(W), \mathrm{Po}(\lambda)) \leq \frac{1-e^{-\lambda}}{\lambda}\Big(\sum_{\alpha\in\Gamma} \pi_\alpha^2 + S\Big), \tag{1.2}$$

where S is given by (1.1), and

$$d_{TV}(\mathcal{L}(W_+), \mathrm{Po}(\lambda_+)) \leq \frac{1-e^{-\lambda_+}}{\lambda_+}\Big(\sum_{\alpha\in\Gamma} (\pi_\alpha^+)^2 + \sum\sum_{\alpha,\beta\in\Gamma,\alpha\neq\beta} |\mathrm{Cov}(I_\alpha^+, I_\beta^+)|\Big), \tag{1.3}$$

a similar result being true for W_-.

Proof Set $X_\alpha^* = -X_\alpha$. Then

$$W = \sum_{\alpha\in\Gamma} I[X_\alpha > t_\alpha^+] + \sum_{\alpha\in\Gamma} I[X_\alpha^* > t_\alpha^-].$$

Theorem 2.L applied to the variables $\{X_\alpha\} \cup \{X_\alpha^*\}$, with two copies of Γ as index set, yields

$$d_{TV}(\mathcal{L}(W), \mathrm{Po}(\lambda)) \leq \frac{1-e^{-\lambda}}{\lambda}\Big(\sum_{\alpha\in\Gamma} (\mathbb{E}I_\alpha^+)^2 + \sum_{\alpha\in\Gamma} (\mathbb{E}I_\alpha^-)^2 + \sum_{\alpha\in\Gamma} (|\mathrm{Cov}(I_\alpha^+, I_\alpha^-)| + |\mathrm{Cov}(I_\alpha^-, I_\alpha^+)|) + S\Big).$$

The first assertion follows because $\mathrm{Cov}(I_\alpha^+, I_\alpha^-) = -\mathbb{E}I_\alpha^+ \mathbb{E}I_\alpha^-$, and thus

$$(\mathbb{E}I_\alpha^+)^2 + (\mathbb{E}I_\alpha^-)^2 + 2|\mathrm{Cov}(I_\alpha^+, I_\alpha^-)| = (\mathbb{E}I_\alpha^+ + \mathbb{E}I_\alpha^-)^2 = \pi_\alpha^2.$$

The second assertion is equivalent to Theorem 2.L. □

Remark 8.1.1 The fixed levels t_α^+, t_α^- in Theorem 8.A may be replaced by random levels that are independent of each other and of $\{X_\alpha\}$. We leave the details to the reader.

Remark 8.1.2 Taking all t_α^+ and t_α^- equal to t, the event $\{W = 0\}$ is the same as $\{\max|X_\alpha| \le t\}$. Consequently Theorem 8.A gives upper bounds for $|\mathbb{P}[\max|X_\alpha| \le t] - e^{-\lambda}|$ and $|\mathbb{P}[\max X_\alpha \le t] - e^{-\lambda_+}|$. Estimates for the distribution of the kth largest value may be derived similarly.

Theorem 8.A is simple to apply numerically in specific examples, as the bound only involves the one- and two-dimensional normal distribution functions. It is also easy to use to obtain asymptotic results. This is illustrated using the following estimates for the covariances. Let Φ and φ denote the standard normal distribution and density functions.

Lemma 8.1.3 *Let (X, Y) have the $N(\binom{0}{0}, \binom{1\,r}{r\,1})$ distribution. If $0 \le r < 1$, then, for any real a and b,*

$$(1-\Phi(a))\Big(1 - \Phi\Big(\frac{b - ra}{\sqrt{1-r^2}}\Big)\Big) \le \mathbb{P}[X > a \text{ and } Y > b]$$
$$\le (1 - \Phi(a))\Big[\Big(1 - \Phi\Big(\frac{b - ra}{\sqrt{1-r^2}}\Big)\Big) + r\frac{\varphi(b)}{\varphi(a)}\Big(1 - \Phi\Big(\frac{a - rb}{\sqrt{1-r^2}}\Big)\Big)\Big].$$

If $-1 < r \le 0$, the inequalities are reversed.

Proof Integrating by parts, we get

$$\begin{aligned}\mathbb{P}[X > a \text{ and } Y > b] &= \int_a^\infty \varphi(x)\Big(1 - \Phi\Big(\frac{b - rx}{\sqrt{1-r^2}}\Big)\Big)dx \\ &= (1 - \Phi(a))\Big(1 - \Phi\Big(\frac{b - ra}{\sqrt{1-r^2}}\Big)\Big) \\ &\quad + \int_a^\infty (1 - \Phi(x))\varphi\Big(\frac{b - rx}{\sqrt{1-r^2}}\Big)\frac{r}{\sqrt{1-r^2}}dx.\end{aligned}$$

If $0 \le r < 1$, the lower bound follows immediately. But since the function $(1 - \Phi(x))/\varphi(x)$ is decreasing, it also follows that

$$\begin{aligned}&\int_a^\infty (1 - \Phi(x))\varphi\Big(\frac{b - rx}{\sqrt{1-r^2}}\Big)\frac{dx}{\sqrt{1-r^2}} \\ &\qquad\le \frac{1 - \Phi(a)}{\varphi(a)}\int_a^\infty \varphi(x)\varphi\Big(\frac{b - rx}{\sqrt{1-r^2}}\Big)\frac{dx}{\sqrt{1-r^2}} \\ &\qquad= \frac{1 - \Phi(a)}{\varphi(a)}\int_a^\infty \varphi(b)\varphi\Big(\frac{x - rb}{\sqrt{1-r^2}}\Big)\frac{dx}{\sqrt{1-r^2}} \\ &\qquad= (1 - \Phi(a))\frac{\varphi(b)}{\varphi(a)}\Big(1 - \Phi\Big(\frac{a - rb}{\sqrt{1-r^2}}\Big)\Big),\end{aligned}$$

giving the upper bound. The same argument works if $-1 < r \le 0$, with the inequalities reversed. □

Lemma 8.1.4 *Let $\{z_n\}$ be a sequence such that $\lambda_n = n(1-\Phi(z_n)) \le K$ for some constant K. Let (X_α, X_β) be $N(\binom{0}{0}, \binom{1r}{r1})$, and let $I_\alpha = I[X_\alpha > z_n]$, $I_\beta = I[X_\beta > z_n]$. Then, for some constants C depending on K only, and for all $n \ge 2$,*

$$\text{(i) } \textit{if} \quad 0 \le r < 1, \quad 0 \le \operatorname{Cov}(I_\alpha, I_\beta) \le C\frac{1}{n^2\sqrt{1-r}}\Big(\frac{n^2}{\log n}\Big)^{\frac{r}{1+r}};$$

$$\text{(ii) } \textit{if} \quad 0 \le r \le 1, \quad 0 \le \operatorname{Cov}(I_\alpha, I_\beta) \le C\frac{r\log n}{n^2}e^{2r\log n};$$

$$\text{(iii) } \textit{if} \ -1 \le r \le 0, \quad 0 \ge \operatorname{Cov}(I_\alpha, I_\beta) \ge -C\frac{|r|\log n}{n^2};$$

$$\text{(iv) } \textit{if} \ -1 \le r \le 0, \quad 0 \ge \operatorname{Cov}(I_\alpha, I_\beta) \ge -Cn^{-2}.$$

Proof For part (i), we may assume that $n > 7K$ and thus $z_n > 1$. Lemma 8.1.3 then yields

$$\begin{aligned}
0 \le \operatorname{Cov}(I_\alpha, I_\beta) \ &\le \ \mathbb{E}(I_\alpha I_\beta) \ \le \ (1-\Phi(z_n))(1+r)\Big(1-\Phi\Big(z_n\sqrt{\frac{1-r}{1+r}}\Big)\Big) \\
&\le (1+r)z_n^{-1}\varphi(z_n)\Big(z_n\sqrt{\frac{1-r}{1+r}}\Big)^{-1}\varphi\Big(z_n\sqrt{\frac{1-r}{1+r}}\Big) \\
&= (1+r)^{3/2}(1-r)^{-1/2}\Big(\frac{\varphi(z_n)}{z_n}\Big)^{1+\frac{1-r}{1+r}} z_n^{-\frac{2r}{1+r}}(2\pi)^{-\frac{r}{1+r}} \\
&\le C(1-r)^{-1/2}\Big(\frac{1}{n}\Big)^{1+\frac{1-r}{1+r}}(\log n)^{-\frac{r}{1+r}}.
\end{aligned}$$

In the last inequality we use the fact that the next to last term is a decreasing function of z_n; hence we may assume that $1-\Phi(z_n) = K/n$ and thus $z_n \sim \sqrt{2\log n}$ and $\varphi(z_n)/z_n \sim K/n$.

For part (ii), we may assume that $z_n > 0$. Then, by Lemma 8.1.3,

$$\begin{aligned}
&\operatorname{Cov}(I_\alpha, I_\beta) \\
&\quad \le (1-\Phi(z_n))\Big[(1+r)\Big(\Phi(z_n)-\Phi\Big(z_n\sqrt{\frac{1-r}{1+r}}\Big)\Big)+r(1-\Phi(z_n))\Big] \\
&\quad \le (1+r)\frac{\varphi(z_n)}{z_n}\Big(z_n-\sqrt{\frac{1-r}{1+r}}z_n\Big)\varphi\Big(z_n\sqrt{\frac{1-r}{1+r}}\Big)+r(1-\Phi(z_n))^2 \\
&\quad \le 2r\varphi(z_n)\varphi\Big(z_n\sqrt{\frac{1-r}{1+r}}\Big)+C\frac{r}{n^2} \\
&\quad = 2rz_n^{\frac{2}{1+r}}\Big(\frac{\varphi(z_n)}{z_n}\Big)^{1+\frac{1-r}{1+r}}(2\pi)^{-\frac{r}{1+r}}+C\frac{r}{n^2} \\
&\quad \le Cr(\log n)^{\frac{1}{1+r}}n^{-2+\frac{2r}{1+r}} \le C\frac{1}{n^2}r(\log n)e^{2r\log n}.
\end{aligned}$$

For part (iii), we may assume that $z_n > 1$. Then Lemma 8.1.3 yields

$$\begin{aligned}
0 &\le -\mathrm{Cov}\,(I_\alpha, I_\beta) \\
&\le (1-\Phi(z_n))\Big[(1+r)\Big(\Phi\Big(z_n\sqrt{\frac{1-r}{1+r}}\Big) - \Phi(z_n)\Big) - r(1-\Phi(z_n))\Big] \\
&\le \frac{\varphi(z_n)}{z_n}(1+r)\Big(\sqrt{\frac{1-r}{1+r}} - 1\Big) z_n\varphi(z_n) + |r|\Big(\frac{\varphi(z_n)}{z_n}\Big)^2 \\
&\le |r|\Big(\frac{\varphi(z_n)}{z_n}\Big)^2 (z_n^2+1) \le C|r|\frac{\log n}{n^2}.
\end{aligned}$$

Finally, for part (iv), $\mathrm{Cov}\,(I_\alpha, I_\beta) \ge -\mathbb{E}I_\alpha \mathbb{E}I_\beta \ge -(K/n)^2$. □

Remark 8.1.5 Of the estimates above, (ii) and (iii) are sharp to within a constant when $|r|\log n$ is bounded and λ_n is bounded from below, while (i) and (iv) are sharp when both $|r|\log n$ and λ_n are bounded from below, provided $r \le r_0 < 1$ and $z_n \ge 0$. This can be proved using the left inequality in Lemma 8.1.3 and calculations similar to those used above.

As an example, consider a stationary Gaussian process. The following result gives rates of convergence for the number of exceedances; see Leadbetter, Lindgren and Rootzén (1983) Chapter 4, where results concerning the convergence of extremes are proved by different methods.

Theorem 8.B *Let $\{X_k\}$ be a standardized stationary normal sequence with covariances $\{r_k\}$ satisfying $r_k \le A/\log k$, $k \ge 2$, for some constant A. Let $\rho = \max(0, r_1, r_2, \ldots)$. Let λ_n and z_n be real numbers such that $\lambda_n = n(1-\Phi(z_n))$ and define $W_n = \sum_{k=1}^n I[X_k > z_n]$. Suppose that $\lambda_n \le B < \infty$, for some constant B. Then*

$$d_{TV}(\mathcal{L}(W_n), \mathrm{Po}\,(\lambda_n)) = O\Big(n^{-\frac{1-\rho}{1+\rho}}(\log n)^{-\frac{\rho}{1+\rho}} + \frac{\log n}{n}\sum_{k=1}^n |r_k|\Big).$$

Remark 8.1.6 Note that $0 \le \rho < 1$, since $r_k = 1$ for some $k \ge 1$ would imply $r_{mk} = 1$, $m \ge 1$, which contradicts $r_k \le A/\log k$.

Proof Using Theorem 8.A and writing $I_k = I[X_k > z_n]$, we have

$$d_{TV}(\mathcal{L}(W_n), \mathrm{Po}\,(\lambda_n)) \le \frac{\lambda_n^2}{n} + 2n\sum_{k=1}^{n-1} |\mathrm{Cov}\,(I_0, I_k)|.$$

If $\rho = 0$, that is $r_k \le 0$ for every $k > 0$, the result follows by Lemma 8.1.4(iii). Let us now assume that $\rho > 0$. Choose $\delta > 0$ with $3\delta < \frac{\rho}{1+\rho}$. We divide the sum $\sum |\mathrm{Cov}\,(I_0, I_k)|$ into four parts.

(i) Since $r_k \leq A/\log k$, only a finite number of k have $r_k > \delta$. From Lemma 8.1.4(i), each of these contributes

$$\mathrm{Cov}\,(I_0, I_k) \leq Cn^{-2}\Big(\frac{n^2}{\log n}\Big)^{r_k/(1+r_k)} \leq Cn^{-2}\Big(\frac{n^2}{\log n}\Big)^{\rho/(1+\rho)},$$

so that their total contribution is $O\big(n^{-2}(n^2/\log n)^{\rho/(1+\rho)}\big)$.

(ii) Next consider the terms with $0 \leq r_k \leq \delta$ and $k < n^\delta$. There are at most n^δ such terms, and by Lemma 8.1.4(i) each contributes

$$\mathrm{Cov}\,(I_0, I_k) \leq Cn^{-2}\Big(\frac{n^2}{\log n}\Big)^{\delta/(1+\delta)} \leq Cn^{-2+2\delta}.$$

Hence their total contribution is at most

$$Cn^{-2+3\delta} = o\bigg(n^{-2}\Big(\frac{n^2}{\log n}\Big)^{\rho/(1+\rho)}\bigg).$$

(iii) For the terms with $r_k \geq 0$ and $k \geq n^\delta$, $0 \leq r_k \leq \frac{A}{\log k} \leq \frac{A}{\delta \log n}$. Hence Lemma 8.1.4(ii) yields $0 \leq \mathrm{Cov}\,(I_0, I_k) \leq Cr_k n^{-2}\log n$ for each such term, and their total contribution is $O(n^{-2}\log n \sum_1^n |r_k|)$.

(iv) The remaining terms are those with $r_k < 0$, and Lemma 8.1.4(iii) shows that their total contribution is $O(n^{-2}\log n \sum_1^n |r_k|)$.

Combining these estimates, we have

$$\frac{\lambda_n^2}{n} + 2n\sum_{k=1}^{n-1} |\mathrm{Cov}\,(I_0, I_k)| = O\bigg(n^{-1}\Big(\frac{n^2}{\log n}\Big)^{\rho/(1+\rho)} + \frac{\log n}{n}\sum_1^n |r_k|\bigg),$$

and the result follows. □

Remark 8.1.7 As a corollary, we see that if $r_k \log k \to 0$ as $k \to \infty$ and $\lambda_n \to \lambda < \infty$, then $W_n \xrightarrow{\mathcal{D}} \mathrm{Po}\,(\lambda)$. In particular, $\mathbb{P}[\max_{k \leq n} X_k \leq z_n] \to e^{-\lambda}$, a result first proved by Berman (1964). Note that $r_k \log k = O(1)$ is not sufficient for this limit result; Mittal and Ylvisaker (1975) have shown that if $r_k \log k \to \gamma > 0$, then W_n converges to a mixture of Poisson distributions: see Leadbetter, Lindgren and Rootzén (1983) Section 6.5.

Remark 8.1.8 The assumption that $r_k \log k$ is bounded above was used only to conclude that $r_k \to 0$ and that $|\{k \leq n : r_k > M/\log n\}| \leq n^\delta$ for some $M < \infty$; it could be replaced for example by $\sum_1^\infty \exp\{-M/|r_k|\} < \infty$, for some M, or $\sum_1^\infty |r_k|^p < \infty$ for some $p < \infty$. It is also easy to obtain limit theorems with explicit rates of convergence for the other conditions on r_k discussed in Leadbetter, Lindgren and Rootzén (1983) Section 4.5.

Remark 8.1.9 In most applications, for example to ARMA processes, where r_k decreases exponentially, $\sum_1^\infty |r_k| < \infty$. In this case it follows from the theorem that

$$d_{TV}(\mathcal{L}(W_n), \mathrm{Po}(\lambda_n)) = O(n^{-\frac{1-\rho}{1+\rho}}(\log n)^{-\frac{\rho}{1+\rho}})$$

when $\rho > 0$, and $d_{TV}(\mathcal{L}(W_n), \mathrm{Po}(\lambda_n)) = O(\frac{\log n}{n})$ when $\rho = 0$. The latter can in fact easily be improved to $o(\frac{\log n}{n})$; and when $\{X_k\}$ is m-dependent and $\rho = 0$ one gets $O(n^{-1})$, using Lemma 8.1.4(iv).

Remark 8.1.10 It follows from Theorem 8.A that if $|r_k| \le A/\log k$, the estimate in Theorem 8.B holds also for $W_n = \sum_{k=1}^n I[|X_k| > z_n]$, with $\lambda_n = 2n(1 - \Phi(z_n))$ and $\rho = \max|r_k|$.

It is clear that the methods used above also apply to non-stationary sequences; this gives results similar to those in Leadbetter, Lindgren and Rootzén (1983) Sections 6.1–6.3, but provides error estimates as well. We may also consider processes with other index sets. As an example, we give a version of Theorem 8.B for stationary processes with multi-dimensional indices. The proof is the same as before, with n replaced by $|B_n|$.

Theorem 8.C *Let $\{X_\mathbf{k} : \mathbf{k} \in \mathbb{Z}^d\}$ be a standardized stationary normal process with covariances $\{r_\mathbf{k}\}$ satisfying* $\sup r_\mathbf{k} \log|\mathbf{k}| < \infty$. *Let $\rho = \max\{0, r_\mathbf{k} : \mathbf{k} \neq 0\}$. Let B_n be subsets of $\mathbb{Z}^d$ and let $\widetilde{B}_n = \{\mathbf{k} - \mathbf{l} : \mathbf{k}, \mathbf{l} \in B_n\}$. Let λ_n and z_n be positive numbers such that $\lambda_n = |B_n|(1 - \Phi(z_n))$ and define $W_n = \sum_{\mathbf{k} \in B_n} I[X_\mathbf{k} > z_n]$. If $\{\lambda_n\}$ is bounded, then*

$$d_{TV}(\mathcal{L}(W_n), \mathrm{Po}(\lambda_n)) = O\Big(|B_n|^{-\frac{1-\rho}{1+\rho}}(\log|B_n|)^{-\frac{\rho}{1+\rho}} + \frac{\log|B_n|}{|B_n|}\sum_{\mathbf{k} \in \widetilde{B}_n} |r_\mathbf{k}|\Big).$$

□

8.2 The scan statistic for Poisson processes

Let $Z_1, \ldots, Z_n$ be independent identically distributed exponential random variables with mean 1. Set $Z_{k+n} = Z_k$ for all k. For $m \ge 2$ and $b > 0$ define

$$I_k = I[Z_k + \cdots + Z_{k+m-1} < b],$$

and introduce the (circular) scan statistic $W = I_1 + \cdots + I_n$. There are many important applications where such random variables appear; see Alm (1983), Janson (1984) and the references therein. The exact distribution of W is complicated, unless n is very small, but provided that

$$\mathbb{P}[Z_1 + \cdots + Z_m < b] = \sum_{k=m}^\infty \frac{b^k}{k!}e^{-b}$$

is small, one could contemplate approximating $\mathcal{L}(W)$ by $\mathrm{Po}(\mathbb{E}W)$.

Theorem 8.D *For the circular scan statistic W,*

$$d_{TV}(\mathcal{L}(W), \mathrm{Po}(\lambda)) \leq (1 - e^{-\lambda})\Big(\frac{\mathrm{Var}\, W}{\lambda} - 1 + \frac{2\lambda}{n}\Big),$$

where $\lambda = \mathbb{E}W$ as usual.

Proof The aim is to use Corollary 2.C.4. The existence of a monotone coupling could be deduced from Theorem 2.G, as in the next section, but we prefer here to exhibit an explicit coupling, assuming, for simplicity, that $2m \leq n$. By symmetry, it suffices to take $\alpha = k = 1$. Set $T_m = Z_1 + \cdots + Z_m$. The random variables $Z_1/T_m, \ldots, Z_m/T_m$ are the successive distances, that is spacings, between $m-1$ points taken at random on the unit interval including 0 and 1; see Chapter 7. The spacings are independent of T_m. Let F and G be the distribution functions of $\mathcal{L}(T_m)$ and $\mathcal{L}(T_m \mid T_m < b)$ respectively. The random variable $H(T_m) = G^{-1}(F(T_m))$ has the distribution $\mathcal{L}(T_m \mid T_m < b)$ and is independent of the spacings. Thus

$$\mathcal{L}(Z_1, \ldots, Z_m \mid T_m < b) = \mathcal{L}\Big(H(T_m)\Big(\frac{Z_1}{T_m}, \ldots, \frac{Z_m}{T_m}\Big)\Big).$$

For $1 < i \leq m$ set

$$J_{i1} = I\Big[(Z_i + \cdots + Z_m)\frac{H(T_m)}{T_m} + Z_{m+1} + \cdots + Z_{m+i-1} < b\Big].$$

As $H(s) \leq s$, we have

$$J_{i1} \geq I[Z_i + \cdots + Z_{m+i-1} < b] = I_i.$$

A similar construction is used for J_{i1} when $n - m < i \leq n$, and for $m < i \leq n - m$ we set $J_{i1} = I_i$. A coupling with $\Gamma_1^+ = \{2, \ldots, n\}$ has thus been constructed. □

In order to make use of Theorem 8.D, estimates of $\mathbb{E}W$ and $\mathrm{Var}\, W$ are needed.

Lemma 8.2.1 *For $2m \leq n$, the variance of W is given by*

$$\frac{\mathrm{Var}\, W}{\lambda} = 1 + 2n \sum_{k=1}^{m-1} \frac{\mathbb{E}I_1 I_{k+1}}{\lambda} - \frac{(2m-1)\lambda}{n}. \tag{2.1}$$

Estimation of the right hand side is facilitated by the observations that

$$(\mathbb{E}I_1)^2 \leq \mathbb{E}I_1 I_{k+1} \leq \mathbb{E}I_1 I_k \leq \mathbb{E}I_1, \tag{2.2}$$

and that

$$\mathbb{E}I_1 I_{k+1} = \mathbb{P}[T_{m-k} + \max(T_k', T_k'') < b], \tag{2.3}$$

where T_{m-k}, T_k' and T_k'' are independent, and distributed as $\Gamma(m-k, 1)$, $\Gamma(k, 1)$ and $\Gamma(k, 1)$ respectively.

Proof The expression for the variance follows from the $(m-1)$-dependence. For $1 \leq k \leq m-1$ and $T_0 = T_0' = T_0'' = 0$, we have

$$\begin{aligned}
\mathbb{E} I_1 I_{k+1} &= \mathbb{P}[Z_1 + \cdots + Z_m < b \text{ and } Z_{k+1} + \cdots + Z_{k+m} < b] \\
&= \mathbb{P}[T_k' + T_{m-k} < b \text{ and } T_{m-k} + T_k'' < b] \\
&= \mathbb{P}[T_{m-k} + \max(T_k', T_k'') < b] \\
&\leq \mathbb{P}[T_{m-k+1} + \max(T_{k-1}', T_{k-1}'') < b] = \mathbb{E} I_1 I_k. \qquad \square
\end{aligned}$$

Lemma 8.2.2 *For $2m \leq n$ and $4b < m$, we have the estimates*

$$\mathbb{E} W = \lambda = n \frac{b^m e^{-b}}{m!} \left(1 + O\left(\frac{b}{m}\right)\right);$$

$$\frac{\operatorname{Var} W}{\lambda} = 1 + \frac{4b}{m+1} + O\left(\left(\frac{b}{m}\right)^2\right); \tag{2.4}$$

$$\frac{\operatorname{Var} W}{\lambda} - 1 \leq \frac{4b}{m+1}\left(1 - \frac{2b}{m}\right)^{-2}\left(1 - \frac{4b}{m}\right)^{-1} - \frac{(2m-1)\lambda}{n}. \tag{2.5}$$

Proof The estimate of the mean is immediate. For the variance, we have

$$\begin{aligned}
\mathbb{E} I_1 I_2 &= \mathbb{P}[T_{m-1} + \max(T_1', T_1'') < b] = \mathbb{P}[T_m + Z_0/2 < b] \\
&= \int_0^b \frac{u^{m-1}}{(m-1)!} e^{-u} \left(1 - e^{-2(b-u)}\right) du \\
&= 2\left(\frac{b^{m+1}}{(m+1)!} e^{-b} + \frac{b^{m+3}}{(m+3)!} e^{-b} + \ldots\right),
\end{aligned}$$

which implies that

$$\frac{\mathbb{E} I_1 I_2}{\mathbb{E} I_1} = \frac{2b}{m+1}\left(1 + O\left(\frac{b}{m}\right)\right).$$

Hence, because of the positive correlations, it follows from (2.1) that

$$\frac{\operatorname{Var} W}{\lambda} \geq 1 + 2\frac{\mathbb{E} I_1 I_2}{\mathbb{E} I_1} - 3\mathbb{E} I_1 = 1 + \frac{4b}{m+1} + O\left(\left(\frac{b}{m}\right)^2\right),$$

and the proof of (2.4) is completed once (2.5) has been established.

Let f_i, F_i be the density and the distribution function of $\Gamma(i, 1)$ respectively. Then, for $1 \le k \le m-1$,

$$
\begin{aligned}
\mathbb{E}I_1 I_{k+1} &= \int_0^b f_{m-k}(b-t)(F_k(t))^2 dt \\
&= e^{-b} \int_0^b \frac{(b-t)^{m-k-1}}{(m-k-1)!} \Big(\sum_{j=k}^{\infty} \frac{t^j}{j!}\Big)^2 e^{-t} dt \\
&\le e^{-b} \int_0^b \sum_{i,j \ge k} \frac{(b-t)^{m-k-1}}{(m-k-1)!} \frac{t^{i+j}}{i!j!} dt \\
&= e^{-b} \sum_{i,j \ge k} \frac{b^{m+i+j-k}}{(m-k-1)!i!j!} \frac{(m-k-1)!(i+j)!}{(m+i+j-k)!} \\
&= e^{-b} \sum_{i,j \ge k} \frac{b^{m+i+j-k}}{(m+i+j-k)!} \frac{(i+j)!}{i!j!} \le e^{-b} \sum_{i,j \ge k} \frac{b^{m+i+j-k}\, 2^{i+j-1}}{(m+k)!m^{i+j-2k}} \\
&= e^{-b} \frac{b^{m+k}}{(m+k)!} 2^{2k-1} \Big(1 - \frac{2b}{m}\Big)^{-2}.
\end{aligned}
$$

Hence it follows that

$$
\begin{aligned}
\sum_{k=1}^{m-1} \mathbb{E}I_1 I_{k+1} &\le \Big(1 - \frac{2b}{m}\Big)^{-2} \sum_{k=1}^{\infty} e^{-b} \frac{b^{m+k}\, 2^{2k-1}}{(m+1)!m^{k-1}} \\
&= 2\Big(1 - \frac{2b}{m}\Big)^{-2} \Big(1 - \frac{4b}{m}\Big)^{-1} e^{-b} \frac{b^{m+1}}{(m+1)!} \\
&\le \frac{2b}{m+1} \Big(1 - \frac{2b}{m}\Big)^{-2} \Big(1 - \frac{4b}{m}\Big)^{-1} \mathbb{E}I_1,
\end{aligned}
$$

which proves (2.5). □

The results of Theorem 8.D and Lemma 8.2.2 can be combined to give the following result.

Corollary 8.D.1 *For the circular scan statistic W, if $4b < m \le n/2$, then*

$$
d_{TV}(\mathcal{L}(W), \mathrm{Po}(\lambda)) \le (1 - e^{-\lambda}) \frac{4b}{m+1} \Big(1 - \frac{2b}{m}\Big)^{-2} \Big(1 - \frac{4b}{m}\Big)^{-1}. \quad (2.6)
$$

□

Remark 8.2.3 If, instead of the cyclic definition used above, we assume that the $\{Z_k\}_{k \ge 1}$ are all independent standard exponential random variables, the variance is slightly smaller. Lemma 8.2.2 and Corollary 8.D.1 hold as stated, except for the variance estimate in (2.4), where $4b/(m+1)$ must be changed to $(1 - \frac{1}{n})4b/(m+1)$.

Remark 8.2.4 If m and $\lambda = \lambda_\infty > 0$ are kept fixed, then by the results above, W converges to $\mathrm{Po}(\lambda_\infty)$ as $n \to \infty$ at a rate $O(b) = O(n^{-1/m})$. Theorem 3.E shows that this rate is sharp: see Remark 3.2.11.

Remark 8.2.5 If $m \to \infty$, $b/m \to c \in (0,1)$ and $\mathbb{E}W \to \lambda_\infty > 0$, it follows from similar estimates that $\mathrm{Var}\, W \not\to \lambda_\infty$. By Theorem 3.B* or Corollary 3.C.1, W can then not be well approximated by any Poisson distribution, but a compound Poisson distribution gives a good approximation: see Theorem 10.N. In Alm (1983) and Janson (1984), very accurate approximations are given in a similar case.

Remark 8.2.6 Results similar to those above can be obtained for indicators defined as $I_k = I[Z_k + \cdots + Z_{k+m-1} > b]$.

8.3 Extremes of associated random variables

Recall from Section 2.2 that the random variables $\{X_\alpha, \alpha \in \Gamma\}$ are *associated* if, for all increasing functions f, g,

$$\mathrm{Cov}\,(f(X_\alpha; \alpha \in \Gamma), g(X_\alpha; \alpha \in \Gamma)) \geq 0.$$

Theorem 2.G then implies that for all families of real numbers $\{z_\alpha\}$ the indicators $I_\alpha = I[X_\alpha > z_\alpha]$, $\alpha \in \Gamma$, are *positively related*. Hence Corollary 2.C.4 yields:

Theorem 8.E *Let $\{X_\alpha; \alpha \in \Gamma\}$ be associated random variables and let $\{z_\alpha; \alpha \in \Gamma\}$ be real numbers. Set*

$$I_\alpha = I[X_\alpha > z_\alpha], \quad \pi_\alpha = \mathbb{E}I_\alpha = \mathbb{P}[X_\alpha > z_\alpha], \quad W = \sum_{\alpha \in \Gamma} I_\alpha, \quad \lambda = \mathbb{E}W.$$

Then

$$d_{TV}(\mathcal{L}(W), \mathrm{Po}\,(\lambda)) \leq (1 - e^{-\lambda})\Big(\frac{\mathrm{Var}\, W}{\lambda} - 1 + 2\frac{\sum \pi_\alpha^2}{\lambda}\Big). \qquad \square$$

Central limit theorems for sums of associated random variables have been studied before; see Birkel (1988) and the references therein. But no general results on extreme value theory for such variables seem to have been obtained previously. Such results follow readily from Theorem 8.E. This will be illustrated for the important special case of moving averages, with non-negative weights, of independent identically distributed random variables. The scan statistic studied in the previous section is a special case of this, after trivial modifications as in Remark 8.2.3.

From now on in this section, let $\{Z_i;\ i \in \mathbb{Z}\}$ be independent identically distributed random variables and let $\{c_i;\ i \in \mathbb{Z}\}$ be non-negative real numbers such that the moving averages

$$X_k = \sum_{i=-\infty}^{\infty} c_i Z_{k-i}, \quad k = 0, 1, \cdots, n,$$

define a stationary sequence; see Leadbetter, Lindgren and Rootzén (1983) Section 3.8. As the weights are non-negative, the X's are increasing functions of the independent Z's, and it follows from Proposition 2.2.2 that the X's are associated. Thus Theorem 8.E implies the following result.

Corollary 8.E.1 *Let W be the number of exceedances above the level z of the moving average defined above, that is $W = \sum_{k=1}^{n} I[X_k > z]$, and set $\lambda = \mathbb{E}W$. Then*

$$d_{TV}(\mathcal{L}(W), \mathrm{Po}(\lambda)) \le (1 - e^{-\lambda})\Big(\frac{\mathrm{Var}\, W}{\lambda} - 1 + 2\mathbb{P}[X_0 > z]\Big). \qquad \square$$

Convergence theorems can also be deduced if $\lambda/n \to 0$ and $\mathrm{Var}\, W/\lambda \to 1$.

For the results to be useful in applications, one has to bound $\mathrm{Var}\, W/\lambda$ and $\mathbb{P}[X_0 > z]$ from above. The next proposition requires only straightforward calculations.

Proposition 8.3.1

$$\begin{aligned}\frac{\mathrm{Var}\, W}{\lambda} &= 1 - \frac{\lambda}{n} + 2\sum_{j=1}^{n} \frac{n-j}{n}(\mathbb{P}[X_j > z \mid X_0 > z] - \mathbb{P}[X_j > z]) \\ &\le 1 + 2\sum_{j=1}^{n}\Big(\frac{\mathbb{P}[X_0 + X_j > 2z]}{\mathbb{P}[X_0 > z]} - \mathbb{P}[X_j > z]\Big). \qquad \square\end{aligned}$$

For estimates of tail probabilities of moving averages, see Rootzén (1987) and the references therein. For special processes it is also possible to estimate or to calculate the probability $\mathbb{P}[X_j > z \mid X_0 > z]$ directly.

Example 8.3.2 Let the Z's be uniformly distributed on the unit interval (0,1) and consider the moving average $X_k = Z_k + Z_{k-1}$, $k = 1, \ldots, n$. Elementary calculation gives $\lambda = n\mathbb{P}\big[Z_k + Z_{k-1} > 2 - \sqrt{2\lambda/n}\big]$ if $n \ge 2\lambda$. Furthermore,

$$\frac{\mathrm{Var}\, W}{\lambda} - 1 = -\frac{\lambda}{n} + \frac{2(n-1)}{n}\Big(\frac{\sqrt{8}}{3}\sqrt{\frac{\lambda}{n}} - \frac{\lambda}{n}\Big) < \frac{4}{3}\sqrt{\frac{2\lambda}{n}},$$

with asymptotic equality if $\lambda = \lambda_\infty$ is kept fixed as $n \to \infty$. Hence, for $\lambda = \lambda_\infty$ as $n \to \infty$,

$$W = \sum_{k=1}^{n} I\big[Z_k + Z_{k-1} > 2 - \sqrt{2\lambda/n}\big] \xrightarrow{\mathcal{D}} \mathrm{Po}(\lambda)$$

with rate of convergence $O(n^{-1/2})$. This rate is sharp, in view of Remark 3.2.11, or from the results of Section 10.4, where it is also shown in

Theorem 10.N that a compound Poisson approximation attains the better rate of $O(n^{-1})$. In particular, it follows that

$$\begin{aligned}|\mathbb{P}[W=0]-e^{-\lambda}|\\ = \left|\mathbb{P}\left[\max_{1\le k\le n}(Z_k+Z_{k-1})\le 2-\sqrt{2\lambda/n}\right]-e^{-\lambda}\right| \le (1-e^{-\lambda})\frac{4}{3}\sqrt{\frac{2\lambda}{n}},\end{aligned}$$

or equivalently, for $y<0$,

$$\left|\mathbb{P}[(n/2)^{1/2}\max_{1\le k\le n}(X_k-2)\le y]-e^{-y^2}\right| \le (1-e^{-y^2})\frac{4\sqrt{2}}{3}\cdot\frac{|y|}{\sqrt{n}}\,.$$

Now consider a sequence $\{\widehat{X}_k\}_{-\infty}^{\infty}$ of independent identically distributed random variables having the same marginal distributions as the X's, that is, triangular on the interval (0,2). It follows from Corollary 8.E.1 that

$$\left|\mathbb{P}[(n/2)^{1/2}(\max_{1\le k\le n}\widehat{X}_k-2)\le y]-e^{-y^2}\right| \le (1-e^{-y^2})\frac{y^2}{n}.$$

Thus $\max_{1\le k\le n}X_k$ and $\max_{1\le k\le n}\widehat{X}_k$ converge, after normalization, to the same limit distribution. In a similar way we could also obtain results for the second largest or other extremes of the moving average sequence. Note, however, that the rates of convergence for the number of exceedances in the two sequences are different, being $n^{-1/2}$ and n^{-1} respectively.

8.4 Patterns and runs

The extreme values in a sequence of moving averages of a 0–1 valued process correspond to long runs of 1's in the original process. The methods of this chapter can also be used to study the occurrence of runs and other patterns directly, a topic already considered in the context of Poisson convergence by von Mises (1921). In this section, we consider a fairly general example: much more on the subject is to be found in Arratia, Goldstein and Gordon (1989, 1990) and in Arratia, Gordon and Waterman (1990).

Let $\{\xi_i, i\ge 0\}$ be independent and identically distributed random variables with values in a set (alphabet) S, and let $p_s=\mathbb{P}[\xi_i=s]$. Let $w=s_1\cdots s_m$ be a fixed string (word) of length m, and let W be the number of occurrences of w in the sequence $\xi_1\cdots\xi_n$. Thus $W=\sum_{i=1}^{n-m+1}I_i$, where $I_i=I[\xi_{i+k-1}=s_k,\ 1\le k\le m]$, and $\mathbb{E}W=(n-m+1)\pi$, where $\pi=\prod_{i=1}^m p_{s_i}$. A Poisson limit theorem for W was proved using the Stein–Chen method by Chryssaphinou and Papastavridis (1988). Here, we wish to estimate the associated error. The sequence (I_i) is $(m-1)$-dependent, and Corollary 2.C.5 could be applied with $\Gamma_i^0=\{j:|i-j|\le m-1\}$. A slightly better result is, however, obtained by using Theorem 2.C in conjunction with an explicit coupling, as follows.

Theorem 8.F *As above, let $W = \sum_{i=1}^{n-m+1} I_i$ be the number of occurrences of the word $w = s_1 \cdots s_m$ in the independent and identically distributed sequence $\xi_1 \cdots \xi_n$. Let π denote $\prod_{i=1}^{m} p_{s_i}$, and let $\lambda = (n-m+1)\pi$ denote $\mathbb{E}W$. Then*

$$\begin{aligned} d_{TV}(\mathcal{L}(W), \mathrm{Po}(\lambda)) & \\ &\leq \lambda^{-1}(1-e^{-\lambda})\Big((n-m+1)\pi^2 + \sum_{i=1}^{n}\sum_{j\neq i} |\mathrm{Cov}(I_i, I_j)|\Big) \\ &\leq \pi + \frac{2}{\pi}\sum_{j=1}^{m-1} |\mathrm{Cov}(I_0, I_j)| \qquad (4.1) \\ &= 2\sum_{j\in A}\prod_{k=1}^{j} p_{s_k} + (2m-1-4|A|)\pi, \end{aligned}$$

where $A = \{j \in \{1, \ldots, m-1\} : s_{j+k} = s_k, 1 \leq k \leq m-j\}$ is the set of periods of w.

Proof Given any i, define $\xi^*_{i+k-1} = s_k$ for $1 \leq k \leq m$, and $\xi^*_l = \xi_l$ otherwise. Then $J_{ji} = I[\xi^*_{j+k-1} = s_k, 1 \leq k \leq m]$ satisfies (2.1.1). Furthermore, $J_{ji} = I_j$ if $|j-i| \geq m$, $J_{ji} \geq I_j$ if $|j-i| \in A$ and $J_{ji} = 0 \leq I_j$ if $1 \leq |j-i| \leq m-1$ but $|j-i| \notin A$. Hence Theorem 2.C may be applied with $\Gamma_i^0 = \emptyset$, and the result follows. □

Remark 8.4.1 The method can also be applied when the ξ_i have different distributions.

The bound in (4.1) is typically small, unless w has a small period. In the latter case, however, clusters of overlapping copies of w occur with rather large probability, and a Poisson approximation is not very good: indeed, it is shown in Chapter 10 that a compound Poisson approximation may be much better (see Example 10.4.2). The extreme case is when w consists of a single letter s repeated m times, and W counts the s-runs of length m. Writing p for p_s, Theorem 8.F then yields

$$d_{TV}(\mathcal{L}(W), \mathrm{Po}(\lambda)) \leq \frac{2p}{1-p}(1-p^{m-1}) - (2m-3)p^m \leq \frac{2p}{1-p}. \qquad (4.2)$$

This case occurs in reliability theory, where it is known as the consecutive k out of n failure system (Chiang and Niu 1981; Papastavridis 1987,1988), here with m for k. ξ_i is the indicator of the failure of the ith component, and the whole system fails if m consecutive components fail, or, equivalently, if $W > 0$.

It is, however, easy to get a much better estimate of $\mathbb{P}[W = 0]$ than that implied by (4.2). Let W' be the number of occurrences of $w' = 011\ldots1$ (with m 1's) in $\xi_0 \cdots \xi_n$. Then

$$|\mathbb{P}[W = 0] - \mathbb{P}[W' = 0]| \leq \mathbb{P}[\xi_0 = \cdots = \xi_m = 1] = p^{m+1} = p\pi, \qquad (4.3)$$

while Theorem 8.F with $\pi' = (1-p)\pi$ yields

$$|\mathbb{P}[W'=0] - \exp\{-(n-m+1)\pi'\}| \\ \le d_{TV}\Big(\mathcal{L}(W'), \mathrm{Po}\,((n-m+1)\pi')\Big) \le (2m+1)\pi', \tag{4.4}$$

giving the following result.

Theorem 8.G *Let W be the number of runs of m 1's in a sequence of n independent* Be (p) *variables. Then*

$$|\mathbb{P}[W=0] - \exp\{-(n-m+1)q\pi| \le (2mq+1)\pi, \tag{4.5}$$

where $q = 1-p$ and $\pi = p^m$. □

Thus the failure probability in the consecutive m out of n system is approximated to an accuracy of order mp^m by $\exp\{-(n-m+1)q\pi\}$, whereas (4.2) only guarantees an accuracy of order p for approximation by $\exp\{-(n-m+1)\pi\}$. Indeed, if $n\to\infty$, $m = o(np)$ and $n\pi \asymp 1$, it follows using Theorem 8.G that

$$\begin{aligned} d_{TV}(\mathcal{L}(W), \mathrm{Po}\,(\lambda)) &\ge |\mathbb{P}[W=0] - \exp\{-(n-m+1)\pi\}| \\ &\ge \exp\{-(n-m+1)q\pi\} - \exp\{-(n-m+1)\pi\} - (2mq+1)\pi \\ &\ge \exp\{-n\pi\}(n-m+1)p\pi - (2mq+1)\pi \asymp p. \end{aligned}$$

Hence the estimates in Theorem 8.F and (4.2) are of the right order.

Note that, except on the event $\{\xi_0 = \cdots = \xi_m = 1\}$, W' is the same as a random variable of interest in its own right when studying runs, the number of runs of at least m consecutive 1's. Inequalities (4.3) and (4.4) then show that the number of such runs has a distribution differing in total variation from $\mathrm{Po}\,(\{n-m+1\}q\pi)$ by at most $(2mq+1)\pi$. Another random variable of interest is W'', the number of runs of exactly m 1's, which can be investigated in similar fashion, starting with the string $011\cdots10$ (with m 1's) and the sequence $\xi_0\cdots\xi_{n+1}$. This leads to the estimate

$$d_{TV}\left(\mathcal{L}(W''), \mathrm{Po}\,(\{n-m+1\}q^2\pi)\right) \le \{(2m-1)q^2+2\}\pi. \tag{4.6}$$

A further random variable, the number of non-overlapping runs of length m (Philippou, Georghiou and Philippou, 1983; Godbole, 1990; Papastavridis, 1990), is often well approximated by the number of runs of length at least m, since they differ only if there is a run of length at least $2m$, an event of probability less than np^{2m}. Its distribution is treated in another way, as an example in the next section.

8.5 Rare sets in Markov chains

Another way of looking at exceedances is to view them as the visits of a process to a rare set: see, for example, Keilson (1979) and Aldous (1989b). If X is a stationary ergodic Markov chain on $\mathbb{Z}$, the analysis becomes particularly amenable to the coupling approach. Suppose that the j-step transition matrix is denoted by $P^{(j)}$ and the stationary distribution by μ, and let W be the number of visits to the set $A \subset \mathbb{Z}$ during the interval $1 \leq j \leq n$, where $\mu(A)$ is thought of as being small: $W = \sum_{i=1}^{n} I_j$ with $I_j = I[X_j \in A]$.

Theorem 8.H *Under the above circumstances,*

$$d_{TV}(\mathcal{L}(W), \mathrm{Po}(\lambda)) \leq (1-e^{-\lambda})\Big[\mu(A) + \frac{2}{\mu(A)} \sum_{r,s \in A} \mu_r \sum_{j \geq 1} |P_{rs}^{(j)} - \mu_s|\Big]. \quad (5.1)$$

Proof In view of Theorem 2.A, it is enough to construct U_i and V_i for each $1 \leq i \leq n$ in such a way that

$$\mathcal{L}(U_i) \stackrel{\mathcal{D}}{=} \mathcal{L}(W); \quad \mathcal{L}(V_i + 1) = \mathcal{L}(W \mid I_i = 1),$$

and

$$\mathbb{E}|U_i - V_i| \leq \mu(A) + \frac{2}{\mu(A)} \sum_{r,s \in A} \mu_r \sum_{j \geq 1} |P_{rs}^{(j)} - \mu_s|.$$

The idea of the proof is as follows. Fix i, and realize a pair of P-chains X' and X'' together in the following way. First, choose (X'_i, X''_i) from a joint distribution with marginals $\mathcal{L}(X'_i) = \mu | A$ and $\mathcal{L}(X''_i) = \mu$. Then construct $\{(X'_{i+j}, X''_{i+j}),\, j \geq 1\}$ using a coupling of the P-chains starting in (X'_i, X''_i), and construct $\{(X'_{i-j}, X''_{i-j}),\, j \geq 1\}$ using a coupling of the $\bar{P}$-chains starting in (X'_i, X''_i), where $\bar{P}$ denotes the transition matrix of the reversed chain:

$$\bar{P}_{rs} = \mu_s P_{sr} / \mu_r. \quad (5.2)$$

Let T^+ and T^- denote the respective coupling times, measured in terms of j, and define

$$U_i = \sum_{j=1}^{n} I[X''_j \in A], \quad V_i + 1 = \sum_{j=1}^{n} I[X'_j \in A].$$

Then U_i and V_i have the required distributions, and

$$|U_i - V_i| \leq I[X''_i \in A] + \sum_{j \geq 1} \Big[I[T^+ > j]\{I[X'_{i+j} \in A] + I[X''_{i+j} \in A]\}$$
$$+ I[T^- > j]\{I[X'_{i-j} \in A] + I[X''_{i-j} \in A]\}\Big],$$

implying that

$$\mathbb{E}|U_i - V_i| \le \mu(A) + \sum_{j\ge 1}\Big[\mathbb{P}[T^+ > j, X'_{i+j} \in A] + \mathbb{P}[T^- > j, X'_{i-j} \in A]$$
$$+ \mathbb{P}[T^+ > j, X''_{i+j} \in A] + \mathbb{P}[T^- > j, X''_{i-j} \in A]\Big]. \quad (5.3)$$

If the P- and $\bar{P}$-chains are constructed using a maximal coupling (Griffeath, 1975), it follows, for instance, that

$$\mathbb{P}[T^+ > j,\ X'_{i+j} \in A] = \sum_{s\in A}\Big[\frac{1}{\mu(A)}\sum_{r\in A}\mu_r P^{(j)}_{rs} - \mu_s\Big]^+$$
$$\le \frac{1}{\mu(A)}\sum_{r\in A}\mu_r \sum_{s\in A}[P^{(j)}_{rs} - \mu_s]^+, \quad (5.4)$$

with the same inequality holding also for the reversed chain, because of (5.2). The theorem is now immediate. □

Remark 8.5.1 It may be convenient to use (5.3) with a simpler coupling, giving a more easily calculated, if weaker, bound than (5.1). If, in particular, the chains X' and X'' are run independently until coupling, an ergodic argument shows that the contributions to (5.3) from the forward and reversed chains are the same, which is also true for the maximal coupling if $A = \{a\}$ is a singleton, since then equality holds in (5.4).

Remark 8.5.2 The argument leading to (5.3) can still be used if X is non-stationary or inhomogeneous, though the formulae become more complicated. Each i now yields a different contribution to the overall error estimate, the distribution of X_i is no longer the same for each i and the reversed chain is less simply expressible.

Example 8.5.3 Consider again the setting of Section 8.4. Let $\xi_1, \ldots, \xi_n$ be independent Be(p) random variables, and define recursively

$$I_i = \prod_{j=0}^{m-1} \xi_{i+j} \prod_{l=i-m+1}^{i-1} (1 - I_l),$$

starting with $I_{2-m} = I_{3-m} = \cdots = I_0 = 0$, so that $W = \sum_{i=1}^{n-m+1} I_i$ counts the number of non-overlapping runs of m consecutive 1's in the sequence $\xi_1, \ldots, \xi_n$. Let $X_j = 0$ if $\xi_j = 0$ and $X_j = r$, $1 \le r \le m$, whenever $\xi_{j-r+1} = \ldots = \xi_j = 1$ and *either* $\xi_{j-r} = 0$ *or* $I_{j-r-m+1} = 1$ *or* $j = r$, so that X_j is zero if $\xi_j = 0$, and otherwise X_j is the number of consecutive 1's preceding time j, modulo m, plus one. Then $(X_j,\ j \ge 1)$ is an ergodic Markov chain with transition matrix

$$P_{r,r+1} = 1 - P_{r0} = p \quad \text{for} \quad 0 \le r \le m-1; \qquad P_{m1} = 1 - P_{m0} = p, \quad (5.5)$$

and with stationary distribution

$$\mu_0 = 1 - p; \qquad \mu_j = (1-p)p^j/(1-p^m),\ 1 \le j \le m,$$

and, with the new definition, $I_i = I[X_{i+m-1} = m]$. Thus W counts the number of times X visits the set $A = \{m\}$ between steps m and n.

First, suppose that X were stationary. Then all that would be needed to apply Theorem 8.H would be an estimate of $\sum_{j\ge1} |P^{(j)}_{mm} - \mu_m|$. Now, for the reversed chain,

$$\bar{P}_{0r} = \mu_r,\ 0 \le r \le m; \quad \bar{P}_{1m} = 1 - \bar{P}_{10} = p^m; \quad \bar{P}_{r,r-1} = 1,\ 2 \le r \le m,$$

and so we may define

$$T^- = 1 + \min\{j : j \ge 1, X'_{i-j} = 0\},$$

and then run X'' backwards and forwards from $X''_{i-T^-} = X'_{i-T^-}$, achieving the correct distributions. Hence

$$\sum_{j\ge1} \mathbb{P}[T^- > j,\ X'_{i-j} \in A] = p^m/(1-p^m)$$

and

$$\sum_{j\ge1} \mathbb{P}[T^- > j,\ X''_{i-j} \in A] = \mu_m \mathbb{E}\{T^- - 1\} = mp^m(1-p)/(1-p^m)^2,$$

which, from Remark 8.5.1 and (5.3), gives

$$\mathbb{E}|U_i - V_i| \le \frac{p^m(1-p)}{1-p^m} + \frac{2p^m}{1-p^m}\left[1 + \frac{m(1-p)}{1-p^m}\right]. \tag{5.6}$$

Hence the same estimate holds for the distance between $\mathcal{L}(\widetilde{W})$ and $\mathrm{Po}(\tilde{\lambda})$, with $\tilde{\lambda} = (n-m+1)p^m(1-p)/(1-p^m)$ and $\widetilde{W} = \sum_{j=m}^n I[X_j = m]$, where X is the stationary chain.

To recover the original model, we need the P-chain X^* for which $X^*_0 = 0$ a.s. instead of the stationary P-chain X. Since $P_{r0} = 1-p$ for each r, the two chains can be coupled in $j \ge 0$ by sending them to zero together. If this coupling time is denoted by T,

$$N^* = \sum_{j\ge0} I[T > j, X^*_j = m] \le N = \sum_{j\ge0} I[T > j, X_j = m] \le N^* + 1,$$

and

$$\mathbb{P}[N > N^*] \le \mathbb{P}[N \ge 1] = \sum_{r\ge1} \mu_r p^{m-r} = mp^m(1-p)/(1-p^m).$$

Hence, taking $W = \sum_{j=m}^{n} I[X_j^* = m]$ to be the number of non-overlapping runs of m consecutive 1's in the sequence $\xi_1, \dots, \xi_n$,

$$d_{TV}(\mathcal{L}(W), \mathcal{L}(\widetilde{W})) \leq mp^m(1-p)/(1-p^m),$$

and so, in view of (5.6),

$$d_{TV}(\mathcal{L}(W), \mathrm{Po}(\tilde{\lambda})) \leq \eta(p,m) = 3p^m\{1 + m(1-p)\}(1-p^m)^{-2}. \quad (5.7)$$

This estimate has the same general appearance as those given for related quantities in (4.5) and (4.6). Approximating W instead by the number of runs of 1's of length at least m, as mentioned in the preceding section, leads to an error estimate of $p^m\{1 + 2m(1-p)\} + np^{2m}$, which is of larger order if $np^m \gg m$.

Example 8.5.4 Let $\{X_j\}_{j=1}^m$ be a collection of m independent two state Markov chains on $\{0,1\}$ with transition matrices $P_j = Q_j + \varepsilon_j R_j$, where $0 \leq \varepsilon_j < 1$,

$$Q_j = \begin{pmatrix} p_j & 1-p_j \\ p_j & 1-p_j \end{pmatrix} \quad \text{and} \quad R_j = \begin{pmatrix} 1-p_j & -(1-p_j) \\ -p_j & p_j \end{pmatrix}.$$

Thus $P_j^r = Q_j + \varepsilon_j^r R_j$, and X_j has its stationary distribution given by $\mathbb{P}[X_j = 0] = 1 - \mathbb{P}[X_j = 1] = p_j$. If $\varepsilon_j = 0$, the sequence $\{X_j(i)\}_{i \geq 1}$ is just a sequence of independent Be(p_j) random variables, but $\varepsilon_j > 0$ introduces a tendency for the zeros and ones to occur in longer runs. Define

$$I_i = \prod_{j=1}^m (1 - X_j(i)) \quad \text{and} \quad \pi = \prod_{j=1}^m p_j,$$

so that $W = \sum_{i=1}^n I_i$ counts how often all of the X_j's are simultaneously zero, and $\lambda = \mathbb{E}W = n\pi$. If all the ε_j are zero, Theorem 2.M gives the estimate

$$d_{TV}(\mathcal{L}(W), \mathrm{Po}(\lambda)) \leq \pi(1 - e^{-\lambda});$$

we now investigate how much the estimate changes if the ε_j's are allowed to be positive.

Applying Theorem 8.H to the equilibrium chain $(X_1, \dots, X_m)$ with $A = \{(0, \dots, 0)\}$, we have $\mu(A) = \pi$ and

$$d_{TV}(\mathcal{L}(W), \mathrm{Po}(\lambda)) \leq (1 - e^{-\lambda})\Big\{\pi + 2\sum_{r \geq 1} \eta_r\Big\},$$

where

$$\eta_r = \Big[\prod_{j=1}^m \{p_j + \varepsilon_j^r(1 - p_j)\} - \pi\Big],$$

using the explicit expression for P_j^r. This estimate simplifies if all the chains have the same structure. For $m = 1$, $\sum_{r\geq 1} \eta_r$ is readily evaluated as $\varepsilon(1-p)/(1-\varepsilon)$, so that good Poisson approximation obtains if both p and ε are small, the larger of p and ε governing the rate. If $m \geq 2$ and $p_j = p$ and $\varepsilon_j = \varepsilon$ for all j, where $0 < p, \varepsilon < 1$, it follows that

$$\eta_r = \sum_{j=1}^{m} \binom{m}{j} (1-p)^j p^{m-j} \varepsilon^{jr} \leq \varepsilon \eta_{r-1},$$

and hence that

$$\sum_{r\geq 1} \eta_r \leq \frac{\eta_1}{1-\varepsilon}$$
$$\leq (1-\varepsilon)^{-1}\{[p+\varepsilon(1-p)]^m \wedge [p+\varepsilon(1-p)]^{m-1} m\varepsilon(1-p)\}. \quad (5.8)$$

Thus, if $m \geq 2$ and $\varepsilon \leq p/\{(m-1)(1-p)\}$,

$$\sum_{r\geq 1} \eta_r \leq [p+\varepsilon(1-p)]^{m-1} \frac{m\varepsilon(1-p)}{1-\varepsilon} \leq ep^{m-1} \frac{m\varepsilon(1-p)}{1-\varepsilon}$$
$$= e\pi \cdot \frac{m\varepsilon(1-p)}{p(1-\varepsilon)} \leq \frac{e\pi m}{m-1} \cdot \frac{1}{1-\varepsilon} \leq \frac{2e\pi}{1-\varepsilon},$$

which, for $\varepsilon \leq p/\{(m-1)(1-p)\}$, is still of order π if π is small. If $\varepsilon > p/\{(m-1)(1-p)\}$, we obtain the estimate

$$\sum_{r\geq 1} \eta_r \leq \frac{[p+\varepsilon(1-p)]^m}{1-\varepsilon} = \frac{\pi}{1-\varepsilon}\left(1 + \frac{\varepsilon(1-p)}{p}\right)^m.$$

The term $2\sum_{r\geq 1} \eta_r \geq 2\eta_1$ dominates π whenever $\varepsilon \gg p/\{(m-1)(1-p)\}$.

9
Independent and dissociated terms

When approximating the distribution of a sum W of independent indicator random variables $(I_\alpha, \alpha \in \Gamma)$, there are many methods available. The simple formula for the characteristic function allows Fourier inversion and complex analytic techniques to be brought into play directly, or as applied by Kerstan (1964). Alternatively, the use of these techniques in an operator framework, as in Le Cam (1960) and Deheuvels and Pfeifer (1988), brings excellent results. That the Stein–Chen method also works well in this context is hardly surprising — independence is mathematically a very powerful tool — but, in view of the many competing possibilities, it is perhaps not to be expected that dramatically better results can be proved.

The real interest in pursuing the case of independence further is to get an idea of what may also be possible in other, more complicated, contexts. After all, if a technique cannot be made to work for independent summands, it is unlikely to perform any better in more difficult circumstances. Thus the main aim of this chapter is to illustrate how much more can be proved, in the case of independent summands, than has been possible in the greater generality of Chapter 2.

The main emphasis is on asymptotic expansions. In Section 9.1, it is shown in Theorem 9.A that it is possible to derive an asymptotic expansion for the distribution of W as a Poisson–Charlier series, analogous to the Edgeworth expansion near the normal limit. What is more, a simple explicit bound for the error in approximating $\mathbb{P}[W \in A]$ by using the Poisson–Charlier series of length l is given, which is valid for all $A \in \mathbb{Z}^+$, and which is sharp enough to show that the series is in fact convergent, provided at least that $\max_{\alpha\in\Gamma} \pi_\alpha < 1/4$. Thus, if this condition is fulfilled, any probability can be computed to any pre-assigned accuracy, merely by taking enough terms in the series. However, if small probabilities are to be estimated to a given relative accuracy, it may require a very large number of terms in the rather complicated expansion, and it becomes more natural to look for alternative expressions. In Corollary 9.B.2, Cramér's method of conjugate distributions is combined with Theorem 9.A(b) to yield asymptotic formulae for tail probabilities, in which the relative error is again shown to be bounded by explicitly calculable quantities. A particular consequence of these formulae is that the $\text{Po}(\lambda)$ approximation is shown in Theorems 9.C and 9.D to have small relative error quite far into the tails. A different approach, which leads to results similar to those of Theorems 9.C and 9.D, and which is potentially more widely applicable, has recently been developed by Chen and Choi (1990).

As noted in Barbour and Eagleson (1987) Remark 2 p. 589, the asymptotic expansion of length two is determined by exactly two adjustable parameters, which can be taken to be the mean and variance of W. Thus W and the binomial distribution with the same mean and variance as W share the same expansion of length two. This shows that the approximation of $\mathcal{L}(W)$ by a binomial distribution can be an order of magnitude better than the Poisson approximation. It is also preferable to the length two expansion, in that the binomial is a well known probability distribution, whereas the expansion of length two is only a signed measure. This suggests trying to prove such an approximation directly, instead of proceeding through the length two expansions, in the hope of obtaining simpler and better error bounds. In Section 9.2, it is shown that Stein's method can be applied directly to establish a binomial approximation, in a fashion little different from the basic Poisson approximation method. The error estimates obtained are of the desired second order of magnitude, and are also reduced by homogeneity in the values of the π_α: if, for instance, the π_α are all equal, the error estimate, as would be hoped, reduces to zero.

In Section 9.3, a particular form of departure from independence is considered, reminiscient of Chen's (1978) finite dependence, which is most commonly encountered in the context of dissociated families of random variables: see Section 2.3. This dependence structure can be exploited most simply using the local approach to Stein's method, though the coupling approach could also be used, and asymptotic expansions can in principle be derived. However, the manipulation becomes very heavy, for lack of an equivalent of the simplifying Lemma 9.1.2 of the independent case, and the argument is not taken beyond the length two expansion. Even here, it is clear that the relative orders of terms to be neglected can vary significantly, depending on the application, so that no such expansion is likely to be universally appropriate.

9.1 Independent summands

Throughout this section, we assume that the indicator random variables $(I_\alpha, \alpha \in \Gamma)$ are independent, and consider the approximation of their sum W. It has already been noted in Theorem 2.M and Corollary 3.D.1 that

$$\min(1,\lambda^{-1})\sum_{\alpha\in\Gamma}\pi_\alpha^2 \geq d_{TV}(\mathcal{L}(W),\mathrm{Po}(\lambda)) \geq \frac{1}{32}\min(1,\lambda^{-1})\sum_{\alpha\in\Gamma}\pi_\alpha^2, \quad (1.1)$$

where $\pi_\alpha = \mathbb{E}I_\alpha$ and $\lambda = \sum_{\alpha\in\Gamma}\pi_\alpha$, demonstrating that, up to a constant, the correct order of accuracy in the approximation of $\mathcal{L}(W)$ by $\mathrm{Po}(\lambda)$ is $\min(1,\lambda^{-1})\sum_{\alpha\in\Gamma}\pi_\alpha^2$. More accurate approximations to $\mathcal{L}(W)$ therefore depend on finding something other than $\mathrm{Po}(\lambda)$ to compare with. In view of (1.1), the distribution $\mathrm{Po}(\lambda)$ is a good first step, and what is needed is to find the small perturbations which are sufficient to compensate the error to a given order of precision. Now, recalling (1.1.13), we have

$$\begin{aligned}\mathbb{P}[W \in A] - \mathrm{Po}(\lambda)\{A\} &= \sum_{\alpha\in\Gamma} \pi_\alpha \mathbb{E}\{g(W+1) - g(W_\alpha+1)\} \\ &= \sum_{\alpha\in\Gamma} \pi_\alpha^2 \mathbb{E}\{g(W_\alpha+2) - g(W_\alpha+1)\}, \quad (1.2)\end{aligned}$$

where $W_\alpha = \sum_{\beta\neq\alpha} I_\beta$ and $g = g_{\lambda,A}$. Since (1.2) is an equality, all that is in principle required is a better estimate of $\mathbb{E}\{g(W_\alpha+2) - g(W_\alpha+1)\}$ than the upper bound Δg used to prove the first half of (1.1).

To obtain this, we start by generalizing (1.2). Let $f : \mathbb{Z}^+ \to \mathbb{R}$ be any bounded function, and let $\theta_\lambda f$ denote the solution g to the equations

$$\lambda g(j) - jg(j-1) = f(j) - \mathrm{Po}(\lambda)\{f\}, \qquad j \geq 0. \quad (1.3)$$

Thus $\theta_\lambda f$ is the solution of the usual Stein equation, shifted one step. By the linearity of (1.3), $(\theta_\lambda f)(m)$, $m \geq 0$, can be written as the linear combination $\sum_{j\geq 0} f(j) g_{\lambda,\{j\}}(m+1)$ of the functions $g_{\lambda,\{j\}}$, evaluated at $m+1$ instead of m. Hence, again by linearity, (1.2) generalizes to

$$\mathbb{E}f(W) - \mathrm{Po}(\lambda)\{f\} = \sum_{\alpha\in\Gamma} \pi_\alpha^2 \mathbb{E}\{(\theta_\lambda f)(W_\alpha+1) - (\theta_\lambda f)(W_\alpha)\}. \quad (1.4)$$

It also follows easily from Lemma 1.1.1 that, if

$$\|f\| = \sup_{j\geq 0} |f(j)| \leq 1,$$

the estimate

$$\sup_{j\geq 0} |(\theta_\lambda f)(j+1) - (\theta_\lambda f)(j)| \leq 2\lambda^{-1}(1-e^{-\lambda}) \quad (1.5)$$

is valid. Thus the error in approximating $\mathbb{E}f(W)$ by $\mathrm{Po}(\lambda)\{f\}$ for any bounded function f can be estimated in the same way as has been used for probabilities.

Returning to (1.2), the quantity to be estimated is expressed in the form $\mathbb{E}f(W_\alpha)$, where $\|f\| \leq \lambda^{-1}(1-e^{-\lambda})$ and W_α is a sum of independent indicators. Thus, using (1.4), $\mathbb{E}\{g_{\lambda,A}(W_\alpha+2) - g_{\lambda,A}(W_\alpha+1)\}$ can be replaced by the corresponding expectation with respect to the $\mathrm{Po}(\lambda - \pi_\alpha)$ distribution, together with an error which, from (1.4), is again a weighted sum of expectations of the form $\mathbb{E}\tilde{f}(\widetilde{W})$, with $\tilde{f}$ a bounded function and $\widetilde{W}$ a sum of independent indicators. This procedure can in principle be iterated any number of times, to produce successively improved approximations to $\mathbb{P}[W \in A]$, but the fact that different Poisson distributions and different sums of indicators appear at each stage makes the detail somewhat intractable. The following observation simplifies the procedure considerably.

Lemma 9.1.1 *Let I be an indicator random variable with expectation π, and let $\nabla^l g$ denote the lth forward difference of g. Then, for any $j \geq 0$,*

$$\left|\mathbb{E}\{Ig(I+j)-\pi g(I+j+1)\} - \sum_{s=1}^{l-1}(-1)^s \pi^{s+1}\mathbb{E}\{\nabla^s g(I+j+1)\}\right|$$
$$= \pi^{l+1}|\nabla^l g(j+1)|.$$

Proof Direct computation yields

$$\sum_{s=1}^{l-1}(-1)^s \pi^{s+1}\mathbb{E}\{\nabla^s g(I+j+1)\}$$
$$= \sum_{s=1}^{l-1}(-1)^s \pi^{s+1}\{\nabla^s g(j+1) + \pi\nabla^{s+1}g(j+1)\}$$
$$= (-1)^{l+1}\pi^{l+1}\nabla^l g(j+1) - \pi^2\nabla g(j+1),$$

and also
$$\mathbb{E}\{Ig(I+j) - \pi g(I+j+1)\} = -\pi^2\nabla g(j+1),$$

from which the lemma is immediate. □

This leads directly to the following result.

Lemma 9.1.2 *For any $g : \mathbb{Z}^+ \to \mathbb{R}$, setting $\lambda_l = \sum_{\alpha\in\Gamma}\pi_\alpha^l$, we have*

$$\varepsilon_l = \left|\mathbb{E}\{\lambda g(W+1) - Wg(W)\} - \sum_{s=1}^{l-1}(-1)^{s+1}\lambda_{s+1}\mathbb{E}\{\nabla^s g(W+1)\}\right|$$
$$\leq \sum_{\alpha\in\Gamma}\pi_\alpha^{l+1}\mathbb{E}|\nabla^l g(W_\alpha+1)| \ \leq \ \lambda_{l+1}\|\nabla^l g\|.$$

If also $\|\nabla^l g\|_1 < \infty$, setting $p^* = \max_{j\geq 0}\mathbb{P}[W=j]$, we have

$$\varepsilon_l \quad \leq \quad 2p^*\lambda_{l+1}\|\nabla^l g\|_1,$$

where $\|\cdot\|_1$ denotes the l_1 norm.

Proof From the independence of I_α and W_α, and since $W = I_\alpha + W_\alpha$, it follows from Lemma 9.1.1 that

$$|\mathbb{E}\{I_\alpha g(W) - \pi_\alpha g(W+1) - \sum_{s=1}^{l-1}(-1)^s\pi_\alpha^{s+1}\mathbb{E}\{\nabla^s g(W+1)\}|$$
$$\leq \pi_\alpha^{l+1}\mathbb{E}|\nabla^l g(W_\alpha+1)| \leq \pi_\alpha^{l+1}\|\nabla^l g\|.$$

The proof of the earlier inequalities is completed by adding over $\alpha \in \Gamma$. For the second part, if $h : \mathbb{Z}^+ \to \mathbb{R}$ satisfies $\|h\|_1 < \infty$, it is easily shown that

$$\mathbb{E}|h(W_\alpha + 1)| \leq \max_{j \geq 0} \mathbb{P}[W_\alpha = j] \|h\|_1.$$

However, it follows from the equations

$$(1 - \pi_\alpha)\mathbb{P}[W_\alpha = j] + \pi_\alpha \mathbb{P}[W_\alpha = j - 1] = \mathbb{P}[W = j], \qquad j \geq 0,$$

that

$$\max_{j \geq 0} \mathbb{P}[W_\alpha = j] \leq p^* \min\{(1 - \pi_\alpha)^{-1}, \pi_\alpha^{-1}\},$$

so that $\max_{j \geq 0} \mathbb{P}[W_\alpha = j] \leq 2p^*$. Hence

$$\sum_{\alpha \in \Gamma} \pi_\alpha^{l+1} \mathbb{E}|\nabla^l g(W_\alpha + 1)| \leq 2p^* \lambda_{l+1} \|\nabla^l g\|_1,$$

as required. □

Remark 9.1.3 If $\|g\| < \infty$, it follows that $\|\nabla^l g\| \leq 2^l \|g\| < \infty$. The same observation is valid with $\|\cdot\|$ replaced by $\|\cdot\|_1$.

Thus, for any bounded f, taking $g(j) = \theta_\lambda f(j-1)$, Lemma 9.1.2 gives the estimate

$$\mathbb{E}f(W) - \mathrm{Po}(\lambda)\{f\} = \sum_{s=1}^{l-1} (-1)^{s+1} \lambda_{s+1} \mathbb{E}\{\nabla^s \theta_\lambda f(W)\} + \varepsilon, \tag{1.6}$$

where

$$|\varepsilon| \leq \lambda_{l+1} \min(\|\nabla^l \theta_\lambda f\|, 2p^* \|\nabla^l \theta_\lambda f\|_1). \tag{1.7}$$

This, for instance, by choosing $f + 1/2$ as the indicator of a set $A \subset \mathbb{Z}^+$, expresses the difference between $\mathbb{P}[W \in A]$ and $\mathrm{Po}(\lambda)\{A\}$ as an asymptotic series in the quantities λ_s, $s \geq 2$: note that $\lambda^{-1}\lambda_s \leq (\max_\alpha \pi_\alpha)^{s-1}$ for all $s \geq 1$, so that the quantities λ_s are geometrically bounded. However, (1.6) as it stands is not useful for computing $\mathbb{E}f(W)$, because the coefficients on the right hand side involve expectations with respect to the unknown distribution $\mathcal{L}(W)$, and so the right hand side of (1.6) must itself be evaluated by using (1.6) again, iterating until all terms of order greater than a prescribed level have been evaluated as expectations with respect to the Poisson distribution $\mathrm{Po}(\lambda)$.

The coefficients in the resulting expansion consist of elements of the form $\mathrm{Po}(\lambda)\{\prod_{j=1}^r (\nabla^{s_j} \theta_\lambda) f\}$, and the bounds on the terms neglected involve quantities of the form $\|\prod_{j=1}^r (\nabla^{s_j} \theta_\lambda) f\|$, where $\sum_{j=1}^r s_j \leq l$. Therefore, in order to obtain a comprehensible expansion and an easily understood error term, some further simplification of these quantities is needed.

This is the subject of the following lemmas. Let $C_n(\lambda; x)$ denote the nth Charlier polynomial,

$$C_n(\lambda; x) = \sum_{r=0}^{n} \binom{n}{r} (-1)^{n-r} \lambda^{-r} x(x-1)\dots(x-r+1). \qquad (1.8)$$

Lemma 9.1.4 *Let Z be a Poisson random variable with distribution $\mathrm{Po}(\lambda)$. Then, for any function $f : \mathbb{Z}^+ \to \mathbb{R}$ for which the expectations exist,*

$$\mathbb{E}\{C_n(\lambda; Z)(\nabla^s f)(Z)\} = \mathbb{E}\{C_{n+s}(\lambda; Z) f(Z)\}, \qquad n, s \geq 0, \qquad (1.9)$$

and

$$\mathbb{E}\{C_n(\lambda; Z)(\theta_\lambda f)(Z)\} = -\frac{1}{n+1}\mathbb{E}\{C_{n+1}(\lambda; Z) f(Z)\}, \qquad n \geq 0. \quad (1.10)$$

Proof In order to establish (1.9), it is enough to consider the case $s = 1$, since it then follows, for $s \geq 2$, that

$$\mathbb{E}\{C_n(\lambda; Z)(\nabla^s f)(Z)\} = \mathbb{E}\{C_{n+1}(\lambda; Z)(\nabla^{s-1} f)(Z)\},$$

and induction on s completes the proof. Now, taking $s = 1$,

$$\begin{aligned}
&\mathbb{E}\{C_n(\lambda; Z)(\nabla f)(Z)\} \\
&= \sum_{j \geq 0} \frac{e^{-\lambda}\lambda^j}{j!} \sum_{r=0}^{n} \binom{n}{r} (-1)^{n-r} \lambda^{-r} (j)_r [f(j+1) - f(j)] \\
&= -\sum_{j \geq 0} \frac{e^{-\lambda}\lambda^j}{j!} \sum_{r=0}^{n} \binom{n}{r} (-1)^{n-r} \lambda^{-r} \left[(j)_r - \lambda^{-1} j (j-1)_r\right] f(j) \\
&= -\sum_{j \geq 0} \frac{e^{-\lambda}\lambda^j}{j!} \sum_{s=0}^{n+1} (j)_s (-1)^{n-s} \lambda^{-s} \left[\binom{n}{s-1} + \binom{n}{s}\right] f(j) \\
&= \mathbb{E}\{C_{n+1}(\lambda; Z) f(Z)\},
\end{aligned}$$

where the notation $(x)_r$ denotes $x(x-1)\dots(x-r+1)$ and impossible combinatorial coefficients are taken to be zero, proving (1.9).

For (1.10), we start with the explicit formula

$$(\theta_\lambda f)(j) = -\lambda^{-j-1} j! \sum_{s>j} \frac{\lambda^s}{s!} [f(s) - \mathrm{Po}(\lambda)\{f\}]$$

for the solution of (1.3): cf. (1.1.18). This gives rise to the formulae

$$\begin{aligned}
&\mathbb{E}\{C_n(\lambda; Z)(\theta_\lambda f)(Z)\} \\
&= -\sum_{j\geq 0} e^{-\lambda}\lambda^{-1} \sum_{s>j} \frac{\lambda^s}{s!}[f(s) - \mathrm{Po}(\lambda)\{f\}] \sum_{r=0}^{n} \binom{n}{r} (-1)^{n-r}\lambda^{-r}(j)_r \\
&= -\sum_{s\geq 1} e^{-\lambda}\frac{\lambda^{s-1}}{s!}[f(s) - \mathrm{Po}(\lambda)\{f\}] \sum_{r=0}^{n} \binom{n}{r} (-1)^{n-r}\lambda^{-r} \sum_{j=0}^{s-1}(j)_r \\
&= -\sum_{s\geq 1} e^{-\lambda}\frac{\lambda^{s-1}}{s!}[f(s) - \mathrm{Po}(\lambda)\{f\}] \sum_{r=0}^{n} \binom{n}{r} (-1)^{n-r}\lambda^{-r}(s)_{r+1}/(r+1) \\
&= -\frac{1}{n+1}\sum_{s\geq 0} e^{-\lambda}\frac{\lambda^{s}}{s!}[f(s) - \mathrm{Po}(\lambda)\{f\}][C_{n+1}(\lambda; s) + (-1)^n] \\
&= -\frac{1}{n+1}\mathbb{E}\{C_{n+1}(\lambda; Z)f(Z)\},
\end{aligned}$$

since, for $n \geq 0$, $\mathbb{E}C_{n+1}(\lambda; Z) = 0$. This proves (1.10). □

As a result of Lemma 9.1.4, we can express any quantity of the form $\mathrm{Po}(\lambda)\{\prod_{j=1}^{r}(\nabla^{s_j}\theta_\lambda)f\}$ alternatively as $\mathcal{Q}\{f\}$, where the signed measure $\mathcal{Q}$, different for different choices of r and $s_1, \ldots, s_r$, has density given by

$$(-1)^r C_{S_r+r}(\lambda; j)\frac{e^{-\lambda}\lambda^j}{j!} / \prod_{m=1}^{r}(S_m + m), \qquad j \geq 0, \tag{1.11}$$

where $S_m = \sum_{k=1}^{m} s_k$. Linear combinations of these signed measures go to make up the asymptotic expansions of Theorem 9.A.

Lemma 9.1.5 *Let $r, s_1, \ldots, s_r \in \mathbb{N}$ be arbitrary, set $l = \sum_{j=1}^{r} s_j$, and let e_λ denote $\lambda^{-1}(1 - e^{-\lambda})$. Then the inequalities*

$$\|\prod_{j=1}^{r}(\nabla^{s_j}\theta_\lambda)h\| \leq 2^l e_\lambda^r \|h\|$$

and

$$\|\prod_{j=1}^{r}(\nabla^{s_j}\theta_\lambda)h\|_1 \leq 2^l e_\lambda^r \|h\|_1$$

hold, for any function $h : \mathbb{Z}^+ \to \mathbb{R}$.

Proof The first inequality follows by applying (1.5) and Remark 9.1.2 repeatedly. For the second, the inequality $\|\nabla^s g\|_1 \leq 2^s\|g\|_1$ for any g is immediate. On the other hand, by linearity, $\nabla\theta_\lambda g = \sum_{j\geq 0} g(j)\nabla\theta_\lambda\delta_j$, where δ_j is the indicator of the set $\{j\}$. Since also $\|\nabla\theta_\lambda\delta_j\|_1 \leq 2e_\lambda$ from the proof of Lemma 1.1.1, it follows that $\|\nabla\theta_\lambda g\|_1 \leq 2e_\lambda\|g\|_1$, and the second inequality of the lemma now follows. □

In order to state the theorem, it is necessary to define the approximating sequence of signed measures. For each $l \geq 1$, let the signed measure Q_l on $\mathbb{Z}^+$ be defined by

$$Q_l(i) = \left\{\frac{e^{-\lambda}\lambda^i}{i!}\right\}\left\{1 + \sum_{s=1}^{l-1}\sum_{[s]}\prod_{j=1}^{s}\left[\frac{1}{r_j!}\left(\frac{(-1)^j\lambda_{j+1}}{j+1}\right)^{r_j}\right]C_{R+s}(\lambda;i)\right\}, \tag{1.12}$$

where $\sum_{[s]}$ denotes the sum over all s-tuples $(r_1, \ldots, r_s) \in (\mathbb{Z}^+)^s$ such that $\sum_{j=1}^{s} jr_j = s$, and $R = \sum_{j=1}^{s} r_j$. Note that this sequence of measures is analogous to the Edgeworth expansion about the normal distribution, with the Poisson density instead of the normal and the Charlier polynomials instead of the Hermite polynomials. Then we can prove the following result.

Theorem 9.A *For any $f : \mathbb{Z}^+ \to \mathbb{R}$ and $l \geq 1$, we have*

$$\mathbb{E}f(W) = \int f dQ_l + \eta_l,$$

with the following estimates.
(a) *If f is bounded,*

$$|\eta_l| \leq 2^{2l-1}e_\lambda\lambda_{l+1}\|f\|.$$

(b) *If $f \in l_1$ and $\lambda_2 e_\lambda \leq 1/8$,*

$$|\eta_l| \leq \sqrt{\frac{2}{e\lambda}}2^{2l}e_\lambda\lambda_{l+1}\|f\|_1.$$

Proof The first step is to use (1.6) iteratively, until all terms involving powers of the π_α lower than $l+1$ appear with coefficients expressed in terms of expectations with respect to the Po(λ) distribution, yielding

$$\mathbb{E}f(W) = \sum_{(l)}\left(\prod_{j=1}^{k}(-1)^{s_j+1}\lambda_{s_j+1}\right)\mathbb{E}\left\{\prod_{j=1}^{k}(\nabla^{s_j}\theta_\lambda)f(Z)\right\} + \eta_l, \tag{1.13}$$

where, using the error estimates in the form appropriate for part (a),

$$|\eta_l| \leq \sum_{(l)}\left\{\prod_{j=1}^{k+1}\lambda_{s_j+1}\right\}\left\|\prod_{j=1}^{k+1}(\nabla^{s_j}\theta_\lambda)f\right\| \tag{1.14}$$

and $\sum_{(l)}$ denotes the sum over

$$\{k \geq 0;\ (s_1, \ldots, s_{k+1}) \in \mathbb{N}^{k+1} : \sum_{j=1}^{k+1} s_j = l\}.$$

Using Lemma 9.1.5 to bound the norm, this implies the estimate

$$|\eta_l| \leq 2^l \sum_{(l)} e_\lambda^{k+1} \Bigl\{ \prod_{j=1}^{k+1} \lambda_{s_j+1} \Bigr\} \|f\|. \tag{1.15}$$

At the cost of some precision, the estimate can be considerably simplified by observing that, for any $s, t \geq 0$,

$$\begin{aligned} \lambda^{-2}\lambda_{s+1}\lambda_{t+1} &= (\sum_{\alpha\in\Gamma} \pi_\alpha \pi_\alpha^s / \sum_{\alpha\in\Gamma} \pi_\alpha)(\sum_{\alpha\in\Gamma} \pi_\alpha \pi_\alpha^t / \sum_{\alpha\in\Gamma} \pi_\alpha) \\ &\leq (\sum_{\alpha\in\Gamma} \pi_\alpha \pi_\alpha^{s+t} / \sum_{\alpha\in\Gamma} \pi_\alpha) = \lambda^{-1}\lambda_{s+t+1}, \end{aligned}$$

using Hölder's inequality. Thus, since $e_\lambda \leq \lambda^{-1}$, the product in (1.15) can be collapsed to give

$$|\eta_l| \leq 2^l e_\lambda \lambda_{l+1} \sum_{(l)} 1 \|f\| = 2^{2l-1} e_\lambda \lambda_{l+1} \|f\|, \tag{1.16}$$

which is the estimate stated in part (a) of the theorem.

It remains to identify the main expression in (1.13) with $\int f\, dQ_l$. From (1.11) and (1.12), this is equivalent to showing that

$$\begin{aligned} &1 + \sum_{s=1}^{l-1} \sum_{[s]} \prod_{j=1}^{s} \Bigl\{ \frac{1}{r_j!} \Bigl(\frac{(-1)^j \lambda_{j+1}}{j+1} \Bigr)^{r_j} \Bigr\} C_{R+s}(\lambda; i) \\ &= \sum_{(l)} \Bigl(\prod_{j=1}^{k} (-1)^{s_j+1} \lambda_{s_j+1} \Bigr) (-1)^k C_{S+k}(\lambda; i) \Big/ \prod_{j=1}^{k} (j + \sum_{r=1}^{j} s_r), \end{aligned} \tag{1.17}$$

where the sums $\sum_{[s]}$ and $\sum_{(l)}$ are as defined following (1.12) and (1.14), $R = \sum_{j=1}^s r_j$ and $S = \sum_{j=1}^k s_j$. To see the equivalence, consider the coefficient of a product of the form $\prod_{i=1}^q \lambda_{t_i+1}$, where $t_1 \leq t_2 \leq \ldots \leq t_q$. Let $T = \sum_{i=1}^q t_i$, and let the *distinct* values taken by the t_i occur in groups of sizes $r_1, \ldots, r_q$, where, when all groups have been exhausted, the remaining r_i are set to zero. We may assume that $T \leq l-1$: otherwise, the term does not appear at all. In the left hand side of (1.17), the coefficient is given by

$$(-1)^T \prod_{i=1}^q (t_i + 1)^{-1} \prod_{i=1}^q (1/r_i!) C_{q+T}.$$

In the right hand side, we have

$$(-1)^T \sum \prod_{i=1}^q (i + \sum_{r=1}^i s_r)^{-1} C_{q+T},$$

where the first sum is taken over all *distinct* rearrangements $s_1, \ldots, s_q$ of $t_1, \ldots, t_q$, of which there are $q!/\prod_{i=1}^q r_i!$. The equivalence of the left and right hand sides now follows from the observation that, for any $a_i > 0$,

$$\sum_{\sigma \in \Sigma_q} \prod_{i=1}^{q} (a_{\sigma(1)} + \ldots + a_{\sigma(i)})^{-1} = \prod_{i=1}^{q} a_i^{-1},$$

as may easily be proved by induction on q. This completes the proof of part (a) of the theorem.

Arguing in exactly the same way, but with the l_1 norm in place of the supremum norm, it follows that

$$\mathbb{E} f(W) = \int f \, dQ_l + \eta_l,$$

where

$$|\eta_l| \leq 2p^* \cdot 2^{2l-1} e_\lambda \lambda_{l+1} \|f\|_1$$

and p^* denotes $\max_{j \geq 0} \mathbb{P}[W = j]$ as before. Applying this result with $l = 1$ and $f = \delta_j$ for any j yields

$$\mathbb{P}[W = j] \leq \mathrm{Po}(\lambda)\{j\} + 4p^* e_\lambda \lambda_2,$$

so that, if $e_\lambda \lambda_2 \leq 1/8$, taking j so that $\mathbb{P}[W = j] = p^*$, it follows that

$$p^* \leq 2 \max_{j \geq 0} \mathrm{Po}(\lambda)\{j\} \leq \sqrt{\frac{2}{e\lambda}}, \tag{1.18}$$

by Proposition A.2.7. This completes the proof of (b). □

The results of Theorem 9.A can naturally be translated into results about probabilities, as follows.

Corollary 9.A.1 *For any $A \subset \mathbb{Z}^+$,*

$$|\mathbb{P}[W \in A] - Q_l\{A\}| \leq 2^{2l-2} e_\lambda \lambda_{l+1}.$$

Proof Take $f(w) = I[w \in A] - 1/2$ in Theorem 9.A(a). □

Corollary 9.A.2 *For any $j \in \mathbb{Z}^+$, if $e_\lambda \lambda_{l+1} \leq 1/8$,*

$$|\mathbb{P}[W = j] - Q_l\{j\}| \leq \sqrt{\frac{2}{e\lambda}} 2^{2l} e_\lambda \lambda_{l+1}.$$

Proof Take $f = \delta_j$ in Theorem 9.A(b). □

The proof of Theorem 9.A(b) does not use Theorem 2.Q, although the argument used to derive the second estimate in Lemma 9.1.2 and in the last part of the proof of Theorem 9.A(b) is very close to that of Theorem 2.Q. However, as mentioned in Section 2.4, the conditional probabilities used in the definition of the error estimates in Theorem 2.Q are awkward to handle, and for independent summands it is advantageous to condition U_α on V_α, rather than the other way round, which is effectively what is done in Lemma 9.1.2. Nor is the constant in Theorem 9.A(b) as small as in the corresponding Theorem 2 of Barbour and Jensen (1989), which, under the assumption $e_\lambda \lambda_2 \le 1/8$, would allow the factor $\sqrt{(2/e)}$ to be replaced by $1/\sqrt{10}$. The principal difference between the arguments lies in the method used to find a bound for $\max_{j\ge 0} \mathbb{P}[W_\alpha = j]$. In Barbour and Jensen (1989), the bound is established directly, as a result in its own right, by a Fourier argument. Here, precision has been lost by using the inequalities $\min(\pi_\alpha^{-1}, (1-\pi_\alpha)^{-1}) \le 2$ in Lemma 9.1.2 and $(1-4e_\lambda\lambda_2)^{-1} \le 2$ when $e_\lambda\lambda_2 \le 1/8$ at the end of the proof of Theorem 9.A(b). Now, from (1.1), good Poisson approximation is only to be expected when $e_\lambda\lambda_2$ is small, and, when this is the case, the inequalities above are far from sharp. More precise versions are not difficult to devise. For instance, if $\pi^* = \max_{\alpha\in\Gamma} \pi_\alpha < 1/3$, the factor 2 in Lemma 9.1.2 can immediately be reduced to $(1-\pi^*)^{-1}$, and thus, since also $e_\lambda\lambda_2 \le \pi^*$, the factor $(1-4e_\lambda\lambda_2)^{-1}$ can be replaced by $\{1-2\pi^*/(1-\pi^*)\}^{-1}$. This leads to an improved estimate

$$|\eta_l| \le \sqrt{\frac{1}{2e\lambda}} 2^{2l-1}(1-3\pi^*)^{-1} e_\lambda \lambda_{l+1} \|f\|_1 \tag{1.19}$$

in place of Theorem 9.A(b), whenever $\pi^* \le 1/3$. If $e_\lambda\lambda_2$ is small but π^* is not, it is more convenient to construct estimates in terms of

$$\lambda_r' = \sum_{i=1}^{n} \pi_\alpha^{r-1} \min\{1, \pi_\alpha/(1-\pi_\alpha)\}.$$

It then follows, by an argument similar to that used for Theorem 9.A(b), noting in particular that $\lambda^{-2}\lambda_s\lambda_t' \le \lambda^{-1}\lambda_{s+t}'$, that

$$\begin{aligned} |\eta_l| &\le \sqrt{\frac{1}{2e\lambda}} 2^l (1-2e_\lambda\lambda_2')^{-1} e_\lambda^{k+1} \sum_{(l)} \left\{ \prod_{j=1}^{k} \lambda_{s_j+1} \right\} \lambda_{s_{k+1}+1}' \|f\|_1 \\ &\le \sqrt{\frac{1}{2e\lambda}} 2^{2l-1} (1-2e_\lambda\lambda_2')^{-1} e_\lambda \lambda_{l+1}' \|f\|_1. \end{aligned} \tag{1.20}$$

In both (1.19) and (1.20), the factor $\sqrt{\frac{1}{2e\lambda}}$ can be replaced by $e^{-\lambda} I_0(\lambda)$, where I_0 denotes the modified Bessel function of order 0: see the proof of

Proposition A.2.7. This improves the estimate a little for large λ, and, with this modification, (1.20) is essentially equivalent to Theorem 2 of Barbour and Jensen (1989), in the limit when $e_\lambda \lambda_2 \to 0$.

In Barbour (1987) and in Barbour and Jensen (1989), the distribution of a sum $\widetilde{W}$ of independent random variables X_α taking values in $\mathbb{Z}^+$, rather than 0, 1, is also considered. Of course, if the probabilities $\mathbb{P}[X_\alpha \geq 2]$ are not considerably smaller than the probabilities $\mathbb{P}[X_\alpha = 1]$, one would not necessarily expect to get good Poisson approximation. This then suggests a very simple procedure. Let $I_\alpha = I[X_\alpha \geq 1]$, $\alpha \in \Gamma$, defining a sequence of independent indicator random variables, to which the methods of this section can be applied. Then, writing $W = \sum_{\alpha\in\Gamma} I_\alpha$ and $\lambda = \mathbb{E}W$ as usual, it follows that

$$\mathbb{P}\Big[\bigcap_{\alpha\in\Gamma}\{X_\alpha = I_\alpha\}\Big] \geq 1 - \sum_{\alpha\in\Gamma}\mathbb{P}[X_\alpha \geq 2],$$

and hence that

$$|\mathbb{P}[\widetilde{W} \in A] - \mathbb{P}[W \in A]| \leq \sum_{\alpha\in\Gamma}\mathbb{P}[X_\alpha \geq 2],$$

implying in particular that

$$d_{TV}(\mathcal{L}(\widetilde{W}), \mathrm{Po}(\lambda)) \leq \sum_{\alpha\in\Gamma}\mathbb{P}[X_\alpha \geq 2] + d_{TV}(\mathcal{L}(W), \mathrm{Po}(\lambda)).$$

Thus, at a cost of $\sum_{\alpha\in\Gamma}\mathbb{P}[X_\alpha \geq 2]$, the results for sums of independent indicators can be carried over to $\widetilde{W}$. However, if for instance the random variables X_α all have the negative binomial distribution $\mathrm{NBi}\,(1,(1-\pi))$, and if $n\pi^2 \asymp 1$, the contribution $\sum_{\alpha\in\Gamma}\mathbb{P}[X_\alpha \geq 2]$ to the error estimate in this approach is of order 1, whereas the distance between the distribution of $\widetilde{W}$, which is $\mathrm{NBi}\,(n,(1-\pi))$, and the Poisson distribution with mean $\mathbb{E}\widetilde{W}$ is in fact of order π. Thus it is reasonable to look for better estimates.

The method used in the above papers is to generalize Lemma 9.1.1, replacing I by any non-negative integer valued random variable X, and obtaining

$$\mathbb{E}\{Xg(X+j) - (\mathbb{E}X)g(X+j+1)\} = \sum_{s=1}^{l-1}\frac{\kappa_{[s+1]}}{s!}\mathbb{E}\{\nabla^s g(X+j+1)\} + \varepsilon_l,$$

where $\kappa_{[s]}$ denotes the sth factorial cumulant of X: the error ε_l is estimated in terms of the factorial moments and cumulants of X, which play the role of the ordinary moments in the error estimates in the Edgeworth expansion. Once this has been accomplished, the argument required to obtain the asymptotic expansion $\mathbb{E}f(W) = \int f\,dQ_l + \eta_l$ proceeds much as before, though Q_l is now defined by

$$Q_l\{i\} = \mathrm{Po}(\lambda)\{i\}\left\{1 + \sum_{s=1}^{l-1}\sum_{[s]}\prod_{j=1}^{s}\left[\frac{1}{r_j!}\left(\frac{\kappa_{[j+1]}}{(j+1)!}\right)^{r_j}\right]C_{R+s}(\lambda;i)\right\},$$

and the estimate of η_l becomes more complicated.

Returning to the case of 0–1 summands, a somewhat different asymptotic expansion for the probabilities $\mathbb{P}[W \in A]$ is given in Deheuvels and Pfeifer (1988). The expansion they obtain, using their development of Le Cam's (1960) operator method, together with the complex analytic techniques of Uspensky (1931) and Shorgin (1977), contains the same terms as in Theorem 9.A(a), but grouped differently. To compare the two expansions asymptotically, consider first the situation where $\lambda \asymp 1$ and $\pi_\alpha = \pi$ for all $\alpha \in \Gamma$. Then the terms present in Q_{l+1} additional to those in Q_l are precisely those with coefficients of order π^l, and the error in Q_{l+1} is estimated to be of order π^{l+1}. By contrast, in the expansion of Deheuvels and Pfeifer, the series of length $l+1$ contains some, but not all, of the terms of order π^l, and a series of length at least $2l$ is needed to encompass all of them: similarly, the error after the series of length $l+1$ is estimated to be of order $\pi^{1+l/2}$, and a series of length $2l+1$ is required to achieve the order of accuracy π^{l+1}. On the other hand, if instead $\lambda \to \infty$, $\lambda_2/\lambda \to 0$ and $\pi^* = \max_{\alpha\in\Gamma}\pi_\alpha$ remains constant, the sequence η_l of error terms in Theorem 9.A(a) is guaranteed to go to zero only as fast as $\lambda^{-1}(4\pi^*)^l$, which is of no great help if $\pi^* \geq 1/4$, whereas the errors in the Deheuvels and Pfeifer expansion decrease like $(\lambda_2/\lambda)^{(l+1)/2}$. Thus, in such limiting situations, the expansion of Deheuvels and Pfeifer is clearly better.

Corollary 9.A.1 provides for the approximation of probabilities of any events associated with W to an accuracy independent of the events considered. However, as in statistics or reliability theory, it is often useful to have approximations to tail probabilities which are also accurate in terms of relative error. This problem, addressed in Barbour and Jensen (1989), can be approached using Cramér's method of conjugate distributions combined with Theorem 9.A. A random variable W_m with the distribution conjugate to that of W but with mean m, for any $1 \leq m \leq |\Gamma|-1$, can be constructed as a sum of independent Bernoulli random variables $(Y_{m\alpha}, \alpha \in \Gamma)$ with

$$\mathbb{P}[Y_{m\alpha} = 1] = \pi_{m\alpha} = \phi_m\pi_\alpha/(1 - \pi_\alpha + \phi_m\pi_\alpha),$$

where ϕ_m denotes the solution to the equation

$$\sum_{\alpha\in\Gamma}\phi\pi_\alpha/(1 - \pi_\alpha + \phi\pi_\alpha) = m. \tag{1.21}$$

For simplicity, we consider only integers m, although, for example, Theorem 9.B is valid for any real m, $0 < m < |\Gamma|$.

Set $\lambda_{mj} = \sum_{\alpha\in\Gamma} \pi^j_{m\alpha}$, and define $\mathcal{Q}_{ml}$ as in (1.12), with $m = \lambda_1$ for λ and λ_{mj} for λ_j, $1 \le j \le l$: thus the sequence of measures $\mathcal{Q}_{ml}$ play the role for W_m that the $\mathcal{Q}_l$ play for W.

Theorem 9.B *For any $i \ge 0$ and for any m, $1 \le m \le |\Gamma| - 1$,*

$$\mathbb{P}[W = i] = \phi_m^{-i} \prod_{\alpha\in\Gamma} (1 - \pi_\alpha + \phi_m \pi_\alpha)[\mathcal{Q}_{ml}\{i\} + \eta],$$

where, if $\lambda_{m2} e_m \le 1/8$, with $e_m = m^{-1}(1 - e^{-m})$ as before,

$$|\eta| \le \sqrt{\frac{2}{em}} 2^{2l} e_m \lambda_{m,l+1}.$$

Proof From the definition of $W_m = \sum_{\alpha\in\Gamma} Y_{m\alpha}$, we have

$$\mathbb{P}[W = i] = \phi_m^{-i} \prod_{\alpha\in\Gamma} (1 - \pi_\alpha + \phi_m \pi_\alpha)\mathbb{P}[W_m = i],$$

and the conditions of Corollary 9.A.2 are satisfied by the random variables $(Y_{m\alpha}, \alpha \in \Gamma)$. □

Corollary 9.B.1 *For any $1 \le m \le |\Gamma| - 1$ such that $\lambda_{m2} e_m \le 1/8$,*

$$\mathbb{P}[W = m] = \phi_m^{-m} \prod_{\alpha\in\Gamma} (1 - \pi_\alpha + \phi_m \pi_\alpha)[\mathcal{Q}_{ml}(m) + \eta_{ml}],$$

where $|\eta_{ml}| \le \sqrt{\frac{2}{em}} 2^{2l} e_m \lambda_{m,l+1}$. □

Remark 9.1.6 By Proposition A.2.9, $\mathcal{Q}_{m1} = \mathrm{Po}(m)\{m\} \asymp m^{-1/2}$, and so, if $m^{-1}\lambda_{m2}$ is small,

$$\mathbb{P}[W = m]/[\phi_m^{-m} \prod_{\alpha\in\Gamma} (1 - \pi_\alpha + \phi_m \pi_\alpha)] \asymp m^{-1/2}$$

also. In such circumstances, the relative error in the approximation given by Corollary 9.B.1 for a fixed l is of order at most $m^{-1}\lambda_{m,l+1}$.

Corollary 9.B.2 *For any $1 \le m \le |\Gamma| - 1$ such that $\lambda_{m2} e_m \le 1/8$,*
(a) *for $m > \lambda$,*

$$\mathbb{P}[W \ge m] = \phi_m^{-m} \prod_{\alpha\in\Gamma} (1 - \pi_\alpha + \phi_m \pi_\alpha)\Big[\sum_{i\ge m} \phi_m^{m-i} \mathcal{Q}_{ml}(i) + \eta\Big];$$

(b) *for $m < \lambda$,*

$$\mathbb{P}[W < m] = \phi_m^{-m} \prod_{\alpha\in\Gamma} (1 - \pi_\alpha + \phi_m \pi_\alpha)\Big[\sum_{i<m} \phi_m^{m-i} \mathcal{Q}_{ml}(i) + \eta\Big],$$

where

$$|\eta| \le \sqrt{\frac{2}{em}} 2^{2l} e_m \lambda_{m,l+1} \{\phi_m / |\phi_m - 1|\}.$$

Proof Use Theorem 9.B, and sum over i. □

Remark 9.1.7 The relative error in the approximation can be seen as follows. Taking the case (a) with $l = 1$, it follows that

$$\mathbb{P}[W \geq m] \Big/ [\phi_m^{-m} \prod_{\alpha \in \Gamma} (1 - \pi_\alpha + \phi_m \pi_\alpha)] = \mathbb{E}\{\phi_m^{m-Z} I[Z \geq m]\} + \eta,$$

where $Z \sim \text{Po}(m)$. If now $\phi_m - 1 \geq m^{-1/2}$ and $m^{-1}\lambda_{m2}$ is sufficiently small, the right hand side of this expression is bounded below by a multiple of $m^{-1/2}\phi_m/(\phi_m - 1)$, so that Corollary 9.B.2 gives a relative error of order $2^{2l}m^{-1}\lambda_{m,l+1}$. For smaller deviations, Corollary 9.A.1 already attains an equivalent relative precision. Case (b) can be treated similarly.

Remark 9.1.8 The estimate of relative error depends on the value of m in a way which is not entirely explicit, because of the factor $\lambda_{m,l+1}$. Clearly, for $m \leq \lambda$, $\lambda_{m,l+1} \leq \lambda_{l+1}$. A simple estimate of $\lambda_{m,l+1}$ for m moderately greater than λ is given in Remark 9.1.10 below.

One drawback to the results of Corollaries 9.B.1 and 9.B.2 is that they are no longer based on approximation throughout the range of m by a single Poisson–Charlier measure. Instead, for each value of m, a new measure has to be computed. It is therefore of interest to know how far into the tails the simple approximation by $\text{Po}(\lambda)$ still maintains small relative error. In order to solve this problem, it is necessary to know more about the solution ϕ_m of (1.21): the results to be used are summarized as follows.

Lemma 9.1.9 *Let ϕ_m denote the solution of (1.21), and set $\psi = \lambda_2/\lambda$. Then, for $\lambda \leq m < \lambda/\psi$,*

(i) $\quad (\lambda^{-1}m - 1) \leq \phi_m - 1 \leq (\lambda^{-1}m - 1)/(1 - \lambda^{-1}m\psi),$

(ii) $\quad 0 \leq m^{-1}\lambda\phi_m - 1 \leq \psi(\lambda^{-1}m - 1)/(1 - \lambda^{-1}m\psi);$

and for $m \leq \lambda$,

(iii) $\quad (1 - \lambda^{-1}m) \leq 1 - \phi_m \leq (1 - \lambda^{-1}m)/(1 - \lambda^{-1}m\psi),$

(iv) $\quad 0 \leq 1 - \lambda\phi_m/m \leq \psi(1 - \lambda^{-1}m)/(1 - \lambda^{-1}m\psi).$

Proof The quantity $\sum_{\alpha \in \Gamma} \phi\pi_\alpha/(1 - \pi_\alpha + \phi\pi_\alpha)$ is increasing in ϕ, and takes the value λ at $\phi = 1$: hence, if $m \geq \lambda$, $\phi_m \geq 1$. Then it follows, using Jensen's inequality, that

$$\lambda\phi_m \geq \sum_{\alpha \in \Gamma} \phi_m\pi_\alpha/(1 - \pi_\alpha + \phi_m\pi_\alpha) = m \geq \lambda\phi_m/(1 + (\phi_m - 1)\psi).$$

Hence, rearranging the inequalities, we have

$$\lambda^{-1}m \leq \phi_m \leq \lambda^{-1}m(1 - \psi)/(1 - \lambda^{-1}m\psi),$$

provided that $m \leq \lambda/\psi$, from which (i) and (ii) are immediate.

If $m \leq \lambda$ and hence $\phi_m \leq 1$, we have instead

$$\lambda\phi_m \leq \sum_{\alpha \in \Gamma} \phi_m\pi_\alpha/(1 - \pi_\alpha + \phi_m\pi_\alpha) = m \leq \lambda\phi_m/(1 + (\phi_m - 1)\psi),$$

from which parts (iii) and (iv) now follow. □

Remark 9.1.10 It follows immediately from part (ii) that, for $\lambda \le m \le \lambda/2\psi$,

$$\lambda_{m,l+1} \le \phi_m^{l+1}\lambda_{l+1} \le \{1 + 2(\lambda^{-1}m - 1)\}^{l+1}\lambda_{l+1}. \tag{1.22}$$

With the help of Lemma 9.1.9, we can demonstrate the following comparisons between the estimates of point probabilities given by Corollary 9.B.1 and the corresponding Poisson probabilities.

Lemma 9.1.11 *Let $\Delta = \lambda^{-1/2}(m-\lambda)$ denote the standardized deviation of m from λ, and write $\psi = \lambda_2/\lambda$. Then*
(a) *if $\lambda \le m < \lambda/\psi$,*

$$\begin{aligned}&\exp\{-\tfrac{1}{2}\psi\Delta^2(1 - \lambda^{-1}m\psi)^{-2}\}\\ &\quad\le \Big[\phi_m^{-m}\prod_{\alpha\in\Gamma}(1 - \pi_\alpha + \phi_m\pi_\alpha)e^{-m}m^m/m!\Big]\Big/\mathrm{Po}(\lambda)\{m\}\\ &\quad\le \exp\{\tfrac{1}{2}\psi\Delta^2(1 - \lambda^{-1}m\psi)^{-2}\};\end{aligned}$$

(b) *if $m \le \lambda$ and $\max_{\alpha\in\Gamma}\pi_\alpha \le \frac{1}{2}(1 - \lambda^{-1}m\psi)/(1 - \lambda^{-1}m)$,*

$$\begin{aligned}&\exp\{-\psi\Delta^2(1 - \lambda^{-1}m\psi)^{-2}\}\\ &\quad\le \Big[\phi_m^{-m}\prod_{\alpha\in\Gamma}(1 - \pi_\alpha + \phi_m\pi_\alpha)e^{-m}m^m/m!\Big]\Big/\mathrm{Po}(\lambda)\{m\}\\ &\quad\le \exp\{\psi^2\Delta^2(1 - \lambda^{-1}m\psi)^{-2}\}.\end{aligned}$$

Proof Writing $\phi = \phi_m$, we use the expression

$$\begin{aligned}&\left(\frac{m}{\lambda\phi}\right)^m e^{\lambda-m}\prod_{\alpha\in\Gamma}\{1 + (\phi - 1)\pi_\alpha\}\\ &= \exp\Big\{\lambda - m + \sum_{\alpha\in\Gamma}\log\{1 + (\phi - 1)\pi_\alpha\} - m\log\{1 + (m^{-1}\lambda\phi - 1)\}\Big\},\end{aligned}$$

together with Lemma 9.1.9 and the inequalities

$$\begin{aligned}x - x^2/2 \le \log(1 + x) \le x, &\qquad x \ge 0;\\ -x - x^2 \le \log(1 - x) \le -x, &\qquad 0 \le x \le 1/2.\end{aligned}$$

For example, if $m \le \lambda$ and $\max_\alpha \pi_\alpha \le \frac{1}{2}(1-\lambda^{-1}m\psi)/(1-\lambda^{-1}m)$, we have $(1 - \phi_m)\pi_\alpha \le 1/2$ for each $\alpha \in \Gamma$ by Lemma 9.1.9(iii), and hence

$$\begin{aligned}&\left(\frac{m}{\lambda\phi}\right)^m e^{\lambda-m}\prod_{\alpha\in\Gamma}\{1 + (\phi - 1)\pi_\alpha\}\\ &\quad\ge \exp\{\lambda - m + \lambda(\phi - 1) - (1 - \phi)^2\lambda_2 - m(m^{-1}\lambda\phi - 1)\}\\ &\quad= \exp\{-\lambda_2(1 - \phi)^2\} \ge \exp\{-\psi\Delta^2(1 - \lambda^{-1}m\psi)^{-2}\},\end{aligned}$$

where the last inequality again uses Lemma 9.1.9(iii). □

Lemma 9.1.11 can be combined with Corollary 9.B.1 to prove the following theorem, which gives the relative accuracy of the Poisson approximation to point probabilities.

Theorem 9.C *Let $\Delta = \lambda^{-1/2}(m-\lambda)$ be as before, and write $\psi = \lambda_2/\lambda$. Then*

(a) *uniformly in m satisfying the inequalities $\lambda \le m \le \lambda/(2\psi)$ and $1+4\Delta^2 \le (16\psi)^{-1}$,*

$$\mathbb{P}[W=m] = \mathrm{Po}(\lambda)\{m\}(1+O(\psi)+O(\psi\Delta^2));$$

(b) *if $\psi \le 1/8$ and $\max_\alpha \pi_\alpha \le 1/2$, then, uniformly in m satisfying $0 \le m \le \lambda$ and $\Delta^2 \le \psi^{-1}$,*

$$\mathbb{P}[W=m] = \mathrm{Po}(\lambda)\{m\}(1+O(\psi)+O(\psi\Delta^2)).$$

Proof In both cases, the conditions are sufficient to enable the application of Lemma 9.1.11, giving

$$\left[\phi_m^{-m}\prod_{\alpha\in\Gamma}(1-\pi_\alpha+\phi_m\pi_\alpha)\frac{e^{-m}m^m}{m!}\right] = \mathrm{Po}(\lambda)\{m\}(1+O(\psi\Delta^2)),$$

uniformly in m. The condition $\lambda_{m2}e_m \le 1/8$, required for the application of Corollary 9.B.1 with $l=1$, is true in case (a) because, from Remark 9.1.10, $\lambda_{m2} \le \{1+2\lambda^{-1/2}\Delta\}^2\lambda_2$, so that

$$\lambda_{m2}\varepsilon_m \le 2(1+4\lambda^{-1}\Delta^2)\lambda_2 e_m \le 2(1+4\Delta^2)\psi.$$

In case (b), we consider only the case $m>0$, since $m=0$ is simpler. Then

$$\lambda_{m2} \le [\phi_m/(1-\tfrac{1}{2}(1-\phi_m))]^2\lambda_2 = \left(\frac{2\phi_m}{1+\phi_m}\right)^2\lambda_2 \le \phi_m\lambda_2 \le m\psi,$$

and thus $\lambda_{m2}e_m \le \psi \le 1/8$ as required. Corollary 9.B.1 with $l=1$ now gives

$$\mathbb{P}[W=m] = \left[\phi_m^{-m}\prod_{\alpha\in\Gamma}(1-\pi_\alpha+\phi_m\pi_\alpha)\frac{e^{-m}m^m}{m!}\right]\{1+O(|\eta_{m1}|/\mathcal{Q}_{m1}(m))\},$$

and the theorem follows because, using Proposition A.2.9,

$$\begin{aligned}|\eta_{m1}|/\mathcal{Q}_{m1}(m) &\le \frac{2^2\lambda_{m2}e_m}{\mathrm{Po}(m)\{m\}}\sqrt{\frac{2}{em}}\\ &\le 4\sqrt{2e}\lambda_{m2}e_m = O\{\psi(1+\Delta^2)\}.\end{aligned}$$

□

We now turn to similar comparisons for tail probabilities.

Lemma 9.1.12 *Let $\phi = \phi_m$ denote the solution of (1.21), and suppose that $\psi\Delta^2 \le 1$.*
(i) *If $m \le \lambda/2\psi$ and $\Delta \ge 1$,*

$$\phi^{-m} \prod_{\alpha\in\Gamma}(1-\pi_\alpha+\phi\pi_\alpha)\sum_{r\ge m}\phi^{m-r}Q_{m1}(r)/\mathrm{Po}(\lambda)\{[m,\infty)\} = 1+\eta,$$

where $|\eta| \le 8\psi(1+\Delta^2)$.
(ii) *If $\psi \le 1/8$ and $-7\lambda^{1/2}/16 \le \Delta \le -1$,*

$$\phi^{-m} \prod_{\alpha\in\Gamma}(1-\pi_\alpha+\phi\pi_\alpha)\sum_{r< m}\phi^{m-r}Q_{m1}(r)/\mathrm{Po}(\lambda)\{[0,m-1]\} = 1+\eta,$$

where $|\eta| \le \psi(4+2\Delta^2)$.

Proof For (i) we use the estimate

$$|\eta| \le \sum_{r\ge m}\mathrm{Po}(\lambda)\{r\}|\varepsilon_r| \Big/ \sum_{r\ge m}\mathrm{Po}(\lambda)\{r\}, \tag{1.23}$$

where

$$\begin{aligned}\varepsilon_r &= \phi^{-r}\prod_{\alpha\in\Gamma}(1-\pi_\alpha+\phi\pi_\alpha)e^{\lambda-m}(m/\lambda)^r - 1\\ &= \left(\frac{m}{\lambda\phi}\right)^{r-m}\cdot\left(\frac{m}{\lambda\phi}\right)^m\prod_{\alpha\in\Gamma}(1-\pi_\alpha+\phi\pi_\alpha)e^{\lambda-m} - 1.\end{aligned}$$

From Lemma 9.1.11, in the range of m considered,

$$\left|\left(\frac{m}{\lambda\phi}\right)^m\prod_{\alpha\in\Gamma}(1-\pi_\alpha+\phi\pi_\alpha)e^{\lambda-m} - 1\right| \le (e^2-1)\psi\Delta^2,$$

and, from Lemma 9.1.9(i), $m \le \lambda\phi$. Hence, from Lemma 9.1.9(ii),

$$|\varepsilon_r| \le (e^2-1)\psi\Delta^2 + (r-m)(m^{-1}\lambda\phi-1) \le (e^2-1)\psi\Delta^2 + 2(r-m)\psi\Delta\lambda^{-1/2}. \tag{1.24}$$

Appealing to Proposition A.2.1(i) and (ii), we find that

$$\mathrm{Po}(\lambda)\{r\}/\mathrm{Po}(\lambda)\{[m,\infty)\} \le 4(1+\lambda^{1/2}\Delta^{-1})^{-1}(1+\lambda^{-1/2}\Delta)^{-(r-m)}, \tag{1.25}$$

and thus, combining (1.23)–(1.25), it follows that $|\eta| \le (e^2-1)\psi\Delta^2 + 8\psi$, which implies part (i).

For (ii), note that the restrictions on Δ and ψ ensure that the condition on $\max_\alpha \pi_\alpha$ in Lemma 9.1.11(b) is automatically satisfied. The argument then follows precisely the same lines as for (i). □

Lemma 9.1.12 can be combined with Corollary 9.B.2 to give an analogue of Theorem 9.C for tail probabilities, showing that, for moderately large deviations, the Poisson approximation still holds with small relative error. The proof is accomplished in the same way as that of Theorem 9.C.

Theorem 9.D *In the notation of Theorem 9.C,*
(a) *uniformly in m satisfying $\Delta \geq 1$, $m \leq \lambda/(2\psi)$ and $1+4\Delta^2 \leq (16\psi)^{-1}$,*

$$\mathbb{P}[W \geq m] = \mathrm{Po}(\lambda)\{[m, \infty)\}\{1 + O(\psi) + O(\psi\Delta^2)\};$$

(b) *if $\psi \leq 1/16$, then, uniformly in m satisfying $-\frac{7\sqrt{\lambda}}{16} \leq \Delta \leq -1$ and $\Delta^2 \leq \psi^{-1}$,*

$$\mathbb{P}[W < m] = \mathrm{Po}(\lambda)\{[0, m-1]\}\{1 + O(\psi) + O(\psi\Delta^2)\}. \qquad \square$$

Remark 9.1.13 If $|\Delta| \leq 1$, Theorem 9.A with $l = 1$ already gives faithful approximation by $\mathrm{Po}(\lambda)$ with relative error as in Theorems 9.C and 9.D.

Remark 9.1.14 The broad conclusion to be drawn from Theorems 9.C and 9.D is that the distribution $\mathrm{Po}(\lambda)$ can be used with small relative error to approximate either $\mathbb{P}[W = m]$, $\mathbb{P}[W \geq m]$ or $\mathbb{P}[W < m]$, provided that $|\Delta| \ll \psi^{-1/2} = (\lambda_2/\lambda)^{-1/2}$, or, equivalently, that $|m - \lambda| \ll \lambda/\sqrt{\lambda_2}$. Note that, in the case of the Poisson approximation to the binomial, the quantity $\psi\Delta^2$, which plays the main role in the relative error estimates in Theorem 9.C, reduces to $n^{-1}(m - np)^2 = O(np^2, m^2n^{-1})$, in agreement with Equation (1.1.1).

9.2 Binomial approximation

When approximating the distribution of a sum of independent indicators, the binomial family is an even more natural choice than the Poisson, since there are two parameters which can be adjusted to improve the fit, and, if the indicators are also identically distributed, the fit is exact. We now show that Stein's method can be applied as well in the binomial context as in the Poisson.

The first step is to derive an analogue of (1.1.10) for the binomial distribution. We start with the equation

$$\alpha_j g(j+1) - \beta_j g(j) = I[j \in A] - c, \qquad 0 \leq j \leq m, \tag{2.1}$$

for positive constants $(\alpha_j, 0 \leq j \leq m-1)$ and $(\beta_j, 1 \leq j \leq m)$, setting $\beta_0 = \alpha_m = 0$: here, $m = \infty$ is also allowed. Let μ denote the probability measure on $\{0, 1, \ldots, m\}$ with

$$\mu\{j\} = \mu_j = C \prod_{r=1}^{j} (\alpha_{r-1}/\beta_r),$$

where C is chosen so that $\mu\{[0,m]\} = 1$: if $m = \infty$, the α's and β's must be such that such a C exists. Then it is easy to check that the μ_j's are summation factors for the equations (2.1), and that if $m < \infty$ the equations are consistent if $c = \mu\{A\}$. For $m = \infty$, $\alpha_j = \lambda$ and $\beta_j = j$, we obtain (1.1.10), and the corresponding μ is the Poisson distribution $\mathrm{Po}(\lambda)$. For $m < \infty$, $\alpha_j = (m-j)p$ and $\beta_j = j(1-p)$, the corresponding μ is the binomial distribution $\mathrm{Bi}(m,p)$. Thus the total variation difference between the distribution of W and $\mathrm{Bi}(m,p)$ can be estimated, if the quantities

$$\mathbb{E}\{(m-W)pg(W+1) - W(1-p)g(W)\} \tag{2.2}$$

can be controlled, for each $g = g_{m,p,A}$ satisfying (2.1) with $\alpha_j = (m-j)p$, $\beta_j = j(1-p)$ and $c = \mathrm{Bi}(m,p)\{A\}$.

To start with, we prove the following analogue of Lemma 1.1.1.

Lemma 9.2.1 *Suppose that the α_j's are non-increasing and the β_j's non-decreasing. Let $g = g_A : \mathbb{Z}^+ \to \mathbb{R}$ satisfy (2.1), where we take $g(0) = g(1)$ and $g(j) = g(m)$ for $j \geq m$. Then*

$$\Delta g = \max_{j\geq 0} |g(j+1) - g(j)| \leq \max_{1\leq j\leq m-1} \min(\alpha_j^{-1}, \beta_j^{-1}).$$

Proof The proof is similar to the last part of Lemma 1.1.1. The function g_A can be explicitly written as

$$\begin{aligned} g_A(j) &= \frac{1}{\beta_j\mu_j}\{\mu\{A\cap U_{j-1}\} - \mu\{A\}\mu\{U_{j-1}\}\} \\ &= -\frac{1}{\alpha_{j-1}\mu_{j-1}}\{\mu\{A\cap U_{j-1}^c\} - \mu\{A\}\mu\{U_{j-1}^c\}\}, \end{aligned}$$

$m \geq j \geq 1$, where $U_j = \{0,1,\ldots,j\}$ as before, and we can express g_A as $\sum_{s\in A} g_{\{s\}}$. Now $g = g_{\{s\}}$ is negative and decreasing in $j \leq s$, and is positive and decreasing in $j > s$, since, for instance, for $2 \leq j \leq s$,

$$\begin{aligned} g(j-1) - g(j) &= \frac{\mu_s}{\beta_j\mu_j}\sum_{r=0}^{j-1}\mu_r - \frac{\mu_s}{\beta_{j-1}\mu_{j-1}}\sum_{r=0}^{j-2}\mu_r \\ &\geq \frac{\mu_s}{\beta_j\mu_j}\sum_{r=1}^{j-1}\mu_r\{1 - \frac{\alpha_{j-1}\beta_j\beta_r}{\beta_j\beta_{j-1}\alpha_{r-1}}\}, \end{aligned}$$

from the definition of μ, and then $\alpha_{j-1} \leq \alpha_{r-1}$ and $\beta_r \leq \beta_{j-1}$. Hence the only positive jump in g is

$$g(s+1) - g(s) = \alpha_s^{-1}\sum_{r=s+1}^{m}\mu_r + \beta_s^{-1}\sum_{r=0}^{s-1}\mu_r,$$

which can be bounded above in two ways, by

$$\alpha_s^{-1}\{\sum_{r=s+1}^{m}\mu_r+\sum_{r=0}^{s-1}\mu_{r+1}\frac{\beta_{r+1}\alpha_s}{\alpha_r\beta_s}\}\le\alpha_s^{-1}\mu\{[1,m]\}\le\alpha_s^{-1},$$

and by

$$\beta_s^{-1}\{\sum_{r=s+1}^{m}\mu_{r-1}\frac{\alpha_{r-1}\beta_s}{\beta_r\alpha_s}+\sum_{r=0}^{s-1}\mu_r\}\le\beta_s^{-1}\mu\{[0,m-1]\}\le\beta_s^{-1}.$$

The rest of the proof is as for Lemma 1.1.1. □

For the binomial distribution $\mathrm{Bi}\,(m,p)$, we have $\alpha_j^{-1}\le p^{-1}$, $\beta_j^{-1}\le(1-p)^{-1}$ and

$$\min(\alpha_j^{-1},\beta_j^{-1})\le m^{-1}\{(m-j)\alpha_j^{-1}+j\beta_j^{-1}\}=[mp(1-p)]^{-1},$$

whenever $1\le j\le m-1$, leading to the estimate

$$\Delta g\le\min\{p^{-1},(1-p)^{-1},[mp(1-p)]^{-1}\}\le[mp(1-p)]^{-1}. \qquad (2.3)$$

We are now able to prove a binomial approximation theorem for W.

Theorem 9.E *Let m be an integer close to λ/ψ, where $\psi=\lambda_2/\lambda$ as before, and let $\varepsilon=(\lambda/\psi)-m$. Suppose that $p=\lambda/m<1$. Then*

$$d_{TV}(\mathcal{L}(W),\mathrm{Bi}\,(m,p))\le(1-p)^{-1}\{4(\lambda^{-1}\lambda_3-\psi^2)+\psi^2|\varepsilon|(\lambda-\varepsilon\psi)^{-1}\}. \quad (2.4)$$

Remark 9.2.2 If λ/ψ is an integer, $m=\lambda/\psi$ makes the second term in braces zero.

Remark 9.2.3 The error is of the same order of magnitude as for the expansion with $l=2$ in Corollary 9.A.1: that is, one order better than for the Poisson approximation. This is not surprising, since there are two adjustable parameters m and p available. If $\pi_\alpha=p$ for all α and $m=|\Gamma|$, the estimate (2.4) reduces to zero. The quantity $\lambda^{-1}\lambda_3-\psi^2$ reflects the heterogeneity of the values of the π_α: it is the variance of the random variable P such that $\mathbb{P}[P=\pi_\alpha]=\lambda^{-1}\pi_\alpha$.

Remark 9.2.4 Ehm (1991) considers approximation with $\mathrm{Bi}\,(m,p)$ with m fixed equal to $|\Gamma|$ and $p=|\Gamma|^{-1}\lambda$, obtaining the inequalities

$$(1/124)\{(npq)^{-1}\wedge 1\}\sum_{\alpha\in\Gamma}(\pi_\alpha-p)^2\le d_{TV}(\mathcal{L}(W),\mathrm{Bi}\,(|\Gamma|,p))$$
$$\le(1-p^{n+1}-q^{n+1})\{(n+1)pq\}^{-1}\sum_{\alpha\in\Gamma}(\pi_\alpha-p)^2,$$

analogous to Corollary 3.D.1. Although the approximation is only first order, since but one of the two parameters is fitted, the upper error estimate is not always inferior to (2.4), because the constant multiplier is smaller.

Proof The aim is to show that (2.2) is small. Using the independence of the summands I_α, it follows easily that

$$\begin{aligned}
&\mathbb{E}\{(m-W)pg(W+1) - W(1-p)g(W)\} \\
&\quad = \sum_{\alpha\in\Gamma} \pi_\alpha \mathbb{E}\{g(W+1) - g(W_\alpha+1) - p[g(W_\alpha+2) - g(W_\alpha+1)]\} \\
&\quad = \sum_{\alpha\in\Gamma} \pi_\alpha(\pi_\alpha - p)(h_\alpha - \bar{h}) + \lambda(\psi - p)\bar{h}, \qquad (2.5)
\end{aligned}$$

where $W_\alpha = \sum_{\beta\neq\alpha} I_\beta$ as before, $h_\alpha = \mathbb{E}\{g(W_\alpha+2) - g(W_\alpha+1)\}$ and $\bar{h} = \lambda^{-1}\sum_{\alpha\in\Gamma}\pi_\alpha h_\alpha$. The estimate

$$\begin{aligned}
|\lambda(\psi-p)\bar{h}| &\leq |\lambda(\psi-p)|[mp(1-p)]^{-1} \\
&= (1-p)^{-1}\left|\psi - \frac{\lambda}{m}\right| = (1-p)^{-1}|\varepsilon|\psi^2(\lambda - \varepsilon\psi)^{-1}
\end{aligned}$$

is now immediate from (2.3).

Now $h_\alpha - \bar{h} = \lambda^{-1}\sum_{\beta\in\Gamma}\pi_\beta(h_\alpha - h_\beta)$, and h_α and h_β are close, since, for any bounded function $f : \mathbb{Z}^+ \to \mathbb{R}$,

$$|\mathbb{E}f(W_\alpha) - \mathbb{E}f(W_\beta)| = |\mathbb{E}\{f(\widetilde{W}_{\alpha\beta} + \tilde{I}_\alpha) - f(\widetilde{W}_{\alpha\beta} + \tilde{I}_\beta)\}|,$$

where

$$\widetilde{W}_{\alpha\beta} \stackrel{\mathcal{D}}{=} \sum_{\gamma\neq\alpha,\beta} I_\gamma, \qquad \tilde{I}_\alpha \stackrel{\mathcal{D}}{=} I_\alpha, \qquad \tilde{I}_\beta \stackrel{\mathcal{D}}{=} I_\beta,$$

the pair $(\tilde{I}_\alpha, \tilde{I}_\beta)$ is independent of $\widetilde{W}_{\alpha\beta}$, and $\tilde{I}_\alpha$ and $\tilde{I}_\beta$ are maximally coupled: hence

$$|\mathbb{E}f(W_\alpha) - \mathbb{E}f(W_\beta)| \leq |\pi_\alpha - \pi_\beta|\Delta f.$$

This implies that $|h_\alpha - h_\beta| \leq 2|\pi_\alpha - \pi_\beta|\Delta g$, and hence that the remaining term in (2.5) can be bounded by

$$\begin{aligned}
\Big|\sum_{\alpha\in\Gamma}\pi_\alpha(\pi_\alpha - p)(h_\alpha - \bar{h})\Big| &= \Big|\sum_{\alpha\in\Gamma}\pi_\alpha(\pi_\alpha - \psi)(h_\alpha - \bar{h})\Big| \\
&\leq 2\Delta g \sum_{\alpha\in\Gamma}\pi_\alpha|\pi_\alpha - \psi|\lambda^{-1}\sum_{\beta\in\Gamma}\pi_\beta|\pi_\alpha - \pi_\beta| \\
&\leq 2\Delta g \sum_{\alpha\in\Gamma}\pi_\alpha\Big(\lambda^{-1}\sum_{\beta\in\Gamma}\pi_\beta|\pi_\alpha - \pi_\beta|\Big)^2 \\
&\leq 2\Delta g \sum_{\alpha\in\Gamma}\pi_\alpha\lambda^{-1}\sum_{\beta\in\Gamma}\pi_\beta(\pi_\alpha - \pi_\beta)^2 \leq 4(1-p)^{-1}(\lambda^{-1}\lambda_3 - \psi^2),
\end{aligned}$$

as required. □

Remark 9.2.5 Consider a collection $(W^{(i)}, 1 \leq i \leq K)$ of independent random variables, each of which is a sum of independent indicators. Let $\mathbf{W}$ denote the vector $(W^{(1)}, \ldots, W^{(K)})$. Then it follows from Corollary 9.A.1 and Proposition A.1.1 that

$$d_{TV}\left(\mathcal{L}(\mathbf{W}), \prod_{i=1}^{K} \mathrm{Po}\,(\lambda^{(i)})\right) \leq \sum_{i=1}^{K} \psi^{(i)}. \tag{2.6}$$

Thus, for identically distributed $W^{(i)}$, one has an estimate of order $K\psi$. Falk and Reiss (1988) have shown that, if W has a binomial distribution $\mathrm{Bi}\,(m,p)$ and $\lambda = mp$, $d_H(\mathcal{L}(W), \mathrm{Po}\,(\lambda)) \leq cp$ for a constant c, where d_H denotes the Hellinger distance. From this and from Proposition A.1.1, it follows that, for K independent copies $W^{(i)}$,

$$d_{TV}\left(\mathcal{L}(\mathbf{W}), \prod_{i=1}^{K} \mathrm{Po}\,(\lambda^{(i)})\right) \leq \sqrt{2} d_H\left(\mathcal{L}(\mathbf{W}), \prod_{i=1}^{K} \mathrm{Po}\,(\lambda)\right) \leq cp\sqrt{K},$$

often an important improvement on the estimate Kp which would follow from (2.6). Theorem 9.E now implies that a similar estimate holds, even if the $W^{(i)}$ are sums of non-identically distributed summands. This is because, using the total-variation inequality, $d_{TV}(\mathcal{L}(\mathbf{W}), \prod_{i=1}^{K} \mathrm{Bi}\,(m^{(i)}, p^{(i)}))$ can be bounded by the sum of the estimates from (2.4), which typically yields a quantity of order Kp^2, where p is some average of the $p^{(i)}$'s. The result of Falk and Reiss can then be used to proceed to the Poisson approximation, if required: the extra error in this step, typically of order $p\sqrt{K}$, is rather larger than the error Kp^2 incurred in the original binomial approximation.

Remark 9.2.6 In principle, this version of Stein's method could also be used for binomial approximation of sums of dependent indicators. For example, much as for Theorem 1.B, suppose that $W = \sum_{\alpha \in \Gamma} I_\alpha$, $\pi_\alpha = \mathbb{E} I_\alpha$ and $mp = \lambda = \sum_{\alpha \in \Gamma} \pi_\alpha$, for some collection of indicators $(I_\alpha, \alpha \in \Gamma)$, and suppose in addition that U_α, V_α and I'_α are random variables satisfying

$$\mathcal{L}(U_\alpha) = \mathcal{L}(W), \quad \mathcal{L}(V_\alpha + 1) = \mathcal{L}(W \mid I_\alpha = 1) \quad \text{and} \quad \mathcal{L}(I'_\alpha) = \mathrm{Be}\,(p),$$

with I'_α and V_α independent. Then

$$d_{TV}(\mathcal{L}(W), \mathrm{Bi}\,(m,p)) \leq \frac{1}{(1-p)\lambda} \sum_{\alpha \in \Gamma} \pi_\alpha \mathbb{E}|U_\alpha - (V_\alpha + I'_\alpha)|.$$

It is conceivable that this inequality could be combined with appropriate couplings to prove some of the results on binomial approximation stated in Chapter 4, though we have not attempted to do so.

9.3 Dissociated summands

In the previous section, the indicators $(I_\alpha, \alpha \in \Gamma)$ were assumed to be independent. Here, we relax the condition slightly, replacing it with the assumption that the indicators are dissociated as in Section 2.3 or, somewhat more generally, that, for each α, there exists $\Gamma_\alpha^i \subset \Gamma$ such that I_α and $\{I_\beta, \beta \in \Gamma_\alpha^i\}$ are independent, and that the $\Gamma_\alpha^s = \Gamma \setminus (\Gamma_\alpha^i \cup \{\alpha\})$ are not too large. Formally, the latter stipulation does not appear as a condition in the following theorem, but is implicit in the form of the estimate

$$\delta_1 = \lambda^{-1}(1 - e^{-\lambda}) \sum_{\alpha \in \Gamma} \{\pi_\alpha^2 + \sum_{\beta \in \Gamma_\alpha^s} (\pi_\alpha \pi_\beta + \mathbb{E} I_\alpha I_\beta)\} \tag{3.1}$$

obtained, which must be small if good Poisson approximation is to be guaranteed.

Theorem 9.F *Under the above assumptions,*

$$\begin{aligned} d_{TV}(\mathcal{L}(W), \mathrm{Po}(\lambda)) &\le \delta_1 \\ &\le \lambda^{-1}(1 - e^{-\lambda}) \sum_{\alpha \in \Gamma} \{\pi_\alpha^2 + \sum_{\beta \in \Gamma_\alpha^s} (2\pi_\alpha \pi_\beta + \rho_{\alpha\beta}^+ \sqrt{\pi_\alpha \pi_\beta})\}, \end{aligned} \tag{3.2}$$

where $\pi_\alpha = \mathbb{E} I_\alpha$, $\lambda = \mathbb{E} W$ *and* $\rho_{\alpha\beta}^+ = \mathrm{Corr}\,(I_\alpha, I_\beta) \vee 0$.

Proof The result is a restatement of Corollary 2.C.5. □

Remark 9.3.1 Suppose that $\pi_\alpha = \pi$, $|\Gamma_\alpha^s| = \gamma$ and $\sum_{\beta \in \Gamma_\alpha^s} \rho_{\alpha\beta}^+ = \rho$ are the same for all α: then the estimate (3.2) does not exceed $\pi(2\gamma + 1) + \rho$. Thus, if $n = |\Gamma|$ is large and $\lambda = n\pi \asymp 1$, good Poisson approximation is obtained, provided that the Γ_α^s are not too large, in the sense that $\gamma \ll n$, and that $\rho \ll 1$, implying that there is no strong correlation between dependent I_α's.

As in Section 9.1, it is reasonable to ask whether the Poisson approximation of Theorem 9.F can be refined to include further terms in an asymptotic expansion. It is clear in principle that this can be done, though the proofs of the results below show how complicated the argument can become. We adopt the approach taken by Barbour and Eagleson (1987), making the simplifying assumptions that $\beta \in \Gamma_\alpha^s \iff \alpha \in \Gamma_\beta^s$ and that $I_\alpha I_\beta$ is independent of $\{I_\gamma, \gamma \in \Gamma_\alpha^i \cap \Gamma_\beta^i\}$, which are obviously true for dissociated variables with the standard choice $\Gamma_\alpha^s = \{\beta : \beta \ne \alpha, \beta \cap \alpha \ne \emptyset\}$.

Let

$$\begin{aligned} &\pi_\alpha = \mathbb{E} I_\alpha; \ \pi_{\alpha\beta} = \mathbb{E} I_\alpha I_\beta; \ \pi_{\alpha\beta\gamma} = \mathbb{E} I_\alpha I_\beta I_\gamma; \\ &\kappa = \mathrm{Var}\, W - \mathbb{E} W = \sum_{\alpha \in \Gamma} \sum_{\beta \in \Gamma_\alpha^s} (\pi_{\alpha\beta} - \pi_\alpha \pi_\beta) - \sum_{\alpha \in \Gamma} \pi_\alpha^2, \end{aligned} \tag{3.3}$$

and define the random variables, writing $\Gamma^s_{\alpha\beta} = \Gamma^s_\beta \setminus (\Gamma^s_\alpha \cup \{\alpha\})$,

$$\begin{aligned} Z_\alpha &= \sum_{\beta\in\Gamma^s_\alpha} I_\beta; \qquad & W_\alpha &= W - Z_\alpha - I_\alpha; \\ Z_{\alpha\beta} &= \sum_{\gamma\in\Gamma^s_{\alpha\beta}} I_\gamma; \qquad & W_{\alpha\beta} &= W_\alpha - Z_{\alpha\beta}; \end{aligned} \tag{3.4}$$

define also the quantity

$$\begin{aligned} \delta_2 = \lambda^{-1}(1-e^{-\lambda})\Big\{ & 2|\kappa|\delta_1 + 2\sum_{\alpha\in\Gamma} \pi_\alpha^3 \\ & + 2\sum_{\alpha\in\Gamma}\sum_{\beta\in\Gamma^s_\alpha} \pi_\alpha(\pi_{\alpha\beta} + \pi_\alpha\pi_\beta + 2|\pi_{\alpha\beta} - \pi_\alpha\pi_\beta|) \\ & + \sum_{\alpha\in\Gamma}\sum_{\substack{\beta,\gamma\in\Gamma^s_\alpha \\ \beta\neq\gamma}} (\pi_\alpha\pi_{\beta\gamma} + 2\pi_\beta\pi_{\alpha\gamma} + 3\pi_{\alpha\beta\gamma} + 4\pi_\beta|\pi_{\alpha\gamma} - \pi_\alpha\pi_\gamma|)\Big\}. \end{aligned} \tag{3.5}$$

Then we are able to prove the following refinement of Theorem 9.F.

Theorem 9.G *For any $A \subset \mathbb{Z}^+$, under the above conditions,*

$$|\mathbb{P}[W \in A] - Q_2(A)| \le \delta_2,$$

where Q_2 is as defined in (1.12), with $-\kappa$ for λ_2:

$$Q_2(i) = \left\{\frac{e^{-\lambda}\lambda^i}{i!}\right\}\left\{1 + \frac{\kappa}{2}C_2(\lambda;i)\right\}.$$

Proof The idea of the proof is the same as in the case of independence. Roughly speaking, the quantity $\mathbb{E}\{\lambda g(W+1) - Wg(W)\}$ for $g = g_{\lambda,A}$ is expressed in terms of a remainder which is second order small, together with the expectation of a function of W multiplying a term which is first order small. The expectation is then evaluated to the required accuracy by using Theorem 9.F to replace W by a Poisson random variable, and the identification of the resulting expression with Q_2 follows from Lemma 9.1.4.

First, note that, because I_α and W_α are independent,

$$\mathbb{E}\{(\pi_\alpha - I_\alpha)g(W_\alpha + 1)\} = 0,$$

and thus, for $g = g_{\lambda,A}$ for any $A \subset \mathbb{Z}^+$,

$$\begin{aligned} \mathbb{E}\{\lambda g(W+1) - Wg(W)\} &= \sum_{\alpha\in\Gamma} \pi_\alpha \mathbb{E}[g(W+1) - g(W_\alpha+1)] \\ &\quad - \sum_{\alpha\in\Gamma} \mathbb{E}[I_\alpha(g(W_\alpha + Z_\alpha + 1) - g(W_\alpha + 1))]. \end{aligned} \tag{3.6}$$

Now, for the first term on the right hand side of (3.6), we have

$$\begin{aligned}\big|g(W+1) - g(W_\alpha+1) - (I_\alpha+Z_\alpha)[g(W_\alpha+2)-g(W_\alpha+1)]\big| \\ \le 2\lambda^{-1}(1-e^{-\lambda})(I_\alpha+Z_\alpha-1)^+,\end{aligned}$$

giving

$$\begin{aligned}\sum_{\alpha\in\Gamma}\pi_\alpha\mathbb{E}[g(W+1)-g(W_\alpha+1)] &= \sum_{\alpha\in\Gamma}\pi_\alpha^2\mathbb{E}[g(W_\alpha+2)-g(W_\alpha+1)] \\ &+ \sum_{\alpha\in\Gamma}\sum_{\beta\in\Gamma_\alpha^s}\pi_\alpha\mathbb{E}\{I_\beta[g(W_\alpha+2)-g(W_\alpha+1)]\}+\varepsilon_1, \quad (3.7)\end{aligned}$$

where

$$\begin{aligned}|\varepsilon_1| &\le 2\lambda^{-1}(1-e^{-\lambda})\sum_{\alpha\in\Gamma}\pi_\alpha\mathbb{E}(I_\alpha+Z_\alpha-1)^+ \\ &\le \eta_1 = \lambda^{-1}(1-e^{-\lambda})\sum_{\alpha\in\Gamma}\pi_\alpha\Bigg(\sum_{\substack{\beta,\gamma\in\Gamma_\alpha^s \\ \beta\ne\gamma}}\pi_{\beta\gamma}+2\sum_{\gamma\in\Gamma_\alpha^s}\pi_{\alpha\gamma}\Bigg), \quad (3.8)\end{aligned}$$

since, for $j\in\mathbb{Z}^+$, $(j-1)^+\le\frac{1}{2}j(j-1)$. Applying the same technique for $\beta\in\Gamma_\alpha^s$,

$$\begin{aligned}\big|\mathbb{E}\{I_\beta[g(W_\alpha+2)-g(W_\alpha+1)]\} - \pi_\beta\mathbb{E}\{g(W+1)-g(W_\alpha+1)\}\big| \\ \le 2\lambda^{-1}(1-e^{-\lambda})\mathbb{E}(I_\beta Z_{\alpha\beta}) \le 2\lambda^{-1}(1-e^{-\lambda})\sum_{\gamma\in\Gamma_{\alpha\beta}^s}\pi_{\beta\gamma}, \quad (3.9)\end{aligned}$$

giving in all the expression

$$\begin{aligned}\sum_{\alpha\in\Gamma}\pi_\alpha\mathbb{E}[g(W+1)-g(W_\alpha+1)] &= \sum_{\alpha\in\Gamma}\pi_\alpha^2\mathbb{E}[g(W_\alpha+2)-g(W_\alpha+1)] \\ &+ \sum_{\alpha\in\Gamma}\sum_{\beta\in\Gamma_\alpha^s}\pi_\alpha\pi_\beta\mathbb{E}[g(W_{\alpha\beta}+2)-g(W_{\alpha\beta}+1)]+\varepsilon_2, \quad (3.10)\end{aligned}$$

where $|\varepsilon_2|\le\eta_1+\eta_2$, and

$$\eta_2 = 2\lambda^{-1}(1-e^{-\lambda})\sum_{\alpha\in\Gamma}\sum_{\beta\in\Gamma_\alpha^s}\sum_{\gamma\in\Gamma_{\alpha\beta}^s}\pi_\alpha\pi_{\beta\gamma}. \quad (3.11)$$

We now turn to the second term on the right hand side of (3.6), and argue in much the same way to obtain an expression to match (3.10). The estimates used are

$$\begin{aligned}\big|I_\alpha[g(W_\alpha+Z_\alpha+1)-g(W_\alpha+1)] - I_\alpha Z_\alpha[g(W_\alpha+2)-g(W_\alpha+1)]\big| \\ \le 2\lambda^{-1}(1-e^{-\lambda})I_\alpha(Z_\alpha-1)^+ \quad (3.12)\end{aligned}$$

and, for $\beta \in \Gamma^s_\alpha$,

$$\left|\mathbb{E}\{I_\alpha I_\beta[g(W_\alpha+2)-g(W_\alpha+1)]\} - \pi_{\alpha\beta}\mathbb{E}\{g(W_{\alpha\beta}+2)-g(W_{\alpha\beta}+1)\}\right| \leq 2\lambda^{-1}(1-e^{-\lambda})\mathbb{E}(I_\alpha I_\beta Z_{\alpha\beta}), \quad (3.13)$$

since $I_\alpha I_\beta$ is independent of $W_{\alpha\beta}$, giving

$$\sum_{\alpha\in\Gamma}\mathbb{E}[I_\alpha(g(W_\alpha+Z_\alpha+1)-g(W_\alpha+1))] = \sum_{\alpha\in\Gamma}\sum_{\beta\in\Gamma^s_\alpha}\pi_{\alpha\beta}\mathbb{E}\{g(W_{\alpha\beta}+2)-g(W_{\alpha\beta}+1)\}+\varepsilon_3, \quad (3.14)$$

where

$$|\varepsilon_3| \leq \eta_3 = \lambda^{-1}(1-e^{-\lambda})\sum_{\alpha\in\Gamma}\sum_{\substack{\beta,\gamma\in\Gamma^s_\alpha\\ \beta\neq\gamma}}\pi_{\alpha\beta\gamma} + 2\lambda^{-1}(1-e^{-\lambda})\sum_{\alpha\in\Gamma}\sum_{\beta\in\Gamma^s_\alpha}\sum_{\gamma\in\Gamma^s_{\alpha\beta}}\pi_{\alpha\beta\gamma}. \quad (3.15)$$

The next step is to observe that

$$\left|[g(W_\alpha+2)-g(W_\alpha+1)]-[g(W+2)-g(W+1)]\right| \leq 2\lambda^{-1}(1-e^{-\lambda})(I_\alpha+Z_\alpha)$$

and that

$$\left|[g(W_{\alpha\beta}+2)-g(W_{\alpha\beta}+1)]-[g(W+2)-g(W+1)]\right| \leq 2\lambda^{-1}(1-e^{-\lambda})(I_\alpha+Z_\alpha+Z_{\alpha\beta}),$$

so that, combining (3.6), (3.10) and (3.14),

$$\mathbb{E}\{\lambda g(W+1)-Wg(W)\} = \left[\sum_{\alpha\in\Gamma}\pi_\alpha^2+\sum_{\alpha\in\Gamma}\sum_{\beta\in\Gamma^s_\alpha}(\pi_\alpha\pi_\beta-\pi_{\alpha\beta})\right]\mathbb{E}\{g(W+2)-g(W+1)\}+\varepsilon_4, \quad (3.16)$$

where $|\varepsilon_4| \leq \eta_1+\eta_2+\eta_3+\eta_4$, and

$$\eta_4 = 2\lambda^{-1}(1-e^{-\lambda})\left[\sum_{\alpha\in\Gamma}\pi_\alpha^2(\pi_\alpha+\sum_{\beta\in\Gamma^s_\alpha}\pi_\beta) + \sum_{\alpha\in\Gamma}\sum_{\beta\in\Gamma^s_\alpha}|\pi_{\alpha\beta}-\pi_\alpha\pi_\beta|(\pi_\alpha+\sum_{\gamma\in\Gamma^s_\alpha}\pi_\gamma+\sum_{\gamma\in\Gamma^s_{\alpha\beta}}\pi_\gamma)\right]. \quad (3.17)$$

Finally, letting Z denote a random variable with distribution $\mathrm{Po}(\lambda)$, and applying Theorem 9.F to the expectation $\mathbb{E}\{g(W+2)-g(W+1)\}$, we conclude that

$$\begin{aligned}\mathbb{P}[W\in A] &= \mathrm{Po}(\lambda)\{A\}-\kappa\mathbb{E}\{g(Z+2)-g(Z+1)\}+\varepsilon_5\\ &= \mathcal{Q}_2(A)+\varepsilon_5,\end{aligned} \quad (3.18)$$

where the last equality uses Lemma 9.1.4, with

$$|\varepsilon_5| \leq \eta_1+\eta_2+\eta_3+\eta_4+2|\kappa|\lambda^{-1}(1-e^{-\lambda})\delta_1.$$

The bounding of this error estimate by δ_2 is straightforward. □

Remark 9.3.2 The signed measure Q_2 has the same first and second moments as W, as may easily be checked.

The estimate δ_2 of the discrepancy is somewhat clumsy, though often easy to calculate in applications. For example, let us consider again the runs studied in Section 8.4.

Example 9.3.3 Suppose that $(Y_j)_{j=1}^{n+k-1}$ are independent random variables such that $\mathbb{P}[Y_j = 1] = p$, and define

$$I_j = I[Y_l = 1, \quad j \leq l \leq j + k - 1], \qquad 1 \leq j \leq n.$$

Then, if $\Gamma = \{\{j, j+1, \ldots, j+k-1\} : 1 \leq j \leq n\}$ and j is used as shorthand for $\alpha = \{j, j+1, \ldots, j+k-1\}$, the random variables $(I_j)_{j=1}^n$ are dissociated. It is immediate that, with $\Gamma_j^s = \{l : 1 \leq |l - j| < k\}$,

$$\pi_j = p^k; \; \pi_{jl} = p^{k+l-j}; \; \pi_{jlm} = p^{k+m-j}, \; 1 \leq j < l < m \leq n \text{ and } j, m \in \Gamma_l^s,$$

and evaluation of the bounds δ_1 and δ_2 yields crude upper estimates of $\delta_1 \leq 4p/(1-p)$ for the accuracy of the Poisson approximation, and of $\delta_2 \leq 80[p/(1-p)]^2$ for the accuracy of Q_2. Note that if p is small, so that δ_1 and δ_2 are small, the leading order contribution to δ_1 comes only from pairs $I_j I_{j+1}$, and, apart from δ_1^2, that to δ_2 comes only from triples $I_j I_{j+1} I_{j+2}$ (provided $k \geq 2$).

Example 9.3.4 A second, more complicated example arises in a simple model of random mating in a two-sex population, where individuals are assumed to mate (in a given time interval) if they are of opposite sex and are close to each other in some space of characteristics: the characteristics may merely define geographical position, but it is also possible to include other information such as the individual's appearance and habits. Formally, we consider individuals described by pairs $\{(Y_i, Z_i)\}_{i=1}^n$, where the Y_i's are independent, identically distributed random elements of a metric space $(\mathcal{Y}, \rho)$, and the Z_i's are independent, identically distributed zero–one random variables with $\mathbb{P}[Z_i = 0] = q$, independent also of the Y_i's. Let $\Gamma = \{\{i, j\} : 1 \leq i \neq j \leq n\}$ and define $I_{ij} = I[\rho(Y_i, Y_j) \leq r; Z_i \neq Z_j]$, for some appropriately chosen threshold r. Then W counts the number of matings in the population.

To simplify the computations, suppose that $q = 1/2$ and that the Y_i's are uniformly distributed on an s-dimensional sphere. Then, letting $\pi = \mathbb{E}I_\alpha$, it follows that $\pi_{\alpha\beta} = \mathbb{E}I_\alpha I_\beta = \pi^2$ whenever $\alpha \neq \beta$, and that $\pi_{\alpha\beta\gamma} = \pi^3$ for α, β and γ distinct, except for the cases $\alpha = \{i, j\}$, $\beta = \{j, l\}$ and $\gamma = \{i, l\}$: here, letting $(\alpha\beta)$ denote this value of γ, it is clear that $I_\alpha I_\beta I_{(\alpha\beta)} = 0$ with probability one, so that $\pi_{\alpha\beta(\alpha\beta)} = 0$. The variables I_α are dissociated and we thus define $\Gamma_\alpha^s = \{\beta : \beta \neq \alpha, \beta \cap \alpha \neq \emptyset\}$, with

$|\Gamma_\alpha^s| = 2(n-2)$. It is then easy to calculate the bounds, with

$$\lambda = \mathbb{E}W = \binom{n}{2}\pi,$$

$$\delta_1 \le 4(n-1)\pi = 8\frac{\lambda}{n}$$

and

$$\delta_2 \le 24(n-1)^2\pi^2 = 96\left(\frac{\lambda}{n}\right)^2,$$

showing that Poisson approximation is good if λ/n is small, and that then approximation by Q_2 is an order of magnitude better.

It is natural to ask why the apparently simpler problem, where we take $I_{ij} = I[\rho(Y_i, Y_j) \le r]$ and W is a measure of clustering, was not analysed first. In this case, the estimates are much the same (with π and λ twice as big), except that $\pi_{\alpha\beta(\alpha\beta)} = c\pi^2$ for some c which need not be small. This does not affect the estimation of δ_1, which is still of order $n\pi$, but it adds a contribution of $6(n-2)c\pi$ to δ_2, so that δ_2 is now only of the same order of magnitude as δ_1. Thus approximation by Q_2 is by no means a universal improvement on the Poisson approximation. Indeed, it suggests that, as soon as independence is relaxed, even only as far as dissociation, it may be more realistic to analyse individual problems rather than to attempt to develop a universal theory.

Counting close pairs is, however, a problem of considerable practical interest, and so we exhibit the next approximation in this case also. We assume the setting of Theorem 9.G, with Γ as in Example 9.3.4, but do not suppose that $\pi_{\alpha\beta(\alpha\beta)}$ is of a smaller order of magnitude than $\pi_{\alpha\beta}$, so that some special treatment of these terms is now needed. Then we have the following modification of Theorem 9.G.

Theorem 9.H *Suppose that* $\Gamma = \{\{i,j\} : 1 \le i \ne j \le n\}$ *and that the* (I_α) *are dissociated. Then, for any* $A \subset \mathbb{Z}^+$,

$$|\mathbb{P}[W \in A] - Q_2'(A)| \le \delta_2',$$

where, with Γ_α^s *as above,*

$$Q_2'(i) = \left\{\frac{e^{-\lambda}\lambda^i}{i!}\right\}\left\{1 + \frac{\kappa}{2}C_2(\lambda; i) + \frac{\kappa'}{6}C_3(\lambda; i)\right\}; \quad \kappa' = \sum_{\alpha\in\Gamma}\sum_{\beta\in\Gamma_\alpha^s} \pi_{\alpha\beta(\alpha\beta)},$$

and

$$\delta_2' = \lambda^{-1}(1-e^{-\lambda})\Big\{ 2|\kappa|\delta_1 + 2\sum_{\alpha\in\Gamma} \pi_\alpha^3$$
$$+ 2\sum_{\alpha\in\Gamma}\sum_{\beta\in\Gamma_\alpha^s} \pi_\alpha(\pi_{\alpha\beta} + \pi_\alpha\pi_\beta + 2|\pi_{\alpha\beta} - \pi_\alpha\pi_\beta|)$$
$$+ \sum_{\alpha\in\Gamma}\sum_{\substack{\beta,\gamma\in\Gamma_\alpha^s \\ \beta\neq\gamma}} (\pi_\alpha\pi_{\beta\gamma} + 2\pi_\beta\pi_{\alpha\gamma} + 4\pi_\beta|\pi_{\alpha\gamma} - \pi_\alpha\pi_\gamma|)$$
$$+ 5\sum_{\alpha\in\Gamma}\sum_{\substack{\beta,\gamma\in\Gamma_\alpha^s \\ \gamma\neq\beta,(\alpha\beta)}} \pi_{\alpha\beta\gamma} + 4\sum_{\alpha\in\Gamma}\sum_{\beta\in\Gamma_\alpha^s} \pi_{\alpha\beta(\alpha\beta)} \sum_{\gamma\in\Gamma_\alpha^s} \pi_\gamma + 2\kappa'\delta_1 \Big\}. \tag{3.19}$$

Proof The proof of Theorem 9.G needs only to be modified where terms involving $\pi_{\alpha\beta(\alpha\beta)}$ occur, that is, in the estimate (3.15) of $|\varepsilon_3|$. Note first that the second sum in (3.15) does not contain any terms of the form $\pi_{\alpha\beta(\alpha\beta)}$, so that its contribution to η_3 in any case does not exceed

$$2\lambda^{-1}(1-e^{-\lambda})\sum_{\alpha\in\Gamma}\sum_{\substack{\beta,\gamma\in\Gamma_\alpha^s \\ \gamma\neq\beta,(\alpha\beta)}} \pi_{\alpha\beta\gamma}.$$

However, the first sum in (3.15) needs to be replaced, using a sharpening of (3.12) to

$$\Big| I_\alpha[g(W_\alpha + Z_\alpha + 1) - g(W_\alpha + 1)] - I_\alpha Z_\alpha[g(W_\alpha + 2) - g(W_\alpha + 1)]$$
$$- \frac{1}{2} I_\alpha \sum_{\beta\in\Gamma_\alpha^s} I_\beta I_{(\alpha\beta)}[g(W_\alpha + 3) - 2g(W_\alpha + 2) + g(W_\alpha + 1)] \Big|$$
$$\leq 2\lambda^{-1}(1-e^{-\lambda}) I_\alpha(Z_\alpha - 1) I\Big[\sum_{\substack{\beta,\gamma\in\Gamma_\alpha^s \\ \gamma\neq\beta,(\alpha\beta)}} I_\beta I_\gamma > 0\Big]$$
$$\leq \lambda^{-1}(1-e^{-\lambda}) \sum_{\substack{\beta,\gamma\in\Gamma_\alpha^s \\ \gamma\neq\beta,(\alpha\beta)}} I_\alpha I_\beta I_\gamma,$$

which now yields a suitably small contribution of

$$\lambda^{-1}(1-e^{-\lambda})\sum_{\alpha\in\Gamma}\sum_{\substack{\beta,\gamma\in\Gamma_\alpha^s \\ \gamma\neq\beta,(\alpha\beta)}} \pi_{\alpha\beta\gamma}$$

to η_3, but also an extra non-negligible term in the expression for $\mathbb{P}[W \in A] - \mathrm{Po}(\lambda)\{A\}$. Now, much as before, we have

$$\Bigg|\mathbb{E}\Big\{\frac{1}{2}\sum_{\alpha\in\Gamma}\sum_{\beta\in\Gamma_\alpha^s} I_\alpha I_\beta I_{(\alpha\beta)}[g(W_\alpha+3)-2g(W_\alpha+2)+g(W_\alpha+1)]\Big\}$$

$$-\frac{1}{2}\sum_{\alpha\in\Gamma}\sum_{\beta\in\Gamma_\alpha^s} \pi_{\alpha\beta(\alpha\beta)}\mathbb{E}[g(W_{\alpha\beta}+3)-2g(W_{\alpha\beta}+2)+g(W_{\alpha\beta}+1)]\Bigg|$$

$$\le 2\lambda^{-1}(1-e^{-\lambda})\sum_{\alpha\in\Gamma}\sum_{\beta\in\Gamma_\alpha^s}\mathbb{E}(I_\alpha I_\beta Z_{\alpha\beta})$$

$$\le 2\lambda^{-1}(1-e^{-\lambda})\sum_{\alpha\in\Gamma}\sum_{\substack{\beta,\gamma\in\Gamma_\alpha^s \\ \gamma\ne\beta,(\alpha\beta)}}\pi_{\alpha\beta\gamma},$$

and

$$\big|\mathbb{E}[g(W_{\alpha\beta}+3)-2g(W_{\alpha\beta}+2)+g(W_{\alpha\beta}+1)]$$

$$-\mathbb{E}[g(W+3)-2g(W+2)+g(W+1)]\big|$$

$$\le 4\lambda^{-1}(1-e^{-\lambda})\Big(\pi_\alpha+\sum_{\gamma\in\Gamma_\alpha^s}\pi_\gamma+\sum_{\gamma\in\Gamma_{\alpha\beta}^s}\pi_\gamma\Big);$$

and then finally, from Theorem 9.F,

$$\big|\mathbb{E}[g(W+3)-2g(W+2)+g(W+1)]$$

$$-\mathbb{E}[g(Z+3)-2g(Z+2)+g(Z+1)]\big| \le 4\lambda^{-1}(1-e^{-\lambda})\delta_1.$$

Collecting these estimates, the inequality

$$\Bigg|\mathbb{P}[W\in A]-\{\mathrm{Po}(\lambda)\{A\}-\kappa\mathbb{E}[g(Z+2)-g(Z+1)]$$

$$-\frac{1}{2}\kappa'\mathbb{E}[g(Z+3)-2g(Z+2)+g(Z+1)]\}\Bigg| \le \delta_2'$$

is obtained, and the theorem follows by applying Lemma 9.1.4. □

In the case of the clustering statistic discussed earlier, where δ_1 and δ_2 are both of order λ/n, we find that $\delta_2' \le 32(1+c)((n-1)\pi)^2 = 128(1+c)(\lambda/n)^2$, where $c \le 1$, so that approximation by Q_2' is indeed an improvement of an order of magnitude upon the Poisson approximation, if λ/n is small. More details on counting close pairs are given in Barbour and Eagleson (1987).

Theorems 9.G and 9.H apply also to the randomly coloured graphs studied in Section 5.3; in fact, the number of adjacent vertices with the same colour is an instance of counting close pairs. As an example, we reconsider the birthday problem.

Example 9.3.5 Colour n points independently with m colours, all having the same probability $p = 1/m$, and let W be the number of pairs with the same colour. Then

$$\mathbb{E}W = \lambda = \binom{n}{2}p; \quad \kappa = -\lambda p \sim -2\lambda^2 n^{-2}; \quad \kappa' = (n)_3 p^2 \sim 4\lambda^2 n^{-1},$$

and $\delta_2' \leq 256\lambda^2 n^{-2}$. In particular, if $\lambda \asymp 1$, it follows that

$$\sup_{A \subset \mathbb{Z}+} \left| \mathbb{P}[W \in A] - \sum_{i \in A} \frac{e^{-\lambda}\lambda^i}{i!}\{1 + \tfrac{\kappa'}{6} C_3(\lambda; i)\} \right| = O(n^{-2}).$$

As a result,

$$d_{TV}(\mathcal{L}(W), \mathrm{Po}\,(\lambda)) = \frac{\kappa'}{12} \sum_{i \geq 0} \frac{e^{-\lambda}\lambda^i}{i!} |C_3(\lambda; i)| + O(n^{-2}) \asymp \kappa' \asymp n^{-1},$$

and thus, as observed more directly in Example 5.3.2, the rate of Poisson convergence as $n \to \infty$ is much slower than $1 - \mathrm{Var}\, W/\mathbb{E}W = p \asymp n^{-2}$.

10
Poisson process approximation

Up to now, attention has been confined to approximating the distribution of a single sum W of indicators I_α by a Poisson random variable. One might, however, wish to say more. For instance, if the underlying indicators form a sequence $I_1, \ldots, I_n$, it may be of interest to test whether the common value π of each $\mathbb{E}I_i$ is less than a given value, in which case $W = \sum_{i=1}^n I_i$ provides an appropriate statistic. On the other hand, it may be suspected that the π_i's actually have a linear trend, in which case a regression statistic $\sum_{i=1}^n iI_i$ would be more suitable. Such a statistic has itself no chance of being Poisson distributed, but, in the spirit of the usual invariance principle, it may be reasonable to suppose that its null distribution can be approximated by a functional of a Poisson process. One economical way of estimating the accuracy of such approximations is first to estimate the distance, in an appropriate metric, between the law of the whole sequence $I_1, \ldots, I_n$ and that of a Poisson process, and then to use smoothness properties of the functionals in question to carry the approximation across to the statistics themselves. Thus it is quite natural to wish to establish rates of approximation for the comparison of point processes, rather than just for the total number of points occurring.

Another reason for wishing to do so comes from the study of the extremes of a sequence of random variables. In this context, it is interesting to know not only when exceptional values occur but also how big they are, and the natural framework for the description of such phenomena is the marked point process of exceedances. It is often useful to know that the exceedance process can be reasonably approximated by a Poisson process on an appropriate space, and finding bounds for the error in such an approximation is therefore of considerable interest. A further reason is provided by multivariate Poisson approximation, as for instance when approximating the joint distribution of the numbers of cycles of lengths 1, 2 and 3 in a random permutation. Since a finite collection of independent Poisson random variables can be thought of as a Poisson process on a finite set, multivariate approximation can be regarded as a special case of Poisson process approximation.

An immediate observation is that, if process approximation can be achieved with respect to the total variation metric, the same accuracy of approximation is valid for the total variation approximation of the distribution of any functional. Such a result therefore leads to enormous simplification. It is thus rather surprising that process analogues of the basic Theorems 1.A, 1.B and 1.C can be proved with almost no extra effort, and that extensions to marked point processes can also be derived. There is, however, an important difference in the rates of approximation obtained, when it comes to large expected numbers λ of events, in that the magic factors λ^{-1} and $\lambda^{-1/2}$ are no longer present in the formulae. Indeed, it can be seen, even in the simplest Bernoulli case, that they cannot be present. Let $I_1, \ldots, I_n$ be a sequence of independent indicators with $\mathbb{E}I_i = p$ for each i, and define a corresponding point process by the finite atomic measure $\sum_{i=1}^n \delta_i I_i$, where δ_i denotes the unit point mass at i. Now compare it with a Poisson process $\sum_{i=1}^n \delta_i Z_i$ with intensity p at each point i, so that the Z_i's are independent $\mathrm{Po}(p)$ random variables. Then the chance of at least one double point occurring in the Poisson process is of order np^2 if np^2 is small, whereas it is zero in the original process: thus $np^2 \wedge 1$ is the true rate for total variation process approximation in this case, and not p as when only the distribution of the total number of points is concerned (the rate cannot be worse in view of Theorem 10.A below).

It has, however, been observed in the preceding chapters that the factor λ^{-1} in particular has been very helpful in extending the range of Poisson approximation, often to its natural limits, and it would be useful to be able to recover it in some form. There are two main ways of going about this. The first is to weaken the metric on the space of distributions of processes. In the Bernoulli example considered above, the Poisson process chosen for comparison is not entirely natural: one would much rather compare the modified process $\Xi_n = \sum_{i=1}^n \delta_{i/n} I_i$ with a Poisson process Ξ of rate $\lambda = np$ on $[0, 1]$. However, this would be a hopeless exercise in the total variation metric, since the Poisson process has no points in the set $n^{-1}\mathbb{Z}$ with probability one, whereas the corresponding probability for the Bernoulli process is $(1-p)^n \sim e^{-\lambda}$. This suggests that the total variation metric is in fact too strong for many practical purposes. Put another way, more in keeping with the Stein approach, one wishes to compare $\mathbb{E}f(\Xi_n)$ with $\mathbb{E}f(\Xi)$, not for all possible functionals f, but only for functionals f belonging to a smaller set, typically those which are not too sensitive to small changes in the positions of the points in a configuration, with respect to some metric on the carrier space. For such sets of functionals f, better estimates may be attainable. Some results in this direction are given in Theorems 10.F and 10.G.

A second approach is to proceed by way of marked point processes. The choice of process $\sum_{i=1}^n \delta_i I_i$ to represent the original sequence is in many contexts natural. In others, a process $\sum_{i=1}^n \delta_{Y_i} I_i$ may be preferable, where the marks $(Y_i, 1 \le i \le n)$ are chosen according some joint distribution to be specified. The choice $Y_i = i$, $1 \le i \le n$, is one extreme possibility, allowing no variation in the y-coordinate, and enabling the index i to be recovered from the position of a point. At the other extreme, the Y_i's could be chosen to be independent $U[0, 1]$ random variables. Then, given that $W = \sum_{i=1}^n I_i = w$, the distribution of the configuration is exactly that of the Poisson process of rate λ on $[0,1]$, given that w points have occurred. Hence, as observed by Michel (1988), the total variation distance between $\mathcal{L}(\sum_{i=1}^n \delta_{Y_i} I_i)$ and $\mathcal{L}(\Xi)$ is exactly the total variation distance $d_{TV}(\mathcal{L}(W), \mathrm{Po}(\lambda))$, which, by Theorem 1.A, has the magic factor λ^{-1}. In this extreme, there is the same variation available for each y-coordinate, and there is no information about the index i contained in the position of a point. Between the two extremes, it is reasonable to hope for improved estimates of total variation distance between the marked point process and a Poisson process, while retaining enough information from the index in the position of a point to make the process approximation of practical use. This is the substance of Theorems 10.H and 10.I.

In Sections 10.3 and 10.4, the results of the first two sections are specialized to yield theorems on multivariate and compound Poisson approximation. The essential idea is that, if the information contained in the marks is only enough to identify a point as belonging to one of rather a few classes, indexed by $1 \le j \le r$, and if the expected number in class j is denoted by λ_j, then bounds on total variation approximation for the joint distribution of the numbers in the different classes by $\prod_{j=1}^r \mathrm{Po}(\lambda_j)$ should benefit if all the λ_j's are large. Theorems 10.K and 10.M show that a magic factor of order at worst $\lambda_{min}^{-1}(1 + \log^+ \lambda_{min})$ can be introduced, as compared with the estimates obtained using Theorems 10.B and 10.E, without altering the assumptions to be made. The final section addresses the question of whether, if the pairs $(I_i, Y_i)_{i=1}^n$ are equidistributed, λ_{min} can be replaced by λ, and the best magic factor recovered. Two examples are given to show how specific problems can be approached, though no general results have as yet been formulated.

A *sine qua non* of the whole enterprise, as far as this book is concerned, is the existence of a Stein's method for Poisson point processes. In Arratia, Goldstein and Gordon (1989), an approach is used which reduces the argument to a succession of one-dimensional estimates, carried through in the usual way. However, it turns out that there is a natural analogue of the Stein–Chen method for the Poisson distribution, once the fundamental equation (1.1.10) has been correctly interpreted. This interpretation, in probabilistic terms, enables the analytic estimates corresponding to those of Lemma 1.1.1 to be accomplished using probabilistic arguments, involving certain couplings. When these arguments are applied in the case of only the total number of points being of interest, the result obtained for Δg is not quite as sharp as that of Lemma 1.1.1, because of the presence of an extra factor of $\log \lambda$ when λ is large, which may well be generally superfluous. However, in the case of $||g||$, the constant obtained is an improvement on that given in Barbour and Eagleson (1983). But perhaps a more important dividend of the extension of the method to processes is that it indicates a way of introducing and applying Stein's method in many other situations and gives a way of understanding why the method works.

10.1 The basic method

The first step in developing Stein's method for Poisson process approximation is to find an analogue of the basic equation (1.1.10); in other words, to understand what is special to the Poisson distribution $\mathrm{Po}(\lambda)$ in the expression $\lambda g(j+1) - jg(j)$. One common connection is clearly that the Poisson density is a summation factor for $\lambda g(j+1) - jg(j)$, which enables one to calculate the explicit solution (1.1.18) of (1.1.10). While this is analytically very convenient, it gives little intuitive feeling for why the method works, or for how multi-dimensional or process versions of the basic equation should be constructed.

A less direct, but more useful connection is obtained by writing the function g as the first backward difference ∇h of a function h: $g(j) = h(j) - h(j-1)$. Then

$$\lambda g(j+1) - jg(j) = \lambda h(j+1) + jh(j-1) - (\lambda + j)h(j) \tag{1.1}$$

is expressed in the form $(\mathcal{A}h)(j)$, where $\mathcal{A}$ is the infinitesimal generator of the Markovian immigration–death process with immigration rate λ and with unit death rate. The connection between $\mathcal{A}$ and $\mathrm{Po}(\lambda)$ is then simply that $\mathrm{Po}(\lambda)$ is the stationary (equilibrium) distribution of the immigration–death process with generator $\mathcal{A}$, and the Stein equation (1.1.10) can be thought of as the equation $\mathcal{A}h = f$, for a particular choice of the function f, where $g = \nabla h$.

Now the equation $\mathcal{A}h = f$, where $\mathcal{A}$ is the generator of a Markov process, can be solved quite generally. The following heuristic discussion indicates how. If Z is an ergodic Markov process on a state space $\mathcal{Z}$ with stationary distribution μ,

$$\mathbb{E}^{\mu} h(Z(t)) = \mathbb{E}^{\mu} h(Z(s)) = \mu(h)$$

for all bounded $h : \mathcal{Z} \to \mathbb{R}$ and for all $s, t \in \mathbb{R}^+$, where $\mathbb{E}^{\mu}$ denotes the expectation with respect to the initial distribution μ, $\mathbb{E}^{\delta_z}$ is as usual abbreviated to $\mathbb{E}^z$ and $\mu(h)$ denotes $\mathbb{E}^{\mu} h(Z(0))$. Thus, taking $s = 0$ and letting $t \to 0$, it should follow that $\mu(\mathcal{A}h) = 0$: that is, that $\mu(\mathcal{A}h) = 0$ for all bounded functions h, perhaps satisfying some further regularity conditions. In a similar vein, if f is a function such that $\mu(f)$ is well defined, the function h, if it exists, defined by

$$h(z) = -\int_0^{\infty} [\mathbb{E}^z f(Z(t)) - \mu(f)]\, dt, \tag{1.2}$$

should satisfy

$$\begin{aligned} u^{-1}&[\mathbb{E}^z h(Z(u)) - h(z)] \\ &= u^{-1}\Big\{-\int_0^{\infty} [\mathbb{E}^z f(Z(t+u)) - \mu(f)]\, dt + \int_0^{\infty} [\mathbb{E}^z f(Z(t)) - \mu(f)]\, dt\Big\} \\ &= u^{-1} \int_0^u [\mathbb{E}^z f(Z(v)) - \mu(f)]\, dv, \end{aligned}$$

which, letting $u \to 0$, would imply that h solves the equation

$$(\mathcal{A}h)(z) = f(z) - \mu(f), \qquad z \in \mathcal{Z}. \tag{1.3}$$

Such arguments can be made precise, and the necessary conditions on the functions h and f established, in great generality, using the theory of Feller processes: see, for example, Ethier and Kurtz (1986) Chapter 1 Section 1. For the immigration–death processes which we need, rather simpler justifications can be given.

From this standpoint, the rationale behind Stein's method can be viewed as follows. Take $f(z) = I[z \in A]$, and let h be the solution (1.2) of (1.3): then $\mu(\mathcal{A}h) = 0$. If, on the other hand, Z' has the equilibrium distribution μ' of a Markov process with generator $\mathcal{A}'$ *which is close to* $\mathcal{A}$,

$$\mathbb{E}(\mathcal{A}h)(Z') \approx \mathbb{E}(\mathcal{A}'h)(Z') = \mu'(\mathcal{A}'h) = 0,$$

implying that

$$\mathbb{P}[Z' \in A] - \mu(A) = \mathbb{E}[f(Z') - \mu(f)] = \mathbb{E}(\mathcal{A}h)(Z') \approx 0.$$

Thus the Stein approach can be expected to work well, whenever the distributions being compared can be realized as (a given function of) the equilibrium distributions of Markov processes with similar generators.

There are, of course, many different processes having a given stationary distribution, and the right choice of a particular process or generator depends on the structure of the problem under consideration. In the Poisson approximation of a sum of indicators, the essential structure is the near independence of the summands. When the summands $(I_\alpha, \alpha \in \Gamma)$ are actually independent, the distribution of their sum W can be represented as the equilibrium distribution of $\sum_{\alpha\in\Gamma} X_\alpha$ in a Markov process X on $\{0,1\}^{|\Gamma|}$, where the coordinates are all independent, and X_α has transition rates given by $q_{01}^\alpha = \pi_\alpha$ and $q_{10}^\alpha = 1 - \pi_\alpha$. In a simple binomial approximation, X is replaced by a similar process X', where the π'_α are all equal to their average value π'. For this process, $W' = \sum_{\alpha\in\Gamma} X'_\alpha$ is itself a Markov process, now on the state-space $\{0, 1, \dots, |\Gamma|\} \subset \mathbb{Z}^+$, with transition rates given by

$$q'_{j,j+1} = \pi'(|\Gamma| - j); \qquad q'_{j,j-1} = j(1 - \pi'), \qquad 0 \le j \le |\Gamma|,$$

and it is clear that this process has a generator very similar to that of the immigration–death process with $\lambda = |\Gamma|\pi'$, provided that π' is small. In this way, the equation (1.3) with $\mathcal{A}$ the generator of an immigration–death process emerges naturally as the Stein equation for the Poisson approximation of a sum of weakly dependent indicators. Note, however, that the above considerations do not preclude other possibilities: in particular, the binomial approximation using X' is not that which was found most effective in Theorem 9.E.

Turning now to the approximation by a Poisson process of a point process generated by a collection of weakly dependent indicators $(I_\alpha, \alpha \in \Gamma)$ and their indices, there is a natural analogue of the one-dimensional argument, suggesting the immigration–death process Z on Γ, with immigration intensity π_α at each $\alpha \in \Gamma$ and with unit per-capita death rate, as the Markov process, with equilibrium distribution μ that of the Poisson process on Γ with intensity $\boldsymbol{\pi} = (\pi_\alpha, \alpha \in \Gamma)$, whose generator $\mathcal{A}$ should yield a useful Stein equation (1.3). If $\xi = \sum_{\alpha\in\Gamma} x_\alpha \delta_\alpha$ denotes a typical element of the space $\mathcal{Z}$ of configurations of point processes over Γ, so that $x_\alpha \in \mathbb{Z}^+$ for each $\alpha \in \Gamma$, the generator $\mathcal{A}$ is defined by

$$(\mathcal{A}h)(\xi) = \sum_{\alpha\in\Gamma} \pi_\alpha[h(\xi+\delta_\alpha) - h(\xi)] + \sum_{\alpha\in\Gamma} x_\alpha[h(\xi-\delta_\alpha) - h(\xi)], \qquad (1.4)$$

for a suitable class of functions h. The corresponding Stein equation, and its solutions for any bounded f, are given by (1.3) and (1.2), with $\mathcal{A}$ given by (1.4), as is implied by the next two slightly more general propositions.

Let Z be an immigration–death process over a measure space $\mathcal{X}$ with immigration intensity measure $\boldsymbol{\lambda}$ and unit per-capita death rate, where $\boldsymbol{\lambda}(\mathcal{X}) = \lambda < \infty$: the possible values of Z at any given time are measures ζ consisting of finite sums of unit atoms (particles) at points of $\mathcal{X}$, and the space of all such finite configurations is denoted by $\mathcal{Z}$. In such a process, particles are born according to a Poisson process with intensity $\boldsymbol{\lambda}$ per unit time, and have lifetimes which are exponentially distributed with unit mean, independently of all other particles and of their own position and time of birth. The configuration space $\mathcal{Z}$ carries a natural σ-field, induced by the σ-field on $\mathcal{X}$, and we henceforth only consider functions on $\mathcal{Z}$ which are measurable with respect to it. Let $\mu = \mathrm{Po}(\boldsymbol{\lambda})$ denote the equilibrium distribution of Z.

Proposition 10.1.1 *For any bounded $f : \mathcal{Z} \to \mathbb{R}$, the function $h : \mathcal{Z} \to \mathbb{R}$ given by*

$$h(\zeta) = -\int_0^\infty [\mathbb{E}^\zeta f(Z(t)) - \mu(f)]\, dt \qquad (1.5)$$

is well defined, and $\sup_{\{\zeta:\zeta(\mathcal{X})=l\}} |h(\zeta)| < \infty$ for each $l \in \mathbb{Z}^+$.

Proof Consider the simple coupling of an immigration–death process Z under $\mathbb{P}^\zeta$ and a similar process $\tilde{Z}$ under $\mathbb{P}^\mu$, taking $Z = Z_0 + D$, $\tilde{Z} = Z_0 + \tilde{D}$, where Z_0, D and $\tilde{D}$ are independent, Z_0 denotes the immigration–death process under $\mathbb{P}^0$ with no initial particles, and D and $\tilde{D}$ are pure death processes with unit per-capita death rate such that $D(0) = \zeta$ and $\tilde{D}(0) \sim \mu$. Then $Z(t) = \tilde{Z}(t)$ for all $t \geq \tau$, where

$$\tau = \inf\{u \geq 0 : D(u) = \tilde{D}(u) = 0\}.$$

Hence

$$\int_0^\infty |\mathbb{E}^\zeta f(Z(t)) - \mu(f)|\, dt = \int_0^\infty |\mathbb{E}^\zeta f(Z(t)) - \mathbb{E}f(\tilde{Z}(t))|\, dt$$
$$\leq 2\int_0^\infty \|f\|\mathbb{P}[\tau > t]\, dt = 2\|f\|\mathbb{E}\tau = 2\|f\|\mathbb{E}\psi(|\zeta| + |\tilde{D}(0)|) < \infty,$$

where $\psi(l) = \sum_{r=1}^{l} 1/r$, and the last inequality follows because $|\tilde{D}(0)|$ has distribution $\mathrm{Po}(\lambda)$. □

Now let

$$(\mathcal{A}h)(\zeta) = \int_{\mathcal{X}} [h(\zeta + \delta_x) - h(\zeta)]\, \boldsymbol{\lambda}(dx) + \int_{\mathcal{X}} [h(\zeta - \delta_x) - h(\zeta)]\, \zeta(dx), \quad (1.6)$$

defined for any function h such that $\sup_{\{\zeta:\zeta(\mathcal{X})=l\}} |h(\zeta)| < \infty$ for each $l \in \mathbb{Z}^+$. It is not difficult to see that, if $\Xi \sim \mathrm{Po}(\boldsymbol{\lambda})$, then

$$\mathbb{E}(\mathcal{A}h)(\Xi) = 0 \quad (1.7)$$

for all functions h that do not grow too fast. This is analogous to (1.1.9) in the random variable case, and the following proposition shows how to solve the analogue of (1.1.10): μ, as before, denotes $\mathrm{Po}(\boldsymbol{\lambda})$.

Proposition 10.1.2 *The function h defined in Proposition 10.1.1 satisfies the equation*

$$(\mathcal{A}h)(\zeta) = f(\zeta) - \mu(f). \quad (1.8)$$

Proof Let $h_t(\zeta) = -\int_0^t [\mathbb{E}^\zeta f(Z(u)) - \mu(f)]\, du$. Then, by considering the first time at which a particle is born or dies, we have

$$h_t(\zeta) = -(f(\zeta) - \mu(f))e^{-q_\zeta t}t + \int_0^t e^{-q_\zeta u}\Big[-q_\zeta u(f(\zeta) - \mu(f))$$
$$+ \int_{\mathcal{X}} h_{t-u}(\zeta + \delta_x)\boldsymbol{\lambda}(dx) + \int_{\mathcal{X}} h_{t-u}(\zeta - \delta_x)\zeta(dx)\Big]\, du,$$

with $q_\zeta = \lambda + |\zeta|$. Letting $t \to \infty$ and using dominated convergence, it follows that

$$h(\zeta) = q_\zeta^{-1}\Big[-\{f(\zeta) - \mu(f)\}$$
$$+ \int_{\mathcal{X}} h(\zeta + \delta_x)\boldsymbol{\lambda}(dx) + \int_{\mathcal{X}} h(\zeta - \delta_x)\zeta(dx)\Big],$$

since, from the proof of Proposition 10.1.1, the functions $h_s(\zeta)$ are uniformly bounded in s for each ζ. The proposition is now immediate. □

In order to use the Stein equation (1.8) to establish Poisson process approximation, it is necessary to have some analogue of Lemma 1.1.1, expressing the smoothness of the solution h given in (1.5) to equation (1.8) in terms of properties of f. This is the substance of the following lemma.

Lemma 10.1.3 *If h is the solution (1.5) to (1.8), where $\mathcal{A}$ is as defined in (1.4), and if $f(\xi) = I[\xi \in A]$,*

$$\text{(i) } \Delta_1 h = \sup_{\xi \in \mathcal{Z}, \alpha \in \Gamma} |h(\xi + \delta_\alpha) - h(\xi)| \leq 1;$$

$$\text{(ii) } \Delta_2 h = \sup_{\xi \in \mathcal{Z}; \alpha, \beta \in \Gamma} |h(\xi + \delta_\alpha + \delta_\beta) - h(\xi + \delta_\alpha) - h(\xi + \delta_\beta) + h(\xi)| \leq 1.$$

Proof For part (i), note that, from the definition (1.5) of h,

$$h(\xi + \delta_\alpha) - h(\xi) = \int_0^\infty [\mathbb{E}^\xi f(Z(t)) - \mathbb{E}^{\xi+\delta_\alpha} f(Z(t))] \, dt,$$

where Z is the immigration–death process on Γ with immigration intensity $(\pi_\beta, \beta \in \Gamma)$ and with unit per-capita death rate. Construct processes Z_1 and Z_2 with distributions $\mathbb{P}^\xi$ and $\mathbb{P}^{\xi+\delta_\alpha}$ together as follows. Let $Z_1 = Z_0 + D$ as for Proposition 10.1.1, where Z_0 has the distribution $\mathbb{P}^0$ and D is a pure death process with unit per-capita death rate and $D(0) = \xi$. Define $Z_2 = Z_1 + \delta_\alpha I[E > t]$, where E is a negative exponential random variable with unit mean which is independent of Z_1: then Z_2 has distribution $\mathbb{P}^{\xi+\delta_\alpha}$, and

$$f(Z_1(t)) - f(Z_2(t)) = I[E > t]\{f(Z_1(t)) - f(Z_1(t) + \delta_\alpha)\}.$$

Hence

$$\mathbb{E}^\xi f(Z(t)) - \mathbb{E}^{\xi+\delta_\alpha} f(Z(t)) = \mathbb{E}[f(Z_1(t)) - f(Z_1(t) + \delta_\alpha)]\mathbb{P}[E > t],$$

and it thus follows that

$$h(\xi + \delta_\alpha) - h(\xi) = \int_0^\infty \mathbb{E}^\xi [f(Z(t)) - f(Z(t) + \delta_\alpha)] e^{-t} \, dt.$$

The estimate $\Delta_1 h \leq 1$ is now immediate.

A similar coupling, introducing two independent negative exponential random variables E_α and E_β with unit means, can be used to link $\mathbb{P}^\xi$, $\mathbb{P}^{\xi+\delta_\alpha}$, $\mathbb{P}^{\xi+\delta_\beta}$ and $\mathbb{P}^{\xi+\delta_\alpha+\delta_\beta}$ by means of processes Z, Z_α, Z_β and $Z_{\alpha\beta}$, where Z has distribution $\mathbb{P}^\xi$,

$$Z_\alpha(t) = Z(t) + \delta_\alpha I[E_\alpha > t]; \qquad Z_\beta(t) = Z(t) + \delta_\beta I[E_\beta > t],$$

and

$$Z_{\alpha\beta}(t) = Z(t) + \delta_\alpha I[E_\alpha > t] + \delta_\beta I[E_\beta > t].$$

Then it follows that

$$\begin{aligned} h(\xi + \delta_\alpha + \delta_\beta) - h(\xi + \delta_\alpha) - h(\xi + \delta_\beta) + h(\xi) \\ = -\int_0^\infty \{\mathbb{E}^{\xi+\delta_\alpha+\delta_\beta} f(Z(t)) - \mathbb{E}^{\xi+\delta_\alpha} f(Z(t)) \\ - \mathbb{E}^{\xi+\delta_\beta} f(Z(t)) + \mathbb{E}^{\xi} f(Z(t))\} \, dt \\ = -\int_0^\infty \mathbb{E}^{\xi} \{f(Z(t) + \delta_\alpha + \delta_\beta) - f(Z(t) + \delta_\alpha) \\ - f(Z(t) + \delta_\beta) + f(Z(t))\} e^{-2t} \, dt, \end{aligned}$$

giving $\Delta_2 h \leq 1$. □

It is now possible to prove simple analogues of Theorems 1.A and 1.B, which bound the error in total variation when approximating the distribution of the process $\Xi = \sum_{\alpha\in\Gamma} \delta_\alpha I_\alpha$ by the Poisson measure $\mathrm{Po}(\boldsymbol{\pi})$ on Γ with intensity $\boldsymbol{\pi} = (\pi_\alpha, \alpha \in \Gamma)$. The first result is essentially Theorem 2 of Arratia, Goldstein and Gordon (1989).

Theorem 10.A *Let $Z_\alpha = \sum_{\beta\in\Gamma_\alpha^s} I_\beta$, $\alpha \in \Gamma$, where Γ_α^s and Γ_α^w are defined as for Theorem 1.A. Then*

$$d_{TV}(\mathcal{L}(\Xi), \mathrm{Po}(\boldsymbol{\pi})) \leq \sum_{\alpha\in\Gamma} \{\pi_\alpha^2 + \pi_\alpha \mathbb{E} Z_\alpha + \mathbb{E}(I_\alpha Z_\alpha) + \eta_\alpha\}, \qquad (1.9)$$

where

$$\eta_\alpha = \mathbb{E}|\mathbb{E}\{I_\alpha|(I_\beta, \beta \in \Gamma_\alpha^w)\} - \pi_\alpha|. \qquad (1.10)$$

Proof The argument is essentially that used for Theorem 1.A. It is enough to estimate $|\mathbb{E}(\mathcal{A}h)(\Xi)|$, where $\mathcal{A}$ is defined by (1.4) and h is given by (1.5) with $f(\xi) = I[\xi \in A]$, for any $A \subset \mathcal{Z}$. Let Ξ_α denote the measure $\sum_{\beta\in\Gamma_\alpha^w} \delta_\beta I_\beta$. Then, as for (1.1.25), but remembering that g corresponds to ∇h, we have

$$|\mathbb{E}\{I_\alpha[h(\Xi - \delta_\alpha) - h(\Xi)] - I_\alpha[h(\Xi_\alpha) - h(\Xi_\alpha + \delta_\alpha)]\}| \leq \Delta_2 h \mathbb{E}(I_\alpha Z_\alpha). \quad (1.11)$$

On the other hand, as for (1.1.26), we have

$$|\pi_\alpha \mathbb{E}\{[h(\Xi_\alpha) - h(\Xi_\alpha + \delta_\alpha)] - [h(\Xi) - h(\Xi + \delta_\alpha)]\}| \leq \Delta_2 h \pi_\alpha(\pi_\alpha + \mathbb{E} Z_\alpha). \quad (1.12)$$

Thus, in order to estimate $|\mathbb{E}(\mathcal{A}h)(\Xi)|$, it remains only to observe that

$$|\mathbb{E}\{(I_\alpha - \pi_\alpha)[h(\Xi_\alpha) - h(\Xi_\alpha + \delta_\alpha)]\}| \leq \eta_\alpha \Delta_1 h. \qquad (1.13)$$

Hence, combining (1.4), (1.11), (1.12) and (1.13),

$$|\mathbb{E}(\mathcal{A}h)(\Xi)| \leq \sum_{\alpha\in\Gamma} [\Delta_2 h(\pi_\alpha^2 + \pi_\alpha \mathbb{E} Z_\alpha + \mathbb{E}(I_\alpha Z_\alpha)) + \eta_\alpha \Delta_1 h],$$

and the theorem follows by Lemma 10.1.3. □

The analogue of Theorem 1.B is equally straightforward.

Theorem 10.B *Suppose that, for each $\alpha \in \Gamma$, random elements Ψ_α and Φ_α of $\mathcal{Z}$ can be realized on the same probability space, in such a way that $\mathcal{L}(\Psi_\alpha) = \mathcal{L}(\Xi)$ and $\mathcal{L}(\Phi_\alpha + \delta_\alpha) = \mathcal{L}(\Xi \mid I_\alpha = 1)$. Then*

$$d_{TV}(\mathcal{L}(\Xi), \mathrm{Po}(\boldsymbol{\pi})) \leq \sum_{\alpha\in\Gamma} \pi_\alpha \mathbb{E} d(\Psi_\alpha, \Phi_\alpha), \tag{1.14}$$

where

$$d(\Psi_\alpha, \Phi_\alpha) = \sum_{\beta\in\Gamma} |\Psi_\alpha\{\beta\} - \Phi_\alpha\{\beta\}|.$$

Proof As for Theorem 1.B, we write

$$\mathbb{E}\{I_\alpha[h(\Xi - \delta_\alpha) - h(\Xi)]\} = \pi_\alpha \mathbb{E}\{h(\Xi - \delta_\alpha) - h(\Xi) \mid I_\alpha = 1\},$$

whence it follows that

$$\mathbb{E}(\mathcal{A}h)(\Xi) = \sum_{\alpha\in\Gamma} \pi_\alpha \mathbb{E}\{[h(\Psi_\alpha + \delta_\alpha) - h(\Psi_\alpha)] + [h(\Phi_\alpha) - h(\Phi_\alpha + \delta_\alpha)]\}.$$

Now, if ψ and ϕ are arbitrary elements of $\mathcal{Z}$, let $\xi = \phi \wedge \psi$ denote the configuration $\sum_{\beta\in\Gamma}(\phi\{\beta\} \wedge \psi\{\beta\})\delta_\beta$: then it follows from the definition of $\Delta_2 h$ that, for any $\alpha \in \Gamma$,

$$|[h(\psi + \delta_\alpha) - h(\psi)] - [h(\xi + \delta_\alpha) - h(\xi)]| \leq d(\psi, \xi)\Delta_2 h$$

and

$$|[h(\phi + \delta_\alpha) - h(\phi)] - [h(\xi + \delta_\alpha) - h(\xi)]| \leq d(\phi, \xi)\Delta_2 h.$$

Hence

$$\begin{aligned} \big|[h(\psi + \delta_\alpha) - h(\psi)] &+ [h(\phi) - h(\phi + \delta_\alpha)]\big| \\ &\leq \Delta_2 h(d(\psi, \xi) + d(\phi, \xi)) = d(\phi, \psi)\Delta_2 h, \end{aligned}$$

and the theorem follows from Lemma 10.1.3 by taking any $A \subset \mathcal{Z}$, and letting h be given by (1.5) with $f(\xi) = I[\xi \in A]$. □

Corollary 10.B.1 *Suppose that random variables $J_{\beta\alpha}$ are defined as in Section 2.1. Then*

$$\begin{aligned} d_{TV}(\mathcal{L}(\Xi), \mathrm{Po}(\boldsymbol{\pi})) &\leq \sum_{\alpha\in\Gamma} \pi_\alpha \mathbb{E}\Big(I_\alpha + \sum_{\beta\neq\alpha} |I_\beta - J_{\beta\alpha}|\Big) \\ &\leq \sum_{\alpha\in\Gamma} \pi_\alpha^2 + \sum_{\alpha\in\Gamma}\sum_{\beta\in\Gamma_\alpha^-} |\mathrm{Cov}(I_\alpha, I_\beta)| \\ &\quad + \sum_{\alpha\in\Gamma}\sum_{\beta\in\Gamma_\alpha^+} \mathrm{Cov}(I_\alpha, I_\beta) + \sum_{\alpha\in\Gamma}\sum_{\beta\in\Gamma_\alpha^0} (\mathbb{E} I_\alpha I_\beta + \pi_\alpha\pi_\beta). \end{aligned}$$

Proof Define $\Psi_\alpha = \Xi = \sum_{\beta\in\Gamma} I_\beta\delta_\beta$ and $\Phi_\alpha = \sum_{\beta\neq\alpha} J_{\beta\alpha}\delta_\beta$, and apply Theorem 10.B. □

Remark 10.1.4 For negatively related indicators $(I_\alpha, \alpha \in \Gamma)$, as in Corollary 2.C.2, this yields

$$d_{TV}(\mathcal{L}(\Xi), \mathrm{Po}(\boldsymbol{\pi})) \leq \lambda - \mathrm{Var}\, W, \tag{1.15}$$

where, as usual, $W = \sum_{\alpha\in\Gamma} I_\alpha$: similarly, for positively related indicators $(I_\alpha, \alpha \in \Gamma)$, as in Corollary 2.C.4,

$$d_{TV}(\mathcal{L}(\Xi), \mathrm{Po}(\boldsymbol{\pi})) \leq \mathrm{Var}\, W - \lambda + 2\sum_{\alpha\in\Gamma} \pi_\alpha^2. \tag{1.16}$$

The estimates (1.9), (1.15) and (1.16) have already been evaluated in many applications in the preceding chapters, when approximating the distribution of $W = \sum_{\alpha\in\Gamma} I_\alpha$, and so process versions of the results can immediately be deduced. However, the estimates obtained for the distribution of W are better for large λ than those for Ξ, because of the factors containing negative powers of λ which are present in Lemma 1.1.1 but not in Lemma 10.1.3: in place of the estimates $\|g\| \leq 1 \wedge \lambda^{-1/2}$ and $\Delta g \leq \lambda^{-1}(1 - e^{-\lambda})$ of Lemma 1.1.1, we only have $\Delta_1 h \leq 1$ and $\Delta_2 h \leq 1$ in Lemma 10.1.3. This should not be surprising, since, if only the distribution of W is at issue, it is enough to make the smoothness estimates for a smaller class of functions h, and so sharper bounds may be attainable: the functions h that are then relevant are those derived through (1.5) from functions f depending on ξ only through $\sum_{\alpha\in\Gamma} x_\alpha$. It is none the less tempting to hope that the results of Theorems 10.A and 10.B can also be improved by λ-factors with negative exponents, but it is not the case. As has been observed in the introduction to the chapter, the order of Poisson approximation in total variation for the process Ξ, even in the simplest case where the $(I_\alpha, \alpha \in \Gamma)$ are independent, is that of Theorems 10.A and 10.B, so that no essential improvement can generally be achieved.

The Poisson approximation Ξ was shown to differ in total variation from the process of independent indicators $\sum_{\alpha\in\Gamma} \delta_\alpha I_\alpha$, by an amount of order $1 \wedge \sum_{\alpha\in\Gamma} \pi_\alpha^2$, because it has a greater tendency to clustering, in the sense that double points can occur. The following example shows that the lack of a tendency towards clustering in the Poisson process can also limit the accuracy of total variation approximation. Suppose that $\Gamma = \{1, 2, \ldots, n\}$, and that the indicators $(I_i, 1 \leq i \leq n)$ are given by $I_i = g(Y_i, Y_{i+1})$ for some 0–1 function g, where the $(Y_i)_{i=1}^{n+1}$ are independent and identically distributed. For Theorem 10.A, it is natural to take $\Gamma_i^s = \{i-1, i+1\}$, with which choice $\eta_i = 0$, and (1.9) reduces to

$$n\pi_1^2 + 2(n-1)\pi_1^2 + 2(n-1)\pi_{12} \sim n\{3\pi_1^2 + 2\pi_{12}\},$$

where π_{12} is used to denote $\mathbb{E} I_1 I_2$. The same estimate is obtained from Corollary 10.B.1 by taking $\Gamma_i^0 = \Gamma_i^s$. Now let

$$A = \{\xi : \sum_{i=0}^{[n/3]-1} x_{3i+1} x_{3i+2} \geq 1\}.$$

The indicators $(I_{3i+1}I_{3i+2})_{i=0}^{[n/3]-1}$ are independent, and each has mean π_{12}, so that, provided that $n\pi_{12} \ll 1$, $\mathbb{P}[\Xi \in A] \approx n\pi_{12}/3$. On the other hand, by the same token, if $n\pi_1^2 \ll 1$, $\mu(A) \approx n\pi_1^2/3$. Thus if, for instance, $\pi_{12} \gg \pi_1^2$, Theorems 10.A and 10.B again give the right order of approximation, and no general improvement through a factor $c(\lambda)$ depending on λ alone can be achieved.

Although no improvement in order can be achieved this way, it might still be that the estimates in Theorems 10.A and 10.B could be reduced, using a factor $c(\lambda)$ such that $\inf_\lambda c(\lambda) > 0$. However, even here, the scope is not very great. If the $(I_\alpha, \alpha \in \Gamma)$ are independent and if $\sum_{\alpha\in\Gamma} \pi_\alpha^2$ is small,

$$\mathrm{Po}\,(\boldsymbol{\pi})(\{\xi : \max_{\alpha\in\Gamma} x_\alpha \geq 2\}) \sim \tfrac{1}{2} \sum_{\alpha\in\Gamma} \pi_\alpha^2,$$

whereas $\mathbb{P}[\max_{\alpha\in\Gamma} I_\alpha \geq 2] = 0$, implying that $c(\lambda)$ cannot be less than one half. What is more, for $\lambda \to 0$, the fact that $d_{TV}(\mathrm{Be}\,(p), \mathrm{Po}\,(p)) \sim p^2$ as $p \to 0$ shows that $\lim_{\lambda\to 0} c(\lambda) = 1$. Thus the results of Theorems 10.A and 10.B for total variation approximation cannot be much improved, without changing the fundamental structure of the estimates (1.9) and (1.14).

The same contrast between the estimates in the one-dimensional and process theorems occurs also when approximating mixed Poisson (Cox) processes by Poisson processes. Suppose that $\Lambda = (\Lambda_\alpha, \alpha \in \Gamma)$ is sampled from a distribution over the space of (finite) intensity measures on Γ, and that, given Λ, Ξ is a Poisson process over Γ with intensity Λ: Ξ then has a mixed Poisson, or Cox, distribution over Γ, which we denote by $\Xi \sim \mathrm{Po}\,(\Lambda)$. Suppose also that $\mathbb{E}\Lambda(\Gamma) < \infty$. Set $\lambda_\alpha = \mathbb{E}\Lambda_\alpha$ and $\boldsymbol{\lambda} = (\lambda_\alpha, \alpha \in \Gamma)$, and define, for each realization Λ,

$$d(\Lambda, \boldsymbol{\lambda}) = \sum_{\alpha\in\Gamma} |\Lambda_\alpha - \lambda_\alpha|.$$

Theorem 10.C *If Ξ has the mixed Poisson distribution* $\mathrm{Po}\,(\Lambda)$ *and $\boldsymbol{\lambda} = \mathbb{E}\Lambda$,*

$$d_{TV}(\mathcal{L}(\Xi), \mathrm{Po}\,(\boldsymbol{\lambda})) \leq \mathbb{E}\{d^2(\Lambda, \boldsymbol{\lambda})\}.$$

Proof As for the one-dimensional Theorem 1.C, for any h that does not grow too fast,

$$\mathbb{E}\{\sum_{\alpha\in\Gamma} \Xi\{\alpha\}[h(\Xi) - h(\Xi - \delta_\alpha)]\} = \mathbb{E}\{\sum_{\alpha\in\Gamma} \Lambda_\alpha \mathbb{E}[h(\Xi + \delta_\alpha) - h(\Xi) \mid \Lambda]\},$$

so that

$$\mathbb{E}(\mathcal{A}h)(\Xi) = \mathbb{E}\{\sum_{\alpha\in\Gamma} (\lambda_\alpha - \Lambda_\alpha)\mathbb{E}[h(\Xi + \delta_\alpha) - h(\Xi) \mid \Lambda]\}.$$

Given Λ, let $Z \sim \mathrm{Po}(\boldsymbol{\lambda})$ and Ξ be realized on a common probability space, by taking three independent Poisson processes Z_j, $0 \le j \le 2$, with intensities given by

$$\lambda_{0\alpha} = \lambda_\alpha \wedge \Lambda_\alpha; \qquad \lambda_{1\alpha} = \Lambda_\alpha - \lambda_{0\alpha}; \qquad \lambda_{2\alpha} = \lambda_\alpha - \lambda_{0\alpha},$$

and setting

$$\Xi = Z_0 + Z_1; \qquad Z = Z_0 + Z_2.$$

Then, as in the proof of Theorem 10.B,

$$\left|\mathbb{E}[h(\Xi + \delta_\alpha) - h(\Xi) \mid \Lambda] - \mathbb{E}[h(Z + \delta_\alpha) - h(Z)]\right| \\ \le \Delta_2 h \sum_{\beta \in \Gamma} \mathbb{E}|Z_{1\beta} - Z_{2\beta}| \le d(\Lambda, \boldsymbol{\lambda}) \Delta_2 h.$$

Hence, since $\Lambda_\alpha - \lambda_\alpha$ is independent of the constant $\mathbb{E}[h(Z + \delta_\alpha) - h(Z)]$ and has mean zero, it follows that

$$|\mathbb{E}(\mathcal{A}h)(\Xi)| \le \Delta_2 h \sum_{\alpha \in \Gamma} \mathbb{E}\{|\Lambda_\alpha - \lambda_\alpha| d(\Lambda, \boldsymbol{\lambda})\} = \Delta_2 h \mathbb{E}\{d^2(\Lambda, \boldsymbol{\lambda})\},$$

and the theorem follows by the usual choice of h. □

Remark 10.1.5 Theorem 10.C remains true, by the same argument, for a mixed Poisson process with finite intensity over a general $\mathcal{X}$, with $d(\Lambda, \boldsymbol{\lambda}) = \int_{\mathcal{X}} |\Lambda(dy) - \boldsymbol{\lambda}(dy)|$, provided that $d(\Lambda, \boldsymbol{\lambda})$ is measurable. This is always the case when $\mathcal{X}$ is a separable metric space. Theorems 10.A and 10.B may similarly be stated in the context of point processes on a general space $\mathcal{X}$: see Barbour and Brown (1990) Theorems 2.4 and 2.6 for details.

If $|\Gamma| = 1$, $\mathbb{E}d^2(\Lambda, \boldsymbol{\lambda})$ reduces to $\mathrm{Var}\,\Lambda$, and the estimate of Theorem 10.C contrasts unfavourably with the estimate $\lambda^{-1}(1 - e^{-\lambda})\mathrm{Var}\,\Lambda$ of Theorem 1.C. However, there is no hope in general of being able to improve the order of the estimate in Theorem 10.C by introducing a factor $c(\lambda)$, where $\lambda = \boldsymbol{\lambda}(\Gamma)$, as the following example shows.

Example 10.1.6 Let $\Gamma = \{1, 2, \ldots, n\}$, and suppose that Λ is defined by

$$\Lambda_j = (1 + J\delta),\ j = 1; \qquad \Lambda_j = 1,\ 2 \le j \le n,$$

where $0 < \delta < 1$ and $\mathbb{P}[J = -1] = \mathbb{P}[J = 1] = 1/2$. Then $\lambda_j = 1$ for all j, $\lambda = n$ and $d(\Lambda, \boldsymbol{\lambda}) = \delta$ with probability one. Thus, from Theorem 10.C, $d_{TV}(\mathcal{L}(\Xi), \mathrm{Po}(\boldsymbol{\lambda})) \le \delta^2$.

On the other hand, $X = \Xi(\{1\})$ is a random variable with the mixed Poisson distribution $\mathrm{Po}(\Lambda)$, where $\mathbb{P}[\Lambda = 1 + \delta] = \mathbb{P}[\Lambda = 1 - \delta] = 1/2$. It thus follows from Example 3.3.1 that

$$d_{TV}(\mathcal{L}(\Xi), \mathrm{Po}(\boldsymbol{\lambda})) \ge d_{TV}(\mathcal{L}(X), \mathrm{Po}(1)) \ge \frac{1}{53}\delta^2,$$

irrespective of the value of λ.

Better estimates can sometimes be obtained by exploiting the structure of the dependence between the components $(\Lambda_\alpha, \alpha \in \Gamma)$. Suppose, for instance, that, for each $\alpha \in \Gamma$, Γ can be partitioned into sets $\{\alpha\}$, Γ_α^s and Γ_α^w, where the random variables $(\Lambda_\beta, \beta \in \Gamma_\alpha^s)$ are thought of as those that can strongly influence Λ_α. Then we have the following result.

Theorem 10.D *Under the above circumstances,*

$$d_{TV}(\mathcal{L}(\Xi), \mathrm{Po}(\boldsymbol{\lambda})) \le \sum_{\alpha\in\Gamma} \{\mathbb{E}(\Lambda_\alpha - \lambda_\alpha)^2 + \sum_{\beta\in\Gamma_\alpha^s} \mathbb{E}|(\Lambda_\alpha - \lambda_\alpha)(\Lambda_\beta - \lambda_\beta)| \\ + \mathbb{E}|\mathbb{E}[\Lambda_\alpha \mid (\Lambda_\beta)_{\beta\in\Gamma_\alpha^w}] - \lambda_\alpha|\}.$$

Proof The proof is very similar to that of Theorem 10.C. However, the conditional expectation $\mathbb{E}[h(\Xi + \delta_\alpha) - h(\Xi) \mid \Lambda]$ is now compared instead with $\mathbb{E}[h(\Xi + \delta_\alpha) - h(\Xi) \mid \Lambda^{(\alpha)}]$, where

$$\Lambda_\beta^{(\alpha)} = \Lambda_\beta,\ \beta \in \Gamma_\alpha^w; \qquad \Lambda_\beta^{(\alpha)} = \lambda_\beta,\ \beta \notin \Gamma_\alpha^w,$$

and a coupling similar to that used in Theorem 10.C then gives

$$|\mathbb{E}[h(\Xi+\delta_\alpha) - h(\Xi) \mid \Lambda] - \mathbb{E}[h(\Xi+\delta_\alpha) - h(\Xi) \mid \Lambda^{(\alpha)}]| \le \Delta_2 h \sum_{\beta\notin\Gamma_\alpha^w} |\Lambda_\beta - \lambda_\beta|.$$

Since also

$$\begin{aligned} &|\mathbb{E}\{(\lambda_\alpha - \Lambda_\alpha)\mathbb{E}[h(\Xi + \delta_\alpha) - h(\Xi) \mid \Lambda^{(\alpha)}]\}| \\ &\quad = |\mathbb{E}\{(\lambda_\alpha - \mathbb{E}[\Lambda_\alpha \mid (\Lambda_\beta)_{\beta\in\Gamma_\alpha^w}])\mathbb{E}[h(\Xi + \delta_\alpha) - h(\Xi) \mid \Lambda^{(\alpha)}]\}| \\ &\quad \le \Delta_1 h \mathbb{E}|\lambda_\alpha - \mathbb{E}[\Lambda_\alpha \mid (\Lambda_\beta)_{\beta\in\Gamma_\alpha^w}]|, \end{aligned}$$

it follows that $|\mathbb{E}(\mathcal{A}h)(\Xi)|$ does not exceed the estimate of the theorem, provided that $\Delta_1 h \le 1$ and $\Delta_2 h \le 1$, which is the case for all h given by (1.5) with $f(\xi) = I[\xi \in A]$, because of Lemma 10.1.3. The theorem now follows. □

If the $(\Lambda_\alpha, \alpha \in \Gamma)$ are independent, the estimate of Theorem 10.D reduces to

$$d_{TV}(\mathcal{L}(\Xi), \mathrm{Po}(\boldsymbol{\lambda})) \le \sum_{\alpha\in\Gamma} \mathbb{E}(\Lambda_\alpha - \lambda_\alpha)^2.$$

On the other hand, by combining Theorem 1.C(ii) and Proposition A.1.1, the estimate

$$d_{TV}(\mathcal{L}(\Xi), \mathrm{Po}(\boldsymbol{\lambda})) \le \sum_{\alpha\in\Gamma} (1 \wedge \lambda_\alpha^{-1})\mathbb{E}(\Lambda_\alpha - \lambda_\alpha)^2$$

is obtained.

Remark 10.1.7 The structure of the estimate is reminiscient of that of Theorem 10.A. A version analogous to Theorem 10.B could also be proved, based on appropriate couplings of intensity measures.

We now turn to the more general context of marked point processes. Suppose that, in addition to the indicators $(I_\alpha, \alpha \in \Gamma)$, there are random variables $(Y_\alpha, \alpha \in \Gamma)$ which take their values in some fixed state space $\mathcal{Y}$, which for reasons of measurability we assume to be metric and separable, and define $\Xi = \sum_{\alpha\in\Gamma} I_\alpha \delta_{Y_\alpha}$, a random element of the space $\mathcal{Z}$ of configurations of (finite) point processes over $\mathcal{Y}$. It suffices that Y_α is defined when $I_\alpha = 1$, since its value when $I_\alpha = 0$ does not affect the configuration. Let the measures $\boldsymbol{\lambda}_\alpha$ and $\boldsymbol{\lambda}$ on $\mathcal{Y}$ be defined by

$$\boldsymbol{\lambda}_\alpha(A) = \mathbb{P}[I_\alpha = 1,\ Y_\alpha \in A]; \quad \boldsymbol{\lambda} = \sum_{\alpha\in\Gamma} \boldsymbol{\lambda}_\alpha.$$

Theorem 10.E *Let $\Xi = \sum_{\alpha\in\Gamma} I_\alpha \delta_{Y_\alpha}$ be as above. Suppose that, for each $\alpha \in \Gamma$ and for $\boldsymbol{\lambda}_\alpha$-almost every $x \in \mathcal{Y}$, random elements $\Psi_{\alpha x}$ and $\Phi_{\alpha x}$ of $\mathcal{Z}$ can be defined on the same probability space, in such a way that $\mathcal{L}(\Psi_{\alpha x}) = \mathcal{L}(\Xi)$ and $\mathcal{L}(\Phi_{\alpha x} + \delta_x) = \mathcal{L}(\Xi \mid I_\alpha = 1 \text{ and } Y_\alpha = x)$. Define*

$$\rho_\alpha(x) = \mathbb{E}d(\Psi_{\alpha x}, \Phi_{\alpha x}) = \mathbb{E}\sum_{y\in\mathcal{Y}} |\Psi_{\alpha x}(y) - \Phi_{\alpha x}(y)|.$$

Then

$$\begin{aligned} d_{TV}(\mathcal{L}(\Xi), \mathrm{Po}(\boldsymbol{\lambda})) &\le \sum_{\alpha\in\Gamma} \pi_\alpha \mathbb{E}\{\rho_\alpha(Y_\alpha) \mid I_\alpha = 1\} \\ &= \sum_{\alpha\in\Gamma} \mathbb{E}\{I_\alpha \rho_\alpha(Y_\alpha)\} = \sum_{\alpha\in\Gamma} \int_{\mathcal{Y}} \rho_\alpha(x)\, \boldsymbol{\lambda}_\alpha(dx). \end{aligned}$$

Proof Let $\mathcal{A}$ and h be as for (1.6). Then, arguing much as usual, it follows that

$$\begin{aligned} &\mathbb{E}\{(\mathcal{A}h)(\Xi)\} \\ &= \mathbb{E}\Big\{\sum_{\alpha\in\Gamma} \int_{\mathcal{Y}} (h(\Xi + \delta_x) - h(\Xi))\, \boldsymbol{\lambda}_\alpha(dx) + \sum_{\alpha\in\Gamma} I_\alpha (h(\Xi - \delta_{Y_\alpha}) - h(\Xi))\Big\} \\ &= \sum_{\alpha\in\Gamma} \int_{\mathcal{Y}} \mathbb{E}\{h(\Psi_{\alpha x} + \delta_x) - h(\Psi_{\alpha x})\} \boldsymbol{\lambda}_\alpha(dx) \\ &\quad + \sum_{\alpha\in\Gamma} \int_{\mathcal{Y}} \mathbb{E}\{h(\Phi_{\alpha x}) - h(\Phi_{\alpha x} + \delta_x)\} \boldsymbol{\lambda}_\alpha(dx) \\ &= \sum_{\alpha\in\Gamma} \int_{\mathcal{Y}} \mathbb{E}\{h(\Psi_{\alpha x} + \delta_x) - h(\Psi_{\alpha x}) - h(\Phi_{\alpha x} + \delta_x) + h(\Phi_{\alpha x})\} \boldsymbol{\lambda}_\alpha(dx), \end{aligned}$$

and thus

$$|\mathbb{E}(\mathcal{A}h)(\Xi)| \leq \sum_{\alpha\in\Gamma} \int_{\mathcal{Y}} \mathbb{E}\Delta_2 h\, d(\Psi_{\alpha x}, \Phi_{\alpha x})\, \boldsymbol{\lambda}_\alpha(dx)$$
$$= \Delta_2 h \sum_{\alpha\in\Gamma} \int_{\mathcal{Y}} \rho_\alpha(x)\, \boldsymbol{\lambda}_\alpha(dx).$$

The proof is completed by applying Lemma 10.1.3. □

10.2 Improved rates

As is shown in the previous section, there is no possibility of improving the order of the bounds on total variation distance given in Theorems 10.A and 10.B by exploiting the fact that λ is large. However, the total variation metric is extremely strong, and is completely unsuitable, for instance, if one wishes to approximate a process on a lattice in $\mathbb{R}^k$ by a Poisson process with a continuous intensity over $\mathbb{R}^k$. There is, therefore, the hope that, by choosing a weaker metric than total variation, one might also be able to introduce factors like λ^{-1} into the bounds on process approximation, and at the same time make the approximation of a discrete process by a continuous process feasible. This is the aim of the present section.

Suppose now that $(\mathcal{Y}, d_0)$ is a metric space, with metric d_0 bounded by 1, and let the indices $\alpha \in \Gamma$ be points in $\mathcal{Y}$. $(\mathcal{Y}, d_0)$ is usually taken to be complete and separable, to avoid measurability problems. Define metrics on the set of configurations over the space $\mathcal{Y}$ and on the set of probability measures over the space $\mathcal{Z}$ of configurations as follows. Let $\mathcal{K}$ denote the set of functions $k : \mathcal{Y} \to \mathbb{R}$ such that

$$s_1(k) = \sup_{y_1\neq y_2\in\mathcal{Y}} |k(y_1) - k(y_2)|/d_0(y_1, y_2) < \infty, \tag{2.1}$$

and define a distance d_1 between finite measures $\boldsymbol{\rho}, \boldsymbol{\sigma}$ over $\mathcal{Y}$ by

$$d_1(\boldsymbol{\rho}, \boldsymbol{\sigma}) = \begin{cases} 1 & \text{if } \boldsymbol{\rho}(\mathcal{Y}) \neq \boldsymbol{\sigma}(\mathcal{Y}), \\ \frac{1}{\boldsymbol{\rho}(\mathcal{Y})} \sup_{k\in\mathcal{K}}\{|\int k\, d\boldsymbol{\rho} - \int k\, d\boldsymbol{\sigma}|/s_1(k)\} & \text{if } \boldsymbol{\rho}(\mathcal{Y}) = \boldsymbol{\sigma}(\mathcal{Y}). \end{cases} \tag{2.2}$$

The space $\mathcal{Z}$ of configurations over $\mathcal{Y}$ consists of elements of the form $\xi = \sum_{i=1}^m \delta_{x_i}$, where $x_1, \ldots, x_m \in \mathcal{Y}$, so that d_1 can be used as a metric over $\mathcal{Z}$. Similarly, let $\mathcal{F}$ denote the set of functions $f : \mathcal{Z} \to \mathbb{R}$ such that

$$s_2(f) = \sup_{\xi_1\neq\xi_2\in\mathcal{Z}} |f(\xi_1) - f(\xi_2)|/d_1(\xi_1, \xi_2) < \infty, \tag{2.3}$$

and define a distance d_2 between probability measures over $\mathcal{Z}$ by

$$d_2(Q, R) = \sup_{f\in\mathcal{F}} \left| \int f\, dQ - \int f\, dR \right| / s_2(f). \tag{2.4}$$

The metric d_1, when restricted to probability measures over $(\mathcal{Y}, d_0)$, is variously known as the Dudley, Fortet–Mourier, Kantorovich $D_{1,1}$ or Wasserstein metric induced by d_0: see Section A.1 of the appendix. In particular, it follows from (A.1.7) that, when considered as a distance between configurations $\xi_1, \xi_2 \in \mathcal{Z}$ with $|\xi_1| = |\xi_2| = n$, $d_1(\xi_1, \xi_2)$ can be interpreted in dual form as

$$\min_{\pi \in S_n} \{n^{-1} \sum_{i=1}^{n} d_0(y_{1i}, y_{2\pi(i)})\},$$

the average distance between the points $(y_{11}, \ldots, y_{1n})$ and $(y_{21}, \ldots, y_{2n})$ of ξ_1 and ξ_2 under the closest matching. The metric d_2 is then the Wasserstein metric induced by d_1 over the probability measures on $\mathcal{Z}$. Note that the distances d_1 and d_2, like d_0, are bounded by 1.

The simplest choice for d_0 is the discrete metric on $\mathcal{Y}$. With this choice, the d_1-distance between two configurations with different numbers of points is, as always, one, and that between two configurations with the same number of points is the proportion of points not common to both configurations, the 'relative variation'. Examples of functions $f \in \mathcal{F}$ are functions such as

$$f(\xi) = |\xi|^{-r} \int_{\mathcal{Y}^r} K(|\xi|; y_1, \ldots, y_r)\, \xi(dy_1) \ldots \xi(dy_r)$$

for suitably bounded kernels K, where $|\xi|$ denotes $\xi(\mathcal{Y})$, or of the form $g(f_1, \ldots, f_s)$, where $g : \mathbb{R}^s \to \mathbb{R}$ is uniformly Lipschitz and $f_1, \ldots, f_s$ are as above, with possibly different values of r. Thus functions of the form $I[|\xi| \in A]$ for $A \subset \mathbb{Z}^+$ are permissible, and a bound on $d_2(Q, R)$ implies the same bound on $Q[|\xi| \in A] - R[|\xi| \in A]$, and hence on the total variation distance between the distributions of $|\xi|$ under Q and R. However, except in trivial cases, functions such as $f(\xi) = I[\xi(B) \in A]$ for some $B \subset \mathcal{Y}$ and $A \subset \mathbb{Z}^+$ have $s_2(f) = \infty$, and restricting $|\xi|$ to its useful range by taking instead $f(\xi) = I[\xi(B) \in A] I[|\xi| \leq 2\lambda]$ still gives $s_2(f) = [2\lambda] \vee 1$. Thus a given bound ε on d_2 only limits the difference between the Q and R probabilities of the corresponding events to being no larger than $(2\lambda \vee 1)\varepsilon$, which introduces an unwelcome λ-factor. Only functions f which, for given $|\xi|$, vary smoothly as a function of the density $|\xi|^{-1}\xi(.)$ have the difference of their expectations under Q and R controlled directly by bounds on d_2.

The relative variation distance d_1 resulting from the choice of the discrete metric d_0 is still too strong, if one wishes to compare an intensity measure concentrated on the points $\{jn^{-1},\ 1 \leq j \leq n\}$ with an intensity absolutely continuous with respect to Lebesgue measure on $[0, 1]$, and the same is true for configurations arising from such intensities. Indeed, for reasons of measurability, the Poisson process with Lebesgue measure on $[0, 1]$ as intensity cannot be defined using the Borel σ-algebra generated by d_0. Thus, when using a Poisson process with continuous intensity to approximate one with a discrete intensity, a choice of d_0 which reflects the natural topology of $\mathcal{Y}$ is better. For instance, if a natural metric d on $\mathcal{Y}$ is given, d_0 can be taken to be $d \wedge 1$. In the case of $[0,1]$ mentioned above, one would take d_0 to be the Euclidean distance. The d_1 distance then reflects an average distance between configurations when the points are optimally paired, and functions $f \in \mathcal{F}$ can be constructed as previously, but with an extra Lipschitz smoothness requirement on the kernels K. Thus a bound on d_2 now leads to useful conclusions for fewer functions f, but can be obtained in circumstances where effective bounds for the metric d_2 based on the discrete metric on $\mathcal{Y}$ cannot be obtained: see Example 10.2.7 below.

The estimates that we obtain for the error in approximating $\mathcal{L}(\Xi)$ by $\mathrm{Po}(\boldsymbol{\lambda})$ with respect to the d_2 metric are in two parts. The first comes from approximating Ξ by the Poisson measure $\mathrm{Po}(\boldsymbol{\pi})$, where $\boldsymbol{\pi} = (\pi_\alpha, \alpha \in \Gamma)$ is as before, and the second from approximating $\mathrm{Po}(\boldsymbol{\pi})$ by $\mathrm{Po}(\boldsymbol{\lambda})$. For the first, the estimate is the same for all d_2 metrics. This is because the argument involves comparison of the values of functions $f : \mathcal{Z} \to \mathbb{R}$ at configurations ξ_1 and ξ_2 which either have different numbers of points, in which case $d_1(\xi_1, \xi_2) = 1$ for all d_1, or where ξ_1 and ξ_2 are identical except for the positions of one or two points, which are then unrestricted in position, so that the choice of d_0 again has no influence on the largest value then possible for $d_1(\xi_1, \xi_2)$. Thus the strongest result for this comparison comes from taking d_0 to be the discrete metric. It is only in the second part, when comparing $\mathrm{Po}(\boldsymbol{\pi})$ with $\mathrm{Po}(\boldsymbol{\lambda})$, that the particular choice of d_0 influences the estimate obtained. The choice of 1 as upper bound is also arbitrary: any other value D could be used, and would appear as a factor multiplying the first part of the estimates.

In order to investigate process approximation in terms of the d_2-metrics, we start with some technical results.

Lemma 10.2.1 *Let $(Z_t)_{t \geq 0}$ be a one-dimensional immigration–death process with immigration rate λ and unit per-capita death rate, such that $\mathbb{P}[Z_0 = k] = 1$. Then*

$$\int_0^\infty e^{-t}\mathbb{E}\{(Z_t+1)^{-1}\}\,dt \leq \left\{\frac{1}{\lambda} + \frac{1}{k+1}\right\}(1 - e^{-\lambda}).$$

Proof Write Z_t as the sum of independent random variables X_t and Y_t, where X_t denotes the number of the initial k individuals still alive at time t, so that $X_t \sim \mathrm{Bi}(k, e^{-t})$, and $Y_t \sim \mathrm{Po}(\lambda(1 - e^{-t}))$ counts the individuals alive at t who arrived after time zero. Thus

$$\mathbb{E}\{(Z_t+1)^{-1}\} \leq \mathbb{E}(X_t+1)^{-1} = \frac{e^t}{k+1}[1 - (1 - e^{-t})^{k+1}]$$

and

$$\mathbb{E}\{(Z_t+1)^{-1}\} \leq \mathbb{E}(Y_t+1)^{-1} = \frac{1}{\lambda(1-e^{-t})}[1 - e^{-\lambda(1-e^{-t})}].$$

Hence, letting t_0 be such that $e^{-t_0} = \lambda/(\lambda + k + 1)$, we have

$$\int_0^\infty e^{-t}\mathbb{E}\{(Z_t + 1)^{-1}\}\,dt$$
$$\leq \int_0^{t_0} \frac{1}{k+1}[1 - (1 - e^{-t_0})^{k+1}]dt + \int_{t_0}^\infty \frac{e^{-t}}{\lambda(1 - e^{-t})}(1 - e^{-\lambda})dt$$
$$\leq (1 - e^{-\lambda})\left\{\frac{1}{k+1}\log(1 + \lambda^{-1}(k+1)) + \frac{1}{\lambda}\log(1 + \lambda(k+1)^{-1})\right\}$$
$$\leq (1 - e^{-\lambda})\left(\frac{1}{\lambda} + \frac{1}{k+1}\right),$$

as required. □

The next lemma is used to control the error incurred by approximating with a Poisson process with intensity $\boldsymbol{\lambda}$ which may not be the same as $\boldsymbol{\pi}$.

Lemma 10.2.2 *Let $\boldsymbol{\lambda}$ be a finite measure over $\mathcal{Y}$, and let $h : \mathcal{Z} \to \mathbb{R}$ be given by (1.5), where $f : \mathcal{Z} \to \mathbb{R}$ is any function in $\mathcal{F}$. Let $\boldsymbol{\pi}$ be another finite measure over $\mathcal{Y}$ with $\boldsymbol{\pi}(\mathcal{Y}) = \boldsymbol{\lambda}(\mathcal{Y}) = \lambda$. Then, for any $\xi \in \mathcal{Z}$,*

$$\left|\int_{\mathcal{Y}} [h(\xi + \delta_y) - h(\xi)](\boldsymbol{\pi}(dy) - \boldsymbol{\lambda}(dy))\right|$$
$$\leq s_2(f)(1 - e^{-\lambda})(1 + \lambda(|\xi| + 1)^{-1})d_1(\boldsymbol{\pi}, \boldsymbol{\lambda}).$$

Proof Let $h_\xi : \mathcal{Y} \to \mathbb{R}$ be given by

$$h_\xi(y) = h(\xi + \delta_y) - h(\xi).$$

Then, from the definition of d_1,

$$\left|\int_{\mathcal{Y}} [h(\xi + \delta_y) - h(\xi)](\boldsymbol{\pi}(dy) - \boldsymbol{\lambda}(dy))\right| \leq s_1(h_\xi)\lambda\, d_1(\boldsymbol{\pi}, \boldsymbol{\lambda}).$$

Now, by taking $(Z_t)_{t\geq 0}$ to be the immigration–death process with $Z_0 = \xi$, and realizing $\mathbb{P}^{\xi+\delta_y}$ and $\mathbb{P}^{\xi+\delta_z}$ together by means of

$$Z_{1t} = Z_t + \delta_y I[E > t]; \qquad Z_{2t} = Z_t + \delta_z I[E > t],$$

where the standard exponential random variable E is independent of Z, we find that

$$|h_\xi(y) - h_\xi(z)| = \left|\int_0^\infty e^{-t}\mathbb{E}^\xi[f(Z_t + \delta_y) - f(Z_t + \delta_z)]dt\right|$$
$$\leq s_2(f)\int_0^\infty e^{-t}\mathbb{E}^\xi d_1(Z_t + \delta_y, Z_t + \delta_z)\,dt$$
$$= s_2(f)d_0(y,z)\int_0^\infty e^{-t}\mathbb{E}^\xi\{(|Z_t| + 1)^{-1}\}dt.$$

Applying Lemma 10.2.1 to the one-dimensional immigration–death process $|Z_t|$, we thus have

$$s_1(h_\xi) \leq s_2(f)(1 - e^{-\lambda})(\lambda^{-1} + (|\xi| + 1)^{-1}),$$

from which the lemma follows. □

The next two lemmas are the analogues of Lemma 10.1.3(i) and (ii). Because the class of functions h under consideration is smaller than that needed for total variation approximation, the smoothness estimates are better.

Lemma 10.2.3 *Under the conditions of Lemma 10.2.2,*

$$\Delta_1 h \leq s_2(f)(1 \wedge 1.65\lambda^{-1/2}).$$

Proof What is required is to estimate $|h(\xi + \delta_y) - h(\xi)|$ for any $\xi \in \mathcal{Z}$, $y \in \mathcal{Y}$, where

$$h(\xi) = -\int_0^\infty [\mathbb{E}^\xi f(Z(t)) - \mu(f)]dt,$$

Z is an immigration–death process over $\mathcal{Y}$ with intensity $\boldsymbol{\lambda}$ and unit per-capita death rate and $\mu = \mathrm{Po}(\boldsymbol{\lambda})$. As for the proof of Lemma 10.1.3(i), construct processes Z_1 and Z_2 with the measures $\mathbb{P}^\xi$ and $\mathbb{P}^{\xi+\delta_y}$ together, by taking independent realizations of a third process Z_0 under $\mathbb{P}^0$, a pure death process X with unit per-capita death rate starting with $X(0) = \xi$, and a standard exponential random variable E, and then setting

$$Z_1(t) = Z_0(t) + X(t); \qquad Z_2(t) = Z_1(t) + \delta_y I[E > t].$$

It then follows that

$$h(\xi + \delta_y) - h(\xi) \qquad (2.5)$$
$$= \int_0^\infty e^{-t} \sum_\eta \mathbb{P}[X(t) = \eta]\mathbb{E}[f(Z_0(t) + \eta) - f(Z_0(t) + \eta + \delta_y)]dt.$$

The inequality $|h(\xi + \delta_y) - h(\xi)| \leq s_2(f)$ is now immediate from (2.2) and (2.3).

For the λ-dependent bound, we may take $\lambda \geq 1$. By conditioning on the value of $|Z_0(t)|$, we have

$$\mathbb{E}[f(Z_0(t) + \eta) - f(Z_0(t) + \eta + \delta_y)] = \mathbb{P}[\,|Z_0(t)| = 0\,]f(\eta)$$
$$+ \sum_{k \geq 0} \Big\{ \mathbb{P}[\,|Z_0(t)| = k + 1\,]\mathbb{E}[f(Z_0(t) + \eta) \mid |Z_0(t)| = k + 1\,]$$
$$- \mathbb{P}[\,|Z_0(t)| = k\,]\mathbb{E}[f(Z_0(t) + \eta + \delta_y) \mid |Z_0(t)| = k\,] \Big\}. \qquad (2.6)$$

The latter sum is simplified by observing that

$$\big|\mathbb{E}[f(Z_0(t) + \eta) \mid |Z_0(t)| = k + 1\,] - \mathbb{E}[f(Z_0(t) + \eta + \delta_y) \mid |Z_0(t)| = k\,]\big|$$
$$= \Big|\lambda^{-1} \int \mathbb{E}[f(Z_0(t) + \eta + \delta_z) - f(Z_0(t) + \eta + \delta_y) \mid |Z_0(t)| = k\,]\boldsymbol{\lambda}(dz)\Big|$$
$$\leq s_2(f)/(k + 1 + |\eta|), \qquad (2.7)$$

since

$$|f(\zeta+\delta_z)-f(\zeta+\delta_y)| \le s_2(f)d_1(\zeta+\delta_z,\zeta+\delta_y)$$
$$\le s_2(f)(|\zeta|+1)^{-1}d_0(z,y) \le s_2(f)(|\zeta|+1)^{-1}.$$

Note also that, since $\frac{1}{2}\{\inf_\zeta f(\zeta)+\sup_\zeta f(\zeta)\}$ may be subtracted from f without altering (2.5), we may take $\sup_\zeta |f(\zeta)| \le \frac{1}{2}s_2(f)$, and hence

$$|\mathbb{E}[f(Z_0(t)+\eta) \mid |Z_0(t)|=k]| \le \tfrac{1}{2}s_2(f). \tag{2.8}$$

Thus, using Proposition A.2.7,

$$|\mathbb{E}[f(Z_0(t)+\eta)-f(Z_0(t)+\eta+\delta_y)]| \le s_2(f)\Big[\mathbb{E}\{(|Z_0(t)|+1)^{-1}\}$$
$$+\tfrac{1}{2}\sum_{k\ge0}\Big|\mathbb{P}[|Z_0(t)|=k+1]-\mathbb{P}[|Z_0(t)|=k]\Big|+\tfrac{1}{2}\mathbb{P}[|Z_0(t)|=0]\Big]$$
$$= s_2(f)\big[\mathbb{E}\{(|Z_0(t)|+1)^{-1}\}+\max_{k\ge0}\mathbb{P}[|Z_0(t)|=k]\big]$$
$$\le s_2(f)\big[\lambda_t^{-1}(1-e^{-\lambda_t})+(2e\lambda_t)^{-1/2}\big], \tag{2.9}$$

where $\lambda_t=\mathbb{E}|Z_0(t)|=\lambda(1-e^{-t})$. Since also a direct estimate yields

$$|\mathbb{E}[f(Z_0(t)+\eta)-f(Z_0(t)+\eta+\delta_y)]| \le s_2(f),$$

we arrive at the formula

$$|h(\xi+\delta_y)-h(\xi)| \le s_2(f)\int_0^\infty e^{-t}\{1\wedge[\lambda_t^{-1}+(2e\lambda_t)^{-1/2}]\}\,dt \tag{2.10}$$
$$\le s_2(f)\left[\int_0^\tau e^{-t}\,dt+\int_\tau^\infty e^{-t}\left\{\frac{1}{\lambda(1-e^{-t})}+\frac{1}{\sqrt{2e\lambda(1-e^{-t})}}\right\}dt\right],$$

where τ is chosen so that $e^{-\tau}=1-\lambda^{-1}$. Computation of the integrals yields the result

$$\lambda^{-1}\left\{1+\log\lambda+\sqrt{\frac{2}{e}}(\sqrt{\lambda}-1)\right\} \le 1.65\lambda^{-1/2},$$

where the last inequality is obtained by calculus, and this implies the lemma. □

Remark 10.2.4 Note that, if f is an indicator function of $|\xi|$ alone, the inequality (2.7) is not needed, since the difference being considered is zero, and also that $s_2(f)=1$ or 0. Hence

$$|\mathbb{E}[f(Z_0(t)+\eta)-f(Z_0(t)+\eta+\delta_y)]| \le \max_{k\ge0}\mathbb{P}[|Z_0(t)|=k]$$

by the argument of Lemma 10.2.3, leading to the estimate

$$\Delta_1 h \leq 1 \wedge \sqrt{\frac{2}{e\lambda}} \approx 1 \wedge 0.858\lambda^{-1/2}. \tag{2.11}$$

However, estimates of $\Delta_1 h$ for functions f which are indicator functions of $|\xi|$ alone are the same as estimates of $\|g\|$, where g satisfies (1.1.10), by means of the relation $g(j+1) = h(\xi + \delta_x) - h(\xi)$ whenever $|\xi| = j$: see also (1.1) above. Hence it follows that the estimate of $\|g\|$ in Lemma 1.1.1 can be sharpened to $1 \wedge \sqrt{\frac{2}{e\lambda}}$.

Lemma 10.2.5 *Under the conditions of Lemma 10.2.2,*

$$\Delta_2 h \leq \left\{1 \wedge \frac{2}{\lambda}\left(1 + 2\log^+\left(\frac{\lambda}{2}\right)\right)\right\} s_2(f).$$

Proof Adapting the proof of Lemma 10.1.3(ii) in much the same way as the proof of Lemma 10.1.3(i) was adapted to prove Lemma 10.2.3, it follows that

$$\begin{aligned} h(\xi + \delta_y + \delta_z) &- h(\xi + \delta_z) - h(\xi + \delta_y) + h(\xi) \\ &= -\int_0^\infty e^{-2t} \sum_\eta \mathbb{P}[X(t) = \eta] \mathbb{E}\{f(Z_0(t) + \eta + \delta_y + \delta_z) \\ &\qquad - f(Z_0(t) + \eta + \delta_y) - f(Z_0(t) + \eta + \delta_z) + f(Z_0(t) + \eta)\}\, dt, \end{aligned} \tag{2.12}$$

from which $\Delta_2 h \leq s_2(f)$ is immediate. For the main estimate, observe that

$$\begin{aligned} &\mathbb{E}[f(Z_0(t) + \eta + \delta_y + \delta_z) - f(Z_0(t) + \eta + \delta_y) \\ &\qquad\qquad - f(Z_0(t) + \eta + \delta_z) + f(Z_0(t) + \eta)] \\ &= \sum_{k \geq -1} \Big\{ \mathbb{P}[|Z_0(t)| = k] \mathbb{E}[f(Z_0(t) + \eta + \delta_y + \delta_z) \mid |Z_0(t)| = k] \\ &\qquad - \mathbb{P}[|Z_0(t)| = k+1] \\ &\qquad \times \mathbb{E}[f(Z_0(t) + \eta + \delta_y) + f(Z_0(t) + \eta + \delta_z) \mid |Z_0(t)| = k+1] \\ &\qquad + \mathbb{P}[|Z_0(t)| = k+2] \mathbb{E}[f(Z_0(t) + \eta) \mid |Z_0(t)| = k+2] \Big\} \\ &\quad + \mathbb{P}[|Z_0(t)| = 0] f(\eta). \end{aligned}$$

Thus, using (2.7) and (2.8), it follows that

$$\begin{aligned}
&\big|\mathbb{E}[f(Z_0(t)+\eta+\delta_y+\delta_z)-f(Z_0(t)+\eta+\delta_y) \\
&\qquad\qquad -f(Z_0(t)+\eta+\delta_z)+f(Z_0(t)+\eta)]\big| \\
&\le \Big|\sum_{k\ge -1}\{\mathbb{P}[|Z_0(t)|=k]-2\mathbb{P}[|Z_0(t)|=k+1]+\mathbb{P}[|Z_0(t)|=k+2]\} \\
&\qquad \times \mathbb{E}[\tfrac{1}{2}f(Z_0(t)+\eta+\delta_y)+\tfrac{1}{2}f(Z_0(t)+\eta+\delta_z)\mid |Z_0(t)|=k+1]\Big| \\
&\quad + s_2(f)\sum_{k\ge -1}\{\mathbb{P}[\,|Z_0(t)|=k\,]+\mathbb{P}[|Z_0(t)|=k+2]\}/(k+2+|\eta|) \\
&\quad + \mathbb{P}[\,|Z_0(t)|=0\,]|f(\eta)| \\
&\le s_2(f)\Big[\tfrac{1}{2}\sum_{k\ge -1}|\mathbb{P}[|Z_0(t)|=k]-2\mathbb{P}[|Z_0(t)|=k+1]+\mathbb{P}[|Z_0(t)|=k+2]| \\
&\qquad + \tfrac{1}{2}\mathbb{P}[\,|Z_0(t)|=0\,] \\
&\qquad + \sum_{k\ge 0}\mathbb{P}[\,|Z_0(t)|=k\,](k+2)^{-1}+\sum_{k\ge 1}\mathbb{P}[\,|Z_0(t)|=k\,]k^{-1}\Big] \\
&\le s_2(f)\Big[\tfrac{1}{2}\sum_{k\ge 0}|\mathbb{P}[|Z_0(t)|=k-2]-2\mathbb{P}[|Z_0(t)|=k-1]+\mathbb{P}[|Z_0(t)|=k]| \\
&\qquad + 3\mathbb{E}\{(|Z_0(t)|+1)^{-1}\}\Big].
\end{aligned}$$

Now, if $Z \sim \mathrm{Po}(\nu)$,

$$\mathbb{P}[Z=k]-2\mathbb{P}[Z=k-1]+\mathbb{P}[Z=k-2]=\mathbb{P}[Z=k]\{(1-\nu^{-1}k)^2-\nu^{-2}k\}$$

for all $k \ge 0$, and hence

$$\begin{aligned}
&\sum_{k\ge 0}|\mathbb{P}[Z=k]-2\mathbb{P}[Z=k-1]+\mathbb{P}[Z=k-2]| \\
&\qquad\qquad \le \mathbb{E}\{(1-\nu^{-1}Z)^2+\nu^{-2}Z\}=2\nu^{-1}.
\end{aligned}$$

Thus

$$\begin{aligned}
&\big|\mathbb{E}[f(Z_0(t)+\eta+\delta_y+\delta_z)-f(Z_0(t)+\eta+\delta_y) \\
&\qquad\qquad - f(Z_0(t)+\eta+\delta_z)+f(Z_0(t)+\eta)]\big| \\
&\qquad\qquad\qquad \le \{2\wedge 4/\lambda(1-e^{-t})\}s_2(f),
\end{aligned}$$

and substitution of this expression into (2.12) leads to the inequality

$$\Delta_2 h \le \Big(\frac{4}{\lambda^2}+\frac{4}{\lambda}\log(\tfrac{\lambda}{2})\Big)s_2(f)\le \frac{2}{\lambda}\Big(1+2\log^+(\tfrac{\lambda}{2})\Big)s_2(f)$$

for all $\lambda \ge 2$, completing the proof of the lemma. □

We are now in a position to prove theorems describing the accuracy, with respect to the new metric, of the approximation of the distribution of $\Xi = \sum_{\alpha\in\Gamma} \delta_\alpha I_\alpha$ by a Poisson measure $\mathrm{Po}(\boldsymbol{\lambda})$ over $\mathcal{Y}$ with intensity $\boldsymbol{\lambda}$ satisfying $\boldsymbol{\lambda}(\mathcal{Y}) = \boldsymbol{\pi}(\mathcal{Y}) = \lambda$.

Theorem 10.F *In the notation of Theorem 10.A,*

$$\begin{aligned} d_2(\mathcal{L}(\Xi), \mathrm{Po}(\boldsymbol{\lambda})) & \\ \le & \left\{1 \wedge \frac{2}{\lambda}\left(1 + 2\log^+\left(\frac{\lambda}{2}\right)\right)\right\} \sum_{\alpha\in\Gamma} (\pi_\alpha^2 + \pi_\alpha \mathbb{E}Z_\alpha + \mathbb{E}(I_\alpha Z_\alpha)) \\ & + \{1 \wedge 1.65\lambda^{-1/2}\} \sum_{\alpha\in\Gamma} \eta_\alpha + d_1(\boldsymbol{\pi}, \boldsymbol{\lambda})(1 - e^{-\lambda})(2 - e^{-\lambda}). \end{aligned}$$

Proof Let f be a function satisfying $s_2(f) < \infty$, and let h be given by (1.5). Then, by the proof of Theorem 10.A, we have

$$|\mathbb{E}(\mathcal{A}h)(\Xi)| \le \Delta_2 h \sum_{\alpha\in\Gamma} (\pi_\alpha^2 + \pi_\alpha \mathbb{E}Z_\alpha + \mathbb{E}(I_\alpha Z_\alpha)) + \Delta_1 h \sum_{\alpha\in\Gamma} \eta_\alpha, \quad (2.13)$$

where $\mathcal{A}$ is defined as in (1.4). We are now able to use Lemmas 10.2.3 and 10.2.5 to bound $\Delta_1 h$ and $\Delta_2 h$, obtaining

$$\begin{aligned} d_2(\mathcal{L}(\Xi), \mathrm{Po}(\boldsymbol{\pi})) \le & \{1 \wedge 1.65\lambda^{-1/2}\} \sum_{\alpha\in\Gamma} \eta_\alpha \\ & + \left\{1 \wedge \frac{2}{\lambda}\left(1 + 2\log^+\left(\frac{\lambda}{2}\right)\right)\right\} \sum_{\alpha\in\Gamma} (\pi_\alpha^2 + \pi_\alpha \mathbb{E}Z_\alpha + \mathbb{E}(I_\alpha Z_\alpha)). \end{aligned}$$

We then use the fact that

$$\begin{aligned} \mathrm{Po}(\boldsymbol{\lambda})(f) - \mathrm{Po}(\boldsymbol{\pi})(f) &= \mathrm{Po}(\boldsymbol{\lambda})(\mathcal{A}h) \\ &= \mathbb{E}\{\int_{\mathcal{Y}} [h(\Xi' + \delta_y) - h(\Xi')](\boldsymbol{\pi}(dy) - \boldsymbol{\lambda}(dy))\}, \end{aligned}$$

where Ξ' has distribution $\mathrm{Po}(\boldsymbol{\lambda})$, which from Lemma 10.2.2 gives

$$d_2(\mathrm{Po}(\boldsymbol{\lambda}), \mathrm{Po}(\boldsymbol{\pi})) \le d_1(\boldsymbol{\pi}, \boldsymbol{\lambda})(1 - e^{-\lambda})(1 + \lambda \sum_{j\ge 0} e^{-\lambda}\lambda^j/(j+1)!).$$

The theorem now follows. □

Theorem 10.G *Under the conditions of Theorem 10.B,*

$$\begin{aligned} d_2(\mathcal{L}(\Xi), \mathrm{Po}(\boldsymbol{\lambda})) \le & \left\{1 \wedge \frac{2}{\lambda}\left(1 + 2\log^+\left(\frac{\lambda}{2}\right)\right)\right\} \sum_{\alpha\in\Gamma} \pi_\alpha \mathbb{E}d(\Psi_\alpha, \Phi_\alpha) \\ & + d_1(\boldsymbol{\pi}, \boldsymbol{\lambda})(1 - e^{-\lambda})(2 - e^{-\lambda}). \end{aligned}$$

Proof The proof of Theorem 10.B is adapted in the same way that the proof of Theorem 10.A was adapted to prove Theorem 10.F. □

Remark 10.2.6 It may well be that the factor $\log^+ \lambda$ is in fact superfluous in Theorems 10.F and 10.G.

Example 10.2.7 If $\Gamma = \{n^{-1}i : 1 \le i \le rn\}$ and $(I_\alpha, \alpha \in \Gamma)$ are independent with $\pi_\alpha = p$ for each α, it follows from Theorems 10.F and 10.G that, taking $d_0(x,y) = |x-y| \wedge 1$,

$$d_2(\mathcal{L}(\Xi), \mathrm{Po}(\boldsymbol{\lambda})) \le 2(1 + 2\log^+(rnp/2))p + 1/(2n),$$

where $\boldsymbol{\lambda}$ is np times Lebesgue measure on $[0,r]$. Note that the bound depends on r only through the logarithmic factor.

Remark 10.2.8 For the approximation of mixed Poisson processes, one can modify the proofs of Theorems 10.C and 10.D in a similar fashion to provide corresponding error estimates in the d_2-metric.

Theorems 10.F and 10.G show that the magic λ-factors can be reintroduced, at the expense of weakening the metric from total variation to a Wasserstein metric. An alternative approach is to try to retain the total variation distance, by instead perturbing the positions of the points in such a way as to reduce the total variation distance of the process from a Poisson point process. This can be accomplished by using marks. Suppose that there is a mark Y_α associated with each $\alpha \in \Gamma$, and that the Y_α are sampled independently of each other and of the $(I_\beta, \beta \in \Gamma)$ from distributions F_α on a mark space $\mathcal{Y}$. A configuration

$$\xi = \sum_{\alpha \in \Gamma} x_\alpha \delta_{y_\alpha} \in \mathcal{Z} \tag{2.14}$$

is then constructed from a realization $(x_\alpha, y_\alpha)_{\alpha\in\Gamma}$ of the random elements $(I_\alpha, Y_\alpha)_{\alpha\in\Gamma}$. The previous setting is recovered if Y_α is taken to be α with probability one. Define $\boldsymbol{\pi} = \sum_{\alpha\in\Gamma} \pi_\alpha F_\alpha$, a measure with total mass $\lambda = \sum_{\alpha\in\Gamma} \pi_\alpha$ on $\mathcal{Y}$. The aim is now to show that, if the distributions of the marks are not all mutually singular, as was the case previously when the index α itself was used as the mark, the total variation distance between $\mathcal{L}(\Xi)$ and $\mu = \mathrm{Po}(\boldsymbol{\pi})$ may be less than the estimates of Theorems 10.A and 10.B.

In Theorems 10.F and 10.G, the improvements in the estimates over those of Theorems 10.A and 10.B came from factors involving the mean λ of the total number of points, which we always take to be positive. Here, the expressions are somewhat different, but the flavour is similar. The results given in Theorems 10.H and 10.I are reminiscient of those obtained by Barbour (1988) for independent summands, and involve a blend of the approach used there and that used for Theorems 10.F and 10.G.

We begin by proving two lemmas, analogous to Lemma 10.1.3(i) and (ii). For any $f : \mathcal{Z} \to [0,1]$, let h be given by (1.5) with $\boldsymbol{\lambda} = \boldsymbol{\pi}$, so that h

satisfies Equation (1.8) with

$$(\mathcal{A}h)(\xi) = \int_{\mathcal{Y}} [h(\xi + \delta_y) - h(\xi)]\pi(dy) + \int_{\mathcal{Y}} [h(\xi - \delta_y) - h(\xi)]\xi(dy). \quad (2.15)$$

Define $\hat{\pi}$ to be the probability measure $\lambda^{-1}\pi$ on $\mathcal{Y}$.

Lemma 10.2.9 *Let F be a probability distribution on $\mathcal{Y}$, and define*

$$\|F'\| = \sup_{y \in \mathcal{Y}} (dF/d\hat{\pi})(y).$$

Then, for all $f : \mathcal{Z} \to [0,1]$ and all $\xi \in \mathcal{Z}$,

$$\left| \int_{\mathcal{Y}} F(dy)[h(\xi + \delta_y) - h(\xi)] \right| \leq \min\{1, \lambda^{-1/2}\sqrt{2\|F'\|/e}\}.$$

Proof The proof once again involves dissecting the immigration–death process Z with generator $\mathcal{A}$ and equilibrium distribution μ, used in the definition (1.5) of h. As in the proof of Lemma 10.2.3, we have

$$h(\xi+\delta_y) - h(\xi) = -\int_0^\infty \mathbb{E}^\xi\{f(X(t) + Z_0(t) + \delta_y) - f(X(t) + Z_0(t))\}e^{-t}\,dt, \quad (2.16)$$

where $X(t)$ denotes the configuration of those of the original particles in ξ still alive at time t, and $Z_0(t)$ has a Poisson distribution on $\mathcal{Y}$ with intensity $\pi(1 - e^{-t})$ and is independent of $X(t)$. Define $\sigma = \lambda/\|F'\|$, and write $Z_0(t) = Y_1(t) + Y_2(t)$, where $Y_1(t)$ and $Y_2(t)$ are independent, and such that $Y_1(t) \sim \mathrm{Po}(\sigma F(1 - e^{-t}))$ and $Y_2(t) \sim \mathrm{Po}((\pi - \sigma F)(1 - e^{-t}))$. Note that $\pi - \sigma F$ is a positive measure, from the definition of σ. Then, defining $\tilde{f} : \mathcal{Z} \times \mathbb{N} \to [-\frac{1}{2}, \frac{1}{2}]$ by

$$\tilde{f}(\xi, n) = \int F(dy_1) \ldots \int F(dy_n)[f(\xi + \delta_{y_1} + \ldots + \delta_{y_n}) - 1/2], \quad (2.17)$$

we have

$$\int_{\mathcal{Y}} F(dy)[h(\xi + \delta_y) - h(\xi)] \quad (2.18)$$

$$= -\int_0^\infty e^{-t}\mathbb{E}^\xi\{\tilde{f}(X(t) + Y_2(t), N_1(t) + 1) - \tilde{f}(X(t) + Y_2(t), N_1(t))\}\,dt,$$

where $N_1(t) = |Y_1(t)|$. Now, given that $X(t) + Y_2(t) = \eta$, the expectation is just

$$-\tilde{f}(\eta, 0)\mathrm{Po}(\sigma_t)\{0\} - \sum_{r \geq 1}\{\mathrm{Po}(\sigma_t)\{r\} - \mathrm{Po}(\sigma_t)\{r-1\}\}\tilde{f}(\eta, r), \quad (2.19)$$

where $\sigma_t = \sigma(1-e^{-t})$, since $N_1(t)$ is independent of $X(t)$ and $Y_2(t)$, and is hence bounded in modulus by

$$\max_{r\geq 0} \mathrm{Po}(\sigma_t)\{r\} \leq \min\{1, (2e\sigma_t)^{-1/2}\}.$$

Thus

$$\left|\int_{\mathcal{Y}} F(dy)[h(\xi+\delta_y) - h(\xi)]\right|$$
$$\leq \int_0^\infty e^{-t}\min\{1, [2e\sigma(1-e^{-t})]^{-1/2}\}\,dt \leq \min\{1, \sqrt{2/e\sigma}\}. \quad \square$$

Lemma 10.2.10 *Let F and G be probability distributions on $\mathcal{Y}$. Then, for all $\xi \in \mathcal{Z}$,*

$$\left|\iint F(dy)G(dz)[h(\xi+\delta_y+\delta_z) - h(\xi+\delta_y) - h(\xi+\delta_z) + h(\xi)]\right|$$
$$\leq \min\left\{\frac{\sqrt{\|F'\|.\|G'\|}}{e\lambda}\left\{1 + 2\log^+\left(\frac{\lambda e}{\sqrt{\|F'\|.\|G'\|}}\right)\right\},\right.$$
$$\left.\frac{8}{3}\sqrt{\frac{\|F'\|}{e\lambda}},\ \frac{8}{3}\sqrt{\frac{\|G'\|}{e\lambda}},\ 1\right\}.$$

Proof Argue as for Lemma 10.2.9, obtaining, in place of (2.18),

$$\iint F(dy)G(dz)[h(\xi+\delta_y+\delta_z) - h(\xi+\delta_y) - h(\xi+\delta_z) + h(\xi)]$$
$$= -\int_0^\infty e^{-2t}\mathbb{E}^\xi\Big\{\bar f(X(t)+Y_3(t), N_1(t)+1, N_2(t)+1)$$
$$+ \bar f(X(t)+Y_3(t), N_1(t), N_2(t)) - \bar f(X(t)+Y_3(t), N_1(t)+1, N_2(t))$$
$$- \bar f(X(t)+Y_3(t), N_1(t), N_2(t)+1)\Big\}\,dt,$$

where $N_1(t)$ and $N_2(t)$ are independent of ξ and of $X(t)+Y_3(t)$, and have distributions $\mathrm{Po}(\sigma_t F)$ and $\mathrm{Po}(\nu_t G)$ respectively, with σ_t and ν_t given by $\sigma_t = (1-e^{-t})\lambda/(2\|F'\|)$ and $\nu_t = (1-e^{-t})\lambda/(2\|G'\|)$, and where

$$\bar f(\xi, m, n) = \int F(dy_1)\ldots\int F(dy_m)$$
$$\int G(dz_1)\ldots\int G(dz_n)\left[f\left(\xi + \sum_{i=1}^m \delta_{y_i} + \sum_{j=1}^n \delta_{z_j}\right) - 1/2\right].$$

Now, given $X(t)+Y_3(t) = \eta$, the expectation is

$$\sum_{r,s\geq 0} \bar f(\eta, r, s)[\mathrm{Po}(\sigma_t)\{r\} - \mathrm{Po}(\sigma_t)\{r-1\}][\mathrm{Po}(\nu_t)\{s\} - \mathrm{Po}(\nu_t)\{s-1\}],$$

which is bounded in modulus by

$$2\max_{r\geq 0}\mathrm{Po}(\sigma_t)\{r\}\max_{s\geq 0}\mathrm{Po}(\nu_t)\{s\} \leq 2\min\{1, (2e\sigma_t)^{-1/2}\}\min\{1, (2e\nu_t)^{-1/2}\},$$

and the lemma follows by elementary computations. $\square$

Remark 10.2.11 If $F = G$, this gives a bound of

$$\lambda^{-1}(\|F'\|/e)\{1 + 2\log^+(\lambda e/\|F'\|)\},$$

as compared with the bound

$$\tfrac{1}{2}\lambda^{-1}\{1 + 2\log^+ 2\lambda\} \int (dF/d\hat{\boldsymbol{\pi}})(y)F(dy) = \tfrac{1}{2}\lambda^{-1}\{1 + 2\log^+ 2\lambda\}\|F'\|_2^2$$

obtainable from Lemma 3 of Barbour (1988) by the natural limiting arguments, where $\|\cdot\|_2$ denotes the $L_2(\hat{\boldsymbol{\pi}})$ norm. The corresponding estimate for $F \neq G$ is

$$\tfrac{1}{2}\lambda^{-1}\{1 + 2\log^+ 2\lambda\}\|F'\|_2 \cdot \|G'\|_2.$$

We are now in a position to prove the following results, quantifying the accuracy of Poisson process approximation of the marked point process $\Xi = \sum_{\alpha\in\Gamma} I_\alpha \delta_{Y_\alpha}$, where the marks Y_α are sampled independently of each other and of the $(I_\beta, \beta \in \Gamma)$ from distributions F_α over $\mathcal{Y}$.

Theorem 10.H *Let Γ_α^s, Γ_α^w, Z_α and η_α be defined as for Theorem 10.A, and set*

$$\Delta_1^{(\alpha)} = \min\{1, (2\|F_\alpha'\|/\lambda e)^{1/2}\}; \qquad \Delta_1 = \max_{\alpha\in\Gamma} \Delta_1^{(\alpha)}; \tag{2.20}$$

$$\begin{aligned} \Delta_2^{(\alpha\beta)} &= \min\Big\{\lambda^{-1}\tau^{(\alpha\beta)}\{1 + 2\log^+(\lambda/\tau^{(\alpha\beta)})\}, \\ &\qquad\qquad \frac{8}{3}\sqrt{\frac{\|F_\alpha'\|}{e\lambda}},\ \frac{8}{3}\sqrt{\frac{\|F_\beta'\|}{e\lambda}},\ 1\Big\} \\ &\le 2\{1 + e^{-1}\log^+(\lambda(\|F_\alpha'\|\cdot\|F_\beta'\|)^{-1/2})\} \\ &\qquad\qquad \times (1 \wedge \{\lambda^{-1}\|F_\alpha'\|\}^{1/2})(1 \wedge \{\lambda^{-1}\|F_\beta'\|\}^{1/2}); \end{aligned} \tag{2.21}$$

$$\Delta_2 = \max_{\alpha,\beta\in\Gamma} \Delta_2^{(\alpha\beta)},$$

where $\tau^{(\alpha\beta)} = e^{-1}\big\{\|F_\alpha'\|\cdot\|F_\beta'\|\big\}^{1/2}$. Then

$$\begin{aligned} d_{TV}(\mathcal{L}(\Xi), \mathrm{Po}(\boldsymbol{\pi})) &\le \sum_{\alpha\in\Gamma}\Big\{\Delta_2^{(\alpha\alpha)}\pi_\alpha^2 + \sum_{\beta\in\Gamma_\alpha^s} \Delta_2^{(\alpha\beta)}(\pi_\alpha\pi_\beta + \mathbb{E}I_\alpha I_\beta) + \Delta_1^{(\alpha)}\eta_\alpha\Big\} \\ &\le \Delta_1 \sum_{\alpha\in\Gamma} \eta_\alpha + \Delta_2 \sum_{\alpha\in\Gamma}(\pi_\alpha^2 + \pi_\alpha \mathbb{E}Z_\alpha + \mathbb{E}(I_\alpha Z_\alpha)). \end{aligned}$$

Remark 10.2.12 If τ is defined to be $e^{-1}\max_{\alpha\in\Gamma}\|F_\alpha'\|$, it is immediate that $\Delta_1 \le \min\{1, \sqrt{2\tau/\lambda}\}$ and $\Delta_2 \le \min\big\{1, \lambda^{-1}\tau(1 + 2\log^+(\lambda/\tau))\big\}$.

Proof The argument is much as for Theorem 10.A. Let $\mathcal{A}$ be given by (2.15), and let f and h be as for Lemma 10.2.9. Then, defining $\Xi_\alpha = \sum_{\beta\in\Gamma_\alpha^w} I_\beta \delta_{Y_\beta}$ and $\Xi'_\alpha = \Xi - I_\alpha\delta_{Y_\alpha}$, it follows that, as in the proof of (1.11),

$$\begin{aligned} &\left|\mathbb{E}\{I_\alpha[h(\Xi-\delta_{Y_\alpha})-h(\Xi)]-I_\alpha[h(\Xi_\alpha)-h(\Xi_\alpha+\delta_{Y_\alpha})]\}\right| \\ &\qquad = \left|\int F_\alpha(dy)\mathbb{E}\{I_\alpha[h(\Xi'_\alpha)-h(\Xi'_\alpha+\delta_y)-h(\Xi_\alpha)+h(\Xi_\alpha+\delta_y)]\}\right|. \end{aligned}$$

In order to apply Lemma 10.2.10, enumerate the elements $\beta\in\Gamma_\alpha^s$ in some order as $\beta(1),\ldots,\beta(k)$, where $k=|\Gamma_\alpha^s|$, and let

$$\Xi_\alpha^{(i)} = \Xi_\alpha + \sum_{j=1}^{i} I_{\beta(j)}\delta_{Y_{\beta(j)}},$$

so that $\Xi_\alpha^{(0)}=\Xi_\alpha$ and $\Xi_\alpha^{(k)}=\Xi'_\alpha$. Then

$$\begin{aligned} &\left|\int F_\alpha(dy)\mathbb{E}\{I_\alpha[h(\Xi'_\alpha)-h(\Xi'_\alpha+\delta_y)-h(\Xi_\alpha)+h(\Xi_\alpha+\delta_y)]\}\right| \\ &= \left|\int F_\alpha(dy)\mathbb{E}\Big\{I_\alpha\sum_{i=1}^{k}[h(\Xi_\alpha^{(i)})-h(\Xi_\alpha^{(i)}+\delta_y) \right. \\ &\qquad \left. -h(\Xi_\alpha^{(i-1)})+h(\Xi_\alpha^{(i-1)}+\delta_y)]\Big\}\right| \\ &= \left|\mathbb{E}\Big\{\sum_{i=1}^{k} I_\alpha I_{\beta(i)}\iint F_\alpha(dy)F_{\beta(i)}(dz) \right. \\ &\qquad \left. \times[h(\Xi_\alpha^{(i-1)}+\delta_z)-h(\Xi_\alpha^{(i-1)}+\delta_z+\delta_y)-h(\Xi_\alpha^{(i-1)})+h(\Xi_\alpha^{(i-1)}+\delta_y)]\Big\}\right| \\ &\le \sum_{i=1}^{k}\mathbb{E}(I_\alpha I_{\beta(i)})\Delta_2^{(\alpha\beta(i))} = \sum_{\beta\in\Gamma_\alpha^s}\Delta_2^{(\alpha\beta)}\mathbb{E}I_\alpha I_\beta, \end{aligned} \tag{2.22}$$

where the inequality follows from Lemma 10.2.10. Arguing in similar fashion, we obtain the inequality

$$\begin{aligned} &\left|\pi_\alpha\int F_\alpha(dy)\mathbb{E}[h(\Xi_\alpha)-h(\Xi_\alpha+\delta_y)]-\pi_\alpha\int F_\alpha(dy)\mathbb{E}[h(\Xi)-h(\Xi+\delta_y)]\right| \\ &\qquad\qquad \le \Delta_2^{(\alpha\alpha)}\pi_\alpha^2+\sum_{\beta\in\Gamma_\alpha^s}\Delta_2^{(\alpha\beta)}\pi_\alpha\pi_\beta, \end{aligned}$$

corresponding to (1.12). Finally, from Lemma 10.2.9, we have the inequality

$$\begin{aligned}&\left|\mathbb{E}I_\alpha[h(\Xi_\alpha)-h(\Xi_\alpha+\delta_{Y_\alpha})]-\pi_\alpha\int F_\alpha(dy)\mathbb{E}[h(\Xi_\alpha)-h(\Xi_\alpha+\delta_y)]\right|\\&\quad=\left|\mathbb{E}\{[\mathbb{E}(I_\alpha\mid(I_\beta:\beta\in\Gamma_\alpha^w))-\pi_\alpha]\int F_\alpha(dy)[h(\Xi_\alpha)-h(\Xi_\alpha+\delta_y)]\}\right|\\&\quad\le\Delta_1^{(\alpha)}\mathbb{E}|\mathbb{E}(I_\alpha\mid(I_\beta:\beta\in\Gamma_\alpha^w))-\pi_\alpha|,\end{aligned}\tag{2.23}$$

corresponding to (1.13), and the theorem follows in the usual way. □

Theorem 10.I *Suppose that, for each $\alpha\in\Gamma$, random elements Ψ_α and Φ_α can be realized on the same probability space, in such a way that $\mathcal{L}(\Psi_\alpha)=\mathcal{L}(\tilde{\Xi})$ and $\mathcal{L}(\Phi_\alpha+\delta_\alpha)=\mathcal{L}(\tilde{\Xi}\mid I_\alpha=1)$, where $\tilde{\Xi}=\sum_{\beta\in\Gamma}I_\beta\delta_\beta$. Let $d(\Psi_\alpha,\Phi_\alpha)$ be as for Theorem 10.B, and define $\Delta_2^{(\alpha\beta)}$, Δ_2 as in Theorem 10.H. Then*

$$\begin{aligned}d_{TV}(\mathcal{L}(\Xi),\mathrm{Po}(\boldsymbol{\pi}))&\le\sum_{\alpha\in\Gamma}\Big\{\Delta_2^{(\alpha\alpha)}\pi_\alpha^2+\sum_{\beta\ne\alpha}\Delta_2^{(\alpha\beta)}\pi_\alpha\mathbb{E}|\Psi_\alpha(\beta)-\Phi_\alpha(\beta)|\Big\}\\&\le\Delta_2\sum_{\alpha\in\Gamma}\pi_\alpha\mathbb{E}d(\Psi_\alpha,\Phi_\alpha).\end{aligned}$$

Proof Take $f:\mathcal{Z}\to[0,1]$, and construct h using (1.5). Then, given Φ_α and Ψ_α, choose just one set of marks $(Y_\alpha,\alpha\in\Gamma)$ independently from the distributions $(F_\alpha,\alpha\in\Gamma)$ to realize configurations $\Psi_{\alpha y}=\sum_{\beta\in\Gamma}\Psi_\alpha(\beta)\delta_{Y_\beta}$ and $\Phi_{\alpha y}=\sum_{\beta\ne\alpha}\Phi_\alpha(\beta)\delta_{Y_\beta}$ in such a way that

$$\mathcal{L}(\Phi_{\alpha y}+\delta_y)=\mathcal{L}(\Xi\mid I_\alpha=1,Y_\alpha=y)\quad\text{and}\quad\mathcal{L}(\Psi_{\alpha y})=\mathcal{L}(\Xi).$$

Note that the processes $\Psi_{\alpha y}=\Psi'_\alpha$ and $\Phi_{\alpha y}=\Phi'_\alpha$ do not involve y in their construction. Then

$$\mathbb{E}I_\alpha[h(\Xi-\delta_{Y_\alpha})-h(\Xi)]=\pi_\alpha\int F_\alpha(dy)\mathbb{E}[h(\Phi'_\alpha)-h(\Phi'_\alpha+\delta_y)],$$

and hence, from (2.15),

$$\begin{aligned}&|\mathbb{E}(\mathcal{A}h)(\Xi)|\\&\quad=\left|\sum_{\alpha\in\Gamma}\pi_\alpha\int F_\alpha(dy)\mathbb{E}[h(\Psi'_\alpha+\delta_y)-h(\Psi'_\alpha)+h(\Phi'_\alpha)-h(\Phi'_\alpha+\delta_y)]\right|.\end{aligned}$$

Enumerate the elements of Γ as $\beta(1),\ldots,\beta(|\Gamma|)$, and let

$$\Theta_\alpha^i=\Phi'_\alpha+\sum_{j=1}^i\{\Psi_\alpha(\beta(j))-\Phi_\alpha(\beta(j))\}\delta_{Y_{\beta(j)}},\quad 0\le i\le|\Gamma|,$$

so that $\Theta_\alpha^0 = \Phi'_\alpha$ and $\Theta_\alpha^{|\Gamma|} = \Psi'_\alpha$. Then, as in the proof of Theorem 10.H,

$$\left|\int F_\alpha(dy)\mathbb{E}[h(\Psi'_\alpha + \delta_y) - h(\Psi'_\alpha) + h(\Phi'_\alpha) - h(\Phi'_\alpha + \delta_y)]\right|$$

$$= \left|\sum_{i=1}^{|\Gamma|} \mathbb{E} \int F_\alpha(dy)[h(\Theta_\alpha^i + \delta_y) - h(\Theta_\alpha^i) + h(\Theta_\alpha^{i-1}) - h(\Theta_\alpha^{i-1} + \delta_y)]\right|$$

$$\le \sum_{i=1}^{|\Gamma|} \mathbb{E}\Big\{ |\Psi_\alpha(\beta(i)) - \Phi_\alpha(\beta(i))| \cdot \Big| \iint F_\alpha(dy) F_{\beta(i)}(dz)$$

$$[h(\Theta_\alpha^{i-1} + \delta_y + \delta_z) - h(\Theta_\alpha^{i-1} + \delta_z) + h(\Theta_\alpha^{i-1}) - h(\Theta_\alpha^{i-1} + \delta_y)]\Big|\Big\}$$

$$\le \sum_{i=1}^{|\Gamma|} \Delta_2^{(\alpha\beta(i))} \mathbb{E}|\Psi_\alpha(\beta(i)) - \Phi_\alpha(\beta(i))| = \sum_{\beta\in\Gamma} \Delta_2^{(\alpha\beta)} \mathbb{E}|\Psi_\alpha(\beta) - \Phi_\alpha(\beta)|,$$

where the last inequality now follows by Lemma 10.2.10. The theorem is now immediate. □

Note that Theorem 10.I yields a bound which is better than the bound in Theorem 10.B by at least a factor of Δ_2. A similar result corresponding to Corollary 10.B.1 follows immediately: the details are left to the reader.

Remark 10.2.13 Two examples are discussed in Section 10.5, where the mark distributions depend on the configuration at other sites. It then becomes much more complicated to apply Lemma 10.2.10.

Example 10.2.14 As an example of the application of Theorems 10.H and 10.I, consider an infinite particle system investigated by Liggett and Port (1988). The system consists of particles which move independently of one another on a countable set S as Markov chains with transition matrix P, and they were interested in conditions under which, given suitable initial configurations, the system would converge to a (mixed) Poisson process on S. Let the initial number of particles at site $x \in S$ be fixed to be $a(x)$, where it is assumed that $\sum_x a(x)P^n(x,y) < \infty$ for all n and y. Take Γ to be the set of particles, indexed by α — that their number is usually infinite is not important here — and fix any finite subset $C \subset S$. Define $I_\alpha = I[Z_\alpha(n) \in C]$, where $Z_\alpha(n)$ denotes the position of particle α at time n, and let $Y_\alpha = Z_\alpha(n)$. Thus, if $Z_\alpha(0) = z$,

$$\pi_\alpha = P^n(z, C), \quad \lambda = \sum_x a(x)P^n(x, C)$$

and

$$\|F'_\alpha\| = \phi(z) = \max_{y\in C} \frac{P^n(z,y)/P^n(z,C)}{\lambda^{-1}\sum_x a(x)P^n(x,y)},$$

and, since the $(I_\alpha, \alpha \in \Gamma)$ are independent, the estimate of Theorem 10.H can be simplified to

$$\begin{aligned} d_{TV}(\mathcal{L}(\Xi), \mathrm{Po}(\boldsymbol{\pi})) &\leq \sum_{\alpha \in \Gamma} \pi_\alpha^2 \Delta_2^{(\alpha\alpha)} \\ &\leq \frac{1}{\lambda e} \sum_z a(z) \{P^n(z, C)\}^2 \phi(z) (1 + 2 \log^+ \{\lambda e / \phi(z)\}), \end{aligned} \tag{2.24}$$

where $\boldsymbol{\pi}$ is the measure $\sum_x a(x) P^n(x, \cdot)$ restricted to C and Ξ is the configuration of particles in C. If the initial configuration a is chosen at random according to some distribution, independently of the subsequent motion, the above estimate still applies for the conditional distribution of Ξ given the initial configuration, and the expectation of the right hand side gives a bound for the total variation distance between the unconditional law of Ξ and that of the mixed Poisson process with intensity $\sum_x a(x) P^n(x, \cdot)$, now random. These results complement those of Liggett and Port, insofar as they address the question of approximation rather than that of convergence. Inequalities from Chapter 3, in particular Theorem 3.A* and Corollary 3.D.1, could also be used to prove lower bounds for Poisson approximation. Note also that the alternative bound given in Remark 10.2.11, which is all that is needed for independent summands, leads to the neater estimate

$$\sum_z a(z) \sum_{y \in C} \frac{\{P^n(z, y)\}^2}{2 \sum_x a(x) P^n(x, y)} (1 + 2 \log^+ 2\lambda),$$

see Barbour (1988) Theorem 1.

Example 10.2.15 This example is used to explore the accuracy of Theorems 10.H and 10.I. Let $\mathcal{Y}$ be the unit circle, and let i correspond to the point $y_i = 2\pi i/n$, $1 \leq i \leq n$. Suppose that each point i has a mark Y_i associated with it, such that F_i is the uniform distribution on $(y_i - \pi\gamma, y_i + \pi\gamma)$, for some $0 < \gamma \leq 1$ such that $n\gamma$ is an integer. Then, if the I_i are independent and $\mathbb{P}[I_i = 1] = p = 1 - \mathbb{P}[I_i = 0]$, Theorem 10.H gives an upper bound of order $\min(1, np^2, p(1 + \log^+(\gamma np))/\gamma)$ for the total variation distance between the law of $\Xi = \sum_{i=1}^n I_i \delta_{Y_i}$ and a homogeneous Poisson process of rate $np/2\pi$ on $\mathcal{Y}$. How good is this bound?

It is possible to construct events which show that the distance must be at least of order $\min(1, np^2, p/\gamma^{1/2})$, as follows. Suppose, to avoid trivial complications, that $1/\gamma = 2k$ is an even integer, and that $4k$ divides n. Partition the circle into $2k$ intervals

$$J_j = [n^{-1}\pi + 2\pi j\gamma, n^{-1}\pi + 2\pi\gamma(j+1)), \quad 0 \leq j < 2k,$$

of length π/k. Then, if $A_0, \ldots, A_{k-1}$ are any events such that A_j depends only on the configuration restricted to J_{2j}, they are independent. Now the distribution of the number N_j of points Y_i with $I_i = 1$ falling in J_{2j} is that of a sum of n/k independent indicators, with probabilities $p_r = p\{1 - |1 - \frac{k}{n}(2r-1)|\}$, $1 \le r \le n/k$, and, applying Corollary 3.D.1, it follows that

$$d_{TV}(\mathcal{L}(N_j), \mathrm{Po}\,(np\gamma)) \ge c_1 p(1 \wedge np/2k).$$

Thus, if $np\gamma \ge \varepsilon$ for some small fixed $\varepsilon > 0$, there is a subset $B \subset \mathbb{Z}^+$ such that, defining

$$q_1 = \mathbb{P}[N_j \in B]; \qquad q_2 = \mathrm{Po}\,(np\gamma)\{B\},$$

the inequality $|q_1 - q_2| \ge c_2 p$ holds. Hence, by considering the distribution of $\sum_j I[N_j \in B]$, the total variation distance between $\mathcal{L}(\Xi)$ and μ is at least

$$d_{TV}(\mathrm{Bi}\,(k, q_1), \mathrm{Bi}\,(k, q_2)) \ge c_3 k^{1/2}|q_1 - q_2| \wedge 1 \asymp \gamma^{-1/2} p \wedge 1.$$

For smaller γ, the set $B = \{2\}$ gives

$$q_2 - q_1 \ge c_4 q_2/n\gamma; \qquad q_2 \asymp (np\gamma)^2,$$

and a similar argument then yields

$$\begin{aligned} d_{TV}(\mathcal{L}(\Xi), \mathrm{Po}\,(\boldsymbol{\pi})) &\ge d_{TV}(\mathrm{Bi}\,(k, q_1), \mathrm{Bi}\,(k, q_2)) \asymp \frac{k|q_1 - q_2|}{(1 + kq_2)^{1/2}} \wedge 1 \\ &\asymp \begin{cases} (np\sqrt{\gamma}/n\gamma) \wedge 1 = p\gamma^{-1/2} \wedge 1 & \text{if } (np)^2\gamma \ge 1, \\ (np\gamma)^2/n\gamma^2 = np^2 & \text{if } (np)^2\gamma \le 1. \end{cases} \end{aligned}$$

Thus, at least when $(np)^2\gamma \le 1$, Theorem 10.H is sharp. However, it seems likely that, for $(np)^2\gamma \ge 1$, the true error is of order $p\gamma^{-1/2} \wedge 1$. It is as yet not known whether Theorem 10.H is typically too weak, except when $\gamma = 1$ or γ is very small, if there is dependence between the I_α.

So far, in this second approach, it has been assumed that the approximating Poisson process has intensity $\boldsymbol{\pi} = \sum_{\alpha \in \Gamma} \pi_\alpha F_\alpha$. If, instead, a Poisson process with intensity measure $\boldsymbol{\lambda}$ is more appropriate, an extra term

$$\left| \int_{\mathcal{Y}} \mathbb{E}[h(\Xi + \delta_y) - h(\Xi)](\boldsymbol{\lambda}(dy) - \boldsymbol{\pi}(dy)) \right|$$

is needed in the estimate of $|\mathbb{E}(\mathcal{A}h)(\Xi)|$. This, using Lemma 10.1.3, adds an extra term $\int |\boldsymbol{\pi} - \boldsymbol{\lambda}|(dy)$ to the bound, which can be construed as allowing for the extra step of approximating $\mathrm{Po}\,(\boldsymbol{\pi})$ by $\mathrm{Po}\,(\boldsymbol{\lambda})$, by introducing in addition an upper estimate of $d_{TV}(\mathrm{Po}\,(\boldsymbol{\pi}), \mathrm{Po}\,(\boldsymbol{\lambda}))$, as in the proof of Theorem 10.F.

In general, the bound

$$d_{TV}(\mathrm{Po}(\boldsymbol{\pi}), \mathrm{Po}(\boldsymbol{\lambda})) \le \int_{\mathcal{Y}} |\boldsymbol{\pi} - \boldsymbol{\lambda}|(dy) \tag{2.25}$$

cannot be much improved, as the following example on $[0, 2+\lambda-\varepsilon/2]$ shows: if

$$\boldsymbol{\pi}(dy)/dy = \begin{cases} 0 & \text{if } y \le 1; \\ \varepsilon/2 & \text{if } 1 \le y \le 2; \\ 1 & \text{if } y \ge 2; \end{cases} \qquad \boldsymbol{\lambda}(dy)/dy = \begin{cases} \varepsilon/2 & \text{if } y \le 1; \\ 0 & \text{if } 1 \le y \le 2; \\ 1 & \text{if } y \ge 2, \end{cases}$$

then $\int |\boldsymbol{\pi} - \boldsymbol{\lambda}|(dy) = \varepsilon$, whereas $d_{TV}(\mathrm{Po}(\boldsymbol{\pi}), \mathrm{Po}(\boldsymbol{\lambda})) = 1 - e^{-\varepsilon/2} \sim \varepsilon/2$. As it happens, if $\boldsymbol{\lambda}(\mathcal{Y}) = \boldsymbol{\pi}(\mathcal{Y})$, the bound given in (2.25) can be halved, and the example shows that no smaller fraction can be achieved. However, in most applications where $\boldsymbol{\lambda}$ is more natural than $\boldsymbol{\pi}$, $\boldsymbol{\lambda}$ is a smooth absolutely continuous perturbation of $\boldsymbol{\pi}$, typically satisfying a condition such as $|(d\boldsymbol{\pi}/d\boldsymbol{\lambda})(y) - 1| \le \theta$, where $\int |\boldsymbol{\pi} - \boldsymbol{\lambda}|(dy) = \varepsilon \asymp \lambda\theta$. In such circumstances, for large λ, a better estimate of $d_{TV}(\mathrm{Po}(\boldsymbol{\pi}), \mathrm{Po}(\boldsymbol{\lambda}))$ than ε can be obtained, introducing a factor of $\lambda^{-1/2}$.

Proposition 10.2.16 *If $\boldsymbol{\pi} \ll \boldsymbol{\lambda}$ and $\lambda \ge 1$, and if*

$$\sup_{y \in \mathcal{Y}} |(d\boldsymbol{\pi}/d\boldsymbol{\lambda})(y) - 1| \le \theta, \tag{2.26}$$

then it follows that

$$d_{TV}(\mathrm{Po}(\boldsymbol{\pi}), \mathrm{Po}(\boldsymbol{\lambda})) \le \theta\sqrt{\lambda}.$$

Proof Using (A.1.14) and Proposition A.1.2, we find that

$$\begin{aligned} d_{TV}(\mathrm{Po}(\boldsymbol{\pi}), \mathrm{Po}(\boldsymbol{\lambda})) &\le \sqrt{2} d_H(\mathrm{Po}(\boldsymbol{\pi}), \mathrm{Po}(\boldsymbol{\lambda})) \le \sqrt{2} d_H(\boldsymbol{\pi}, \boldsymbol{\lambda}) \\ &= \left[\int_{\mathcal{Y}} \left\{ \left(\frac{d\boldsymbol{\pi}}{d\boldsymbol{\lambda}} \right)^{\frac{1}{2}} - 1 \right\}^2 \boldsymbol{\lambda}(dy) \right]^{\frac{1}{2}} \le \left[\int_{\mathcal{Y}} \left(\frac{d\boldsymbol{\pi}}{d\boldsymbol{\lambda}} - 1 \right)^2 \boldsymbol{\lambda}(dy) \right]^{\frac{1}{2}} \le \{\theta^2 \lambda\}^{\frac{1}{2}}, \end{aligned}$$

where the penultimate inequality follows because

$$|x - 1| = |(\sqrt{x} - 1)(\sqrt{x} + 1)| \ge |\sqrt{x} - 1| \quad \text{in} \quad x \ge 0.$$ □

10.3 Multivariate Poisson approximation

Consider again a collection $(I_\alpha, \alpha \in \Gamma)$ of indicator variables, but assume now that a partition $\Gamma = \cup_{j=1}^{r} \Gamma_j$ of Γ is given and that the random variables $W_j = \sum_{\alpha \in \Gamma_j} I_\alpha$ are of interest. The methods of the previous chapters are useful for approximating the distribution of each W_j by $\mathrm{Po}(\lambda_j)$, where

$\lambda_j = \mathbb{E}W_j = \sum_{\alpha\in\Gamma_j}\pi_\alpha$, but we can now also give results on the accuracy of approximation of the *joint* distribution of (W_j) by independent Poisson variables. The first result is a simple consequence of Theorem 10.B. We assume, as in Chapter 2, that there exists a coupling of (I_β) with $(J_{\beta\alpha})$ such that

$$\mathcal{L}((J_{\beta\alpha}), \beta\in\Gamma) = \mathcal{L}((I_\beta), \beta\in\Gamma \mid I_\alpha = 1). \tag{3.1}$$

Theorem 10.J *Let $W_j = \sum_{\alpha\in\Gamma_j} I_\alpha$ and $\lambda_j = \mathbb{E}W_j$ be as above, and assume that $(J_{\beta\alpha})$ satisfies (3.1). Then*

$$d_{TV}\left(\mathcal{L}(\{W_j\}_{j=1}^r), \prod_{j=1}^r \mathrm{Po}(\lambda_j)\right) \le \sum_{\alpha\in\Gamma}\left(\pi_\alpha^2 + \sum_{\beta\ne\alpha}\pi_\alpha\mathbb{E}|J_{\beta\alpha} - I_\beta|\right).$$

Proof Let $\Xi = \Psi_\alpha = \sum I_\beta\delta_\beta$ and $\Phi_\alpha = \sum_{\beta\ne\alpha} J_{\beta\alpha}\delta_\beta$, note that

$$d_{TV}\left(\mathcal{L}(\{W_j\}_{j=1}^r), \prod_{j=1}^r \mathrm{Po}(\lambda_j)\right) \le d_{TV}(\mathcal{L}(\Xi), \mathrm{Po}(\boldsymbol{\pi})),$$

and apply Theorem 10.B.

Corollary 10.J.1 *If there exists a coupling satisfying (3.1) and (2.1.3), then*

$$d_{TV}\left(\mathcal{L}(\{W_j\}_{j=1}^r), \prod_{j=1}^r \mathrm{Po}(\lambda_j)\right)$$
$$\le \sum_{\alpha\in\Gamma}\pi_\alpha^2 + \sum_{\alpha\in\Gamma}\sum_{\beta\in\Gamma_\alpha^-}|\mathrm{Cov}(I_\alpha, I_\beta)|$$
$$+ \sum_{\alpha\in\Gamma}\sum_{\beta\in\Gamma_\alpha^+}\mathrm{Cov}(I_\alpha, I_\beta) + \sum_{\alpha\in\Gamma}\sum_{\beta\in\Gamma_\alpha^0}(\mathbb{E}I_\alpha I_\beta + \pi_\alpha\pi_\beta) \qquad \square$$

We thus obtain the same estimates as in Theorem 2.C and its corollaries, but without the factor $\lambda^{-1}(1-e^{-\lambda})$. In particular, given a sequence of vectors $\{W_j^{(n)}\}_{j=1}^r$ as above such that $\mathbb{E}W_j^{(n)} \to \lambda_j$ for some $\lambda_j < \infty$, joint convergence in distribution of the $W_j^{(n)}$ to independent Poisson $\mathrm{Po}(\lambda_j)$ variables follows in all cases where Theorem 2.C or its corollaries may be used to prove Poisson convergence of $W^{(n)} = \sum_{j=1}^r W_j^{(n)}$.

In order at least partly to reinstate the factor $\lambda^{-1}(1-e^{-\lambda})$ when $\lambda \ge 1$, we use instead Theorem 10.I, choosing the mark space $\mathcal{Y} = \{1, \ldots, r\}$ and $F_\alpha = \delta_j$ (i.e. $Y_\alpha = j$) if $\alpha\in\Gamma_j$. Then $\boldsymbol{\pi} = \sum_{j=1}^r \lambda_j\delta_j$, $\|F'_\alpha\| = \lambda/\lambda_j$ and thus $\lambda^{-1}\tau^{(\alpha\beta)} = e^{-1}(\lambda_j\lambda_k)^{-\frac{1}{2}}$ if $\alpha\in\Gamma_j$ and $\beta\in\Gamma_k$. Furthermore, $\Xi = \sum_j W_j\delta_j$ can be identified with the vector $(W_1, \ldots, W_r)$ and the Poisson process $\mathrm{Po}(\boldsymbol{\pi})$ consists simply of r independent Poisson processes with intensities λ_j. We may thus apply Theorem 10.I with Ψ_α and Φ_α as in the proof of Theorem 10.J, and obtain the following result.

Theorem 10.K *Let $W_j = \sum_{\alpha\in\Gamma_j} I_\alpha$ and $\lambda_j = \mathbb{E}W_j$ be as above, and define*

$$\Delta^{(jk)} = \min\Big(\frac{1+\log^+(e^2\lambda_j\lambda_k)}{e\sqrt{\lambda_j\lambda_k}}, \frac{8}{3}\sqrt{\frac{1}{e\lambda_j}}, \frac{8}{3}\sqrt{\frac{1}{e\lambda_k}}, 1\Big).$$

Suppose that $(J_{\beta\alpha})$ satisfies (3.1). Then

$$d_{TV}\Big(\mathcal{L}(\{W_j\}_{j=1}^r), \prod_{j=1}^r \mathrm{Po}(\lambda_j)\Big)$$
$$\leq \sum_j \Delta^{(jj)} \sum_{\alpha\in\Gamma_j} \pi_\alpha^2 + \sum_j\sum_k \Delta^{(jk)} \sum_{\alpha\in\Gamma_j} \sum_{\substack{\beta\in\Gamma_k\\ \beta\neq\alpha}} \pi_\alpha \mathbb{E}|J_{\beta\alpha} - I_\beta|$$
$$\leq 2(1+e^{-1}\log^+\max\lambda_j)\Big\{\sum_j (1\wedge\lambda_j^{-1}) \sum_{\alpha\in\Gamma_j} \pi_\alpha^2$$
$$+ \sum_j\sum_k (1\wedge\lambda_j^{-\frac{1}{2}})(1\wedge\lambda_k^{-\frac{1}{2}}) \sum_{\alpha\in\Gamma_j} \sum_{\substack{\beta\in\Gamma_k\\ \beta\neq\alpha}} \pi_\alpha \mathbb{E}|J_{\beta\alpha} - I_\alpha|\Big\}.$$

In particular,

$$d_{TV}\Big(\mathcal{L}(\{W_j\}_{j=1}^r), \prod_{j=1}^r \mathrm{Po}(\lambda_j)\Big)$$
$$\leq \frac{1+2\log^+(e\min\lambda_j)}{e\min\lambda_j} \sum_\alpha \Big(\pi_\alpha^2 + \sum_{\beta\neq\alpha} \pi_\alpha \mathbb{E}|J_{\beta\alpha} - I_\beta|\Big). \qquad \square$$

Thus if all the λ_j are equal to $\lambda_j = \lambda/r$, we obtain, after a small calculation, the estimates of Theorem 2.C and its corollaries, multiplied by the factor $\frac{3}{e-1}r(1+\log^+\lambda)$.

Example 10.3.1 *The joint distribution of small cycles in $K_{n,c/n}$.* We take the standard model $K_{n,p}$ of a random graph on n vertices described in Section 5.1, with edge probability $p = n^{-1}c$ for c fixed, and consider for fixed r the joint distribution of $(W_3, W_4, \ldots, W_r)$, where W_j denotes the number of distinct j-cycles appearing in $K_{n,p}$. Here, Γ_j denotes the set of all distinct j-cycles in K_n, and, for $\alpha\in\Gamma_j$, $\pi_\alpha = p^j$, so that, by counting,

$$\lambda_j = \mathbb{E}W_j = \binom{n}{j}\frac{j!}{2j}p^j \sim \tfrac{1}{2}j^{-1}c^j, \quad 3\leq j\leq r. \qquad (3.2)$$

A coupling of (I_β) with $(J_{\beta\alpha})$ can be arranged by first realizing $K_{n,p}$, and with it the I_β, and then forcing all edges in α to be present, giving the $J_{\beta\alpha}$. Since the coupling only involves adding edges, $J_{\beta\alpha}\geq I_\beta$ for all

$\beta \neq \alpha$, and the indicators are positively related. This makes Corollary 10.J.1 particularly easy to apply, since $\Gamma_\alpha^- = \Gamma_\alpha^0 = \emptyset$, and hence

$$d_{TV}\left(\mathcal{L}(\{W_j\}_{j=1}^r), \prod_{j=1}^r \mathrm{Po}(\lambda_j)\right) \leq \sum_{\alpha\in\Gamma} \pi_\alpha^2 + \sum_{\alpha\in\Gamma}\sum_{\beta\in\Gamma_\alpha^+} \mathrm{Cov}(I_\alpha, I_\beta). \quad (3.3)$$

Furthermore, $\mathrm{Cov}(I_\alpha, I_\beta) = 0$ unless α and β have at least two vertices in common. If α is a fixed j-cycle and $3 \leq l \leq r$, the number of l-cycles β sharing m vertices with α, $2 \leq m \leq l \wedge j$, is $\binom{j}{m}\binom{n-j}{l-m}\frac{l!}{2l}$. For each such $\beta \neq \alpha$, the covariance $\mathrm{Cov}(I_\alpha, I_\beta)$ is at most $p^{j+l-m+1}$, though this estimate is for the most part very crude. Combining these facts with (3.3) gives an estimate

$$\begin{aligned} &d_{TV}\left(\mathcal{L}(\{W_j\}_{j=1}^r), \prod_{j=1}^r \mathrm{Po}(\lambda_j)\right) \\ &\leq \tfrac{1}{2}\sum_{j=3}^r j^{-1}n^j\left\{p^{2j} + \tfrac{1}{2}\sum_{l=3}^r\sum_{m=2}^{l\wedge j}\binom{j}{m}\frac{(l-1)!}{(l-m)!}n^{l-m}p^{j+l-m+1}\right\} \\ &= O(n^{-1}), \end{aligned}$$

for c and r fixed as $n \to \infty$.

10.4 Compound Poisson approximation

We have occasionally noted that a strong tendency to clustering makes the Poisson approximation less good, and that approximation with a compound Poisson distribution seems better. One way to achieve such results is through the Poisson process approximations treated in this chapter. This idea is also exploited in Arratia, Goldstein and Gordon (1990) Section 3.

Recall that if μ is a finite measure on $\mathbb{R}\setminus\{0\}$ (or, more generally, a measure such that $\int(|x|\wedge 1)d\mu < \infty$), the compound Poisson distribution $\mathrm{CP}(\mu)$ is the distribution with characteristic function $\exp(\int_{-\infty}^{\infty}(e^{itx}-1)d\mu(x))$; a variable with this distribution may be constructed as $\int_{-\infty}^{\infty} x\,d\Xi$, where Ξ is a Poisson process on $\mathbb{R}$ with intensity μ. μ is often given as a measure on the whole of $\mathbb{R}$, in which case we ignore $\mu\{0\}$. We are mainly concerned with the case of a discrete measure $\mu = \sum \mu\{a_k\}\delta_{a_k}$, where the construction reduces to $\sum a_k Z_k$, with the Z_k independent $\mathrm{Po}(\mu\{a_k\})$ random variables. Such distributions occur most naturally if we want to approximate sums $\sum X_\alpha$ of random variables that may take values other than 0 and 1.

Our first result is a corollary to Theorem 10.E.

Theorem 10.L *Let $W = \sum_{\alpha\in\Gamma} X_\alpha$, where the X_α are real valued random variables with distributions F_α, and let $\boldsymbol{\lambda} = \sum_{\alpha\in\Gamma} F_\alpha$. Suppose that,* *for each $\alpha \in \Gamma$ and (almost) every $x \neq 0$, random variables $(Y_{\beta\alpha x})_{\beta \in \Gamma}$ can be defined on the same probability space as the (X_β), in such a way that $\mathcal{L}(\{Y_{\beta\alpha x}\}_\beta) = \mathcal{L}(\{X_\beta\}_\beta \mid X_\alpha = x)$. Then*

$$d_{TV}(\mathcal{L}(W), \mathrm{CP}(\boldsymbol{\lambda})) \leq \sum_{\alpha \in \Gamma} \Big(\mathbb{P}[X_\alpha \neq 0]^2 + \mathbb{E}\{\rho'_\alpha(X_\alpha) \cdot I[X_\alpha \neq 0]\} \Big),$$

where

$$\begin{aligned} \rho'_\alpha(x) &= \sum_{\beta \neq \alpha} \Big(\mathbb{P}[Y_{\beta\alpha x} \neq X_\beta \neq 0] + \mathbb{P}[0 \neq Y_{\beta\alpha x} \neq X_\beta] \Big) \\ &\leq 2 \sum_{\beta \neq \alpha} \mathbb{P}[Y_{\beta\alpha x} \neq X_\beta]. \end{aligned}$$

Proof Let $I_\alpha = I[X_\alpha \neq 0]$ and consider the point process $\Xi = \sum_\alpha I_\alpha \delta_{X_\alpha}$ on $\mathbb{R} \setminus \{0\}$. Then $W = \int x d\Xi(x)$, and thus

$$d_{TV}(\mathcal{L}(W), \mathrm{CP}(\boldsymbol{\lambda})) \leq d_{TV}(\mathcal{L}(\Xi), \mathrm{Po}(\boldsymbol{\lambda})).$$

Define $\Psi_{\alpha x} = \Xi$ and $\Phi_{\alpha x} = \sum_{\beta \neq \alpha} I[Y_{\beta\alpha x} \neq 0] \delta_{Y_{\beta\alpha x}}$. The assumptions of Theorem 10.E are fulfilled, and the result follows because

$$\begin{aligned} \rho_\alpha(x) &= \mathbb{E} d(\Psi_{\alpha x}, \Phi_{\alpha x}) \\ &\leq \mathbb{E} I[X_\alpha \neq 0] + \mathbb{E} \sum_{\beta \neq \alpha} \Big(I[X_\beta \notin \{0, Y_{\beta\alpha x}\}] + I[Y_{\beta\alpha x} \notin \{0, X_\beta\}] \Big) \\ &= \mathbb{P}[X_\alpha \neq 0] + \rho'_\alpha(x). \end{aligned}$$

□

A 'local' version follows as a corollary.

Corollary 10.L.1 *Let $W = \sum_{\alpha \in \Gamma} X_\alpha$ and $\boldsymbol{\lambda} = \sum_{\alpha \in \Gamma} F_\alpha$ be as above, and suppose that for each α there is a partition $\Gamma = \{\alpha\} \cup \Gamma^s_\alpha \cup \Gamma^i_a$, such that X_α and $(X_\beta, \beta \in \Gamma^i_\alpha)$ are independent. Then*

$$\begin{aligned} d_{TV}(\mathcal{L}(W), \mathrm{CP}(\boldsymbol{\lambda})) \leq \sum_{\alpha \in \Gamma} \Big(& \mathbb{P}[X_\alpha \neq 0]^2 \\ & + \sum_{\beta \in \Gamma^s_\alpha} (\mathbb{P}[X_\beta \neq 0 \neq X_\alpha] + \mathbb{P}[X_\beta \neq 0] \mathbb{P}[X_\alpha \neq 0]) \Big). \end{aligned}$$

Proof For each α, construct a coupling such that $Y_{\beta\alpha x} = X_\beta$ for all $\beta \in \Gamma^i_\alpha$ and all x. Then

$$\rho'_\alpha(x) \leq \sum_{\beta \in \Gamma^s_\alpha} \Big(\mathbb{P}[Y_{\beta\alpha x} \neq 0] + \mathbb{P}[X_\beta \neq 0] \Big) = \sum_{\beta \in \Gamma^s_\alpha} (\varphi_{\beta\alpha}(x) + \pi_\beta),$$

where

$$\varphi_{\beta\alpha}(x) = \mathbb{P}[Y_{\beta\alpha x} \neq 0] = \mathbb{P}[X_\beta \neq 0 \mid X_\alpha = x].$$

Hence

$$\begin{aligned}\mathbb{E}\{\varphi_{\beta\alpha}(X_\alpha) \cdot I[X_\alpha \neq 0]\} &= \mathbb{E}\Big(\mathbb{E}(I[X_\beta \neq 0] \mid X_\alpha) I[X_\alpha \neq 0]\Big) \\ &= \mathbb{P}[X_\beta \neq 0 \ \text{ and } \ X_\alpha \neq 0],\end{aligned}$$

and

$$\mathbb{E}\{\rho'_\alpha(X_\alpha) \cdot I[X_\alpha \neq 0]\} \leq \sum_{\beta \in \Gamma^s_\alpha} \Big(\mathbb{P}[X_\beta \neq 0 \neq X_\alpha] + \mathbb{P}[X_\beta \neq 0]\mathbb{P}[X_\alpha \neq 0]\Big).$$

□

If the variables X_α are discrete, we may obtain better bounds by using Theorem 10.K. For simplicity, we consider only the case $X_\alpha \in \mathbb{Z}^+$.

Theorem 10.M *Suppose that the assumptions of Theorem 10.L hold and that the X_α take their values in $\mathbb{Z}^+$. Then, if $\lambda_j = \boldsymbol{\lambda}\{j\} = \sum_{\alpha\in\Gamma} \mathbb{P}[X_\alpha = j]$,*

$$\begin{aligned}&d_{TV}(\mathcal{L}(W), \mathrm{CP}\,(\boldsymbol{\lambda})) \\ &\quad \leq 2(1 + e^{-1}\log^+ \max_j \lambda_j)\Bigg[\sum_{\alpha\in\Gamma}\Big(\sum_j (1 \wedge \lambda_j^{-\frac{1}{2}})\mathbb{P}[X_\alpha = j]\Big)^2 \\ &\quad + \sum_{\alpha\in\Gamma}\sum_{\beta\neq\alpha}\sum_j\sum_k (1 \wedge \lambda_j^{-\frac{1}{2}})(1 \wedge \lambda_k^{-\frac{1}{2}})\mathbb{P}[X_\alpha = j] \\ &\qquad \times \Big(\mathbb{P}[Y_{\beta\alpha j} \neq X_\beta = k] + \mathbb{P}[X_\beta \neq Y_{\beta\alpha j} = k]\Big)\Bigg]\end{aligned}$$

Proof Define $I_{\alpha j} = I[X_\alpha = j]$, $j = 1, 2, \ldots$, and $W_j = \sum_\alpha I_{\alpha j}$. Then $\mathbb{E}W_j = \lambda_j$. Furthermore, $W = \sum_{j\geq 1} jW_j$, and thus

$$d_{TV}(\mathcal{L}(W), \mathrm{CP}\,(\boldsymbol{\lambda})) \leq d_{TV}\Big(\mathcal{L}(\{W_j\}_{j\geq 1}), \prod_{j=1}^{\infty} \mathrm{Po}\,(\lambda_j)\Big).$$

Set $J_{\beta k\alpha j} = I(Y_{\beta\alpha j} = k)$. Theorem 10.K applies with $\Gamma_j = \Gamma \times \{j\}$, together with a truncation argument if $\boldsymbol{\lambda}$ assigns positive mass to infinitely many values j, and yields

$$
\begin{aligned}
& d_{TV}(\mathcal{L}(W), \mathrm{CP}(\boldsymbol{\lambda})) \\
& \quad \le 2(1 + e^{-1}\log^+ \max_j \lambda_j)\Big\{ \sum_j (1 \wedge \lambda_j^{-1}) \sum_{\alpha\in\Gamma} \mathbb{P}[X_\alpha = j]^2 \\
& \qquad + \sum_j \sum_k (1 \wedge \lambda_j^{-\frac{1}{2}})(1 \wedge \lambda_k^{-\frac{1}{2}}) \\
& \qquad \times \sum_{\alpha\in\Gamma} \Big[\sum_{\beta\ne\alpha} \mathbb{P}[X_\alpha = j]\mathbb{P}[X_\beta \ne Y_{\beta\alpha j} \text{ and } X_\beta \text{ or } Y_{\beta\alpha j} = k] \\
& \qquad + I[j \ne k]\mathbb{P}[X_\alpha = j]\mathbb{P}[X_\alpha \ne Y_{\alpha\alpha j} \text{ and } X_\alpha \text{ or } Y_{\alpha\alpha j} = k]\Big]\Big\},
\end{aligned}
$$

and the result follows because $Y_{\alpha\alpha j} = j$, so the last term may be combined with the first sum. □

Theorems 10.L and 10.M can sometimes be applied to sums of indicators after a suitable transformation, as in the following theorem.

Theorem 10.N *Let $(I_i)_{i=-\infty}^{\infty}$ be a stationary sequence of indicator variables and let $W = \sum_{i=1}^n I_i$. Assume that the I_i are m-dependent. Then*

$$
d_{TV}(\mathcal{L}(W), \mathrm{CP}(\boldsymbol{\lambda})) \le (11m + 1)n(\mathbb{E}I_0)^2 + 2\sum_{j=1}^m j\mathbb{E}I_0 I_j,
$$

where $\boldsymbol{\lambda} = \sum_{k=1}^{m+1} \lambda_k \delta_k$,

$$
\lambda_k = n\mathbb{P}\Big[I_{-m} = \cdots = I_{-1} = 0,\ I_0 = 1 \text{ and } \sum_{i=0}^m I_i = k\Big]
$$

Proof Define $X_i = I[I_i = 1,\ I_{i-m} = \cdots = I_{i-1} = 0]\sum_{j=i}^{i+m} I_j$ and $W^* = \sum_{i=1}^n X_i$. W^* differs from W principally because some of the I_j may not be counted: we use the fact that the I_j, $1 \le j \le m$ are always counted if $I_i = 0$, $-m+1 \le i \le 0$, and I_j, $m+1 \le j \le n$ is counted whenever $I_i = 0$, $j - 2m \le i \le j - m - 1$. It is, however, also possible for some of the I_j, $n+1 \le j \le n+m$ to be counted in W^* as well. This leads to the inequality

$$
\begin{aligned}
|W - W^*| \le & \sum_{j=1}^m I\big[I_j = 1 \text{ and } I_l = 1 \quad \text{for some} \quad l \in [-(m-1), 0]\big] \\
& + \sum_{j=m+1}^n I\big[I_j = 1 \text{ and } I_l = 1 \quad \text{for some} \quad l \in [j - 2m, j - m - 1]\big] \\
& + \sum_{j=n+1}^{n+m} I\big[I_j = 1 \text{ and } I_l = 1 \quad \text{for some} \quad l \in [j - m, n]\big].
\end{aligned}
$$

Thus, writing $\pi = \mathbb{E}I_0$ and $\pi_{0,j} = \mathbb{E}I_0 I_j$, it follows that

$$\begin{aligned} d_{TV}(\mathcal{L}(W),\mathcal{L}(W^*)) &\le \mathbb{E}|W - W^*| \\ &\le \sum_{j=1}^{m}\sum_{i=0}^{m-1} \pi_{0,i+j} + (n-m)\sum_{l=m+1}^{2m} \pi_{0,l} + \sum_{j=1}^{m}\sum_{i=0}^{m-j} \pi_{0,i+j} \\ &\le 2\sum_{j=1}^{m} j\pi_{0,j} + n\sum_{l=m+1}^{2m} \pi_{0,l} = 2\sum_{j=1}^{m} j\pi_{0,j} + mn\pi^2. \end{aligned} \tag{4.1}$$

Since X_i is independent of $\{X_j\}$, $|j-i| > 3m$, we can apply Corollary 10.L.1 and obtain

$$\begin{aligned} &d_{TV}(\mathcal{L}(W^*), \mathrm{CP}\,(\boldsymbol{\lambda})) \\ &\quad\le \sum_i \Big(\mathbb{P}[X_i \ne 0]^2 \\ &\qquad + \sum_{0<|j-i|\le 3m} \{\mathbb{P}[X_j \ne 0 \ne X_i] + \mathbb{P}[X_j \ne 0]\mathbb{P}[X_i \ne 0]\}\Big) \\ &\quad\le n(1+6m)\pi^2 + \sum_i \sum_{0<|j-i|\le 3m} \mathbb{P}[X_j \ne 0 \ne X_i]. \end{aligned} \tag{4.2}$$

Now the construction implies that $\mathbb{P}[X_j \ne 0 \ne X_i] = 0$ if $|j-i| \le m$, and $\mathbb{P}[X_j \ne 0 \ne X_i] \le \mathbb{P}[I_j = I_i = 1] = \pi^2$ if $|j-i| > m$. Hence the result follows by combining (4.1) and (4.2). □

Remark 10.4.1 Since $|\sum k\lambda_k - n\mathbb{E}I_0| = |\mathbb{E}(W^* - W)| \le \mathbb{E}|W^* - W|$, we may modify λ_1, for example, in such a way that $\sum k\lambda_k$ becomes equal to $\mathbb{E}W$, without increasing the error to more than $(12m+1)n(\mathbb{E}I_0)^2 + 4\sum_{j=1}^{m} j\mathbb{E}I_0 I_j$.

Corollary 10.N.1 *With notation as above,*

$$|\mathbb{P}[W=0] - \exp(-n\mathbb{P}[I_0 = 1,\ I_1 = \ldots = I_m = 0])| \le (11m+1)n(\mathbb{E}I_0)^2 + 2\sum_{1}^{m} j\mathbb{E}I_0 I_j.$$

Proof If $Z = \sum kZ_k \sim \mathrm{CP}\,(\boldsymbol{\lambda})$, with the Z_k independent $\mathrm{Po}(\lambda_k)$ random variables, then

$$\mathbb{P}[Z=0] = \prod_{k\ge 1} \mathbb{P}[Z_k = 0] = \exp\Big(-\sum_{k\ge 1} \lambda_k\Big).$$

Furthermore,

$$\begin{aligned} \sum_{k\ge 1} \lambda_k &= n\mathbb{P}[I_{-m} = \ldots = I_{-1} = 0,\ I_0 = 1] \\ &= n\mathbb{P}[I_0 = 1,\ I_1 = \ldots = I_m = 0], \end{aligned}$$

the latter equality following by stationarity. □

Consequently, if we define a measure of the tendency to clustering by $\delta = \mathbb{P}[I_0 = 1 \text{ and } \sum_{i=1}^{m} I_i \neq 0]/m\pi^2$, then $\mathbb{E}W - \sum_{k\geq 1}\lambda_k = n\delta m\pi^2$ and $\mathbb{E}I_0 I_j = \mathbb{E}I_j I_0 \leq \delta m\pi^2$ for $1 \leq j \leq m$, so that, from Corollary 10.N.1,

$$\begin{aligned} d_{TV}(\mathcal{L}(W),\text{Po}(\mathbb{E}W)) &\geq |\mathbb{P}[W=0] - e^{-\mathbf{E}W}| \\ &\geq \exp\{-\sum_{k\geq 1}\lambda_k\} - e^{-\mathbf{E}W} - (11m+1)n\pi^2 - 2\sum_{j=1}^{m} j\mathbb{E}I_0 I_j \\ &\geq \Big(\mathbb{E}W - \sum_{k\geq 1}\lambda_k\Big)e^{-\mathbf{E}W} - (11m+1)n\pi^2 - m(m+1)\delta m\pi^2 \\ &= (ne^{-\mathbf{E}W} - m(m+1))\delta m\pi^2 - (11m+1)n\pi^2. \end{aligned}$$

If m is fixed, $\mathbb{E}W$ stays bounded and $\delta \gg 1$, this shows that the bound $d_{TV}(\mathcal{L}(W), \text{Po}(\mathbb{E}W)) = O(\delta n m\pi^2)$ for Poisson approximation that is obtained, for example, from Corollary 2.C.5 is of the right order of magnitude, and that the compound Poisson approximation of Theorem 10.N, of order $(n + m^2\delta)m\pi^2$, is much more accurate.

Example 10.4.2 *The number of occurrences of a word with small period.* We suppose, as in Section 8.4, that $\{\xi_i,\ i \geq 0\}$ are independent and identically distributed letters from an alphabet S, and that W counts the occurrences of the word $w = s_1 \dots s_m$ in the sequence $\xi_1 \dots \xi_{n+m-1}$. Let $A = \{j \in \{1, 2, \dots, m-1\} :\ s_{j+k} = s_k,\ 1 \leq k \leq m - j\}$ denote the set of periods of w. The accuracy of Poisson approximation for $\mathcal{L}(W)$ is given in Theorem 8.F, where it is also observed that, if w has small periods, a compound Poisson approximation may be better. Here, we use Theorem 10.N to demonstrate this.

The random variable W is just $\sum_{i=1}^{n} I_i$, where the indicator random variables $I_i = I[\xi_{i+k-1} = s_k,\ i \leq k \leq m]$ are stationary and $(m-1)$-dependent. If p_s denotes the probability of the letter s and if $\pi = \mathbb{E}I_i = \prod_{k=1}^{m} p_{s_k}$, it follows that

$$\mathbb{E}(I_0 I_j) = \pi \prod_{k=1}^{j} p_{s_k}, \qquad j \in A,$$

and that $\mathbb{E}(I_0 I_j) = 0$ for $j \in \{1, 2, \dots, m-1\} \setminus A$. Thus, if the $(\lambda_k)_{k=1}^{m}$ are defined as in Theorem 10.N,

$$d_{TV}(\mathcal{L}(W), \text{CP}(\boldsymbol{\lambda})) \leq (11m+1)n\pi^2 + 2\pi\sum_{j\in A} j \prod_{k=1}^{j} p_{s_k}. \tag{4.3}$$

Comparison of this estimate with that of Theorem 8.F shows that a better rate is frequently obtained, as when, for example, $\lambda = n\pi = O(1)$, $m =$

$O(1)$ and $\sum_{j\in A}\prod_{k=1}^{j} p_{s_k}$ is the largest term in (8.4.1). In the extreme case of runs of the same letter, where $s_1 = \cdots = s_m$ and $A = \{1,\ldots,m-1\}$, the inequality (8.4.1) gives an estimate which exceeds $2p(1-p^{m-1})$ (with $p = p_{s_i}$) for the accuracy of the Poisson $\mathrm{Po}(\lambda)$ approximation, whereas (4.3) gives an upper bound of

$$(12\lambda m + 2m^2 p)p^m \tag{4.4}$$

for the accuracy of the $\mathrm{CP}(\boldsymbol{\lambda})$ approximation, which is almost always much better. Note that, in this case, $\lambda_k = n(1-p)^2 p^{m+k-1}$ when $k \le m-1$, $\lambda_m = n(1-p)p^{2m-1}$ and $\sum k\lambda_k = \lambda - n\pi^2$. A much more complicated argument is given below in Example 10.5.1, which shows that the factor λ can be improved upon.

10.5 Further developments

The results of Sections 10.3 and 10.4 are based either on Theorems 10.B and 10.E, in which case a large λ does not improve the estimates at all, or on Theorem 10.I, but then with a very concentrated distribution for each F_α, so that a large λ does not improve the estimates as much as might be hoped: instead of λ^{-1}, factors like $(\lambda_j\lambda_k)^{-1/2}$ appear, where $\sum_j \lambda_j = \lambda$. None the less, if the marks give little or no information as to which of the I_α are non-zero, it is plausible that the factor λ^{-1} could perhaps be re-introduced. A big step in achieving this aim would be to find a version of Theorem 10.I in which dependent Y_α's were allowed, or, equivalently, to frame useful conditions under which the quantity $|\mathbb{E}(\mathcal{A}h)(\Xi)|$ for a point process Ξ can be reduced to a form in which Lemma 10.2.10 can be applied. Although we have not yet managed to find a satisfactory general formulation, the following two examples show how it is sometimes possible in specific applications. We begin by further developing the example of Section 8.4.

Example 10.5.1 *The joint distribution of long runs.* Let $\xi_1,\ldots,\xi_n$ be independent $\mathrm{Be}(p)$ random variables, and let W_j denote the number of runs of 1's of exact length j. Assume, to avoid edge effects, that indices of the form $i + ln$ are identified with i whenever $1 \le i \le n$ and $l \in \mathbb{Z}$, and define

$$W_j = \sum_{i=1}^{n} I[\xi_i = 0,\ \xi_{i+1} = \cdots = \xi_{i+j} = 1,\ \xi_{i+j+1} = 0].$$

Then, arguing very much as for (8.4.6), it follows that, if $n > 2j+2$,

$$d_{TV}(\mathcal{L}(W_j), \mathrm{Po}(nq^2p^j)) \le \{(2j-1)q+2\}qp^j, \tag{5.1}$$

where $q = 1 - p$. Here, we prove a multivariate extension, that

$$d_{TV}(\mathcal{L}(\{W_j\}_{j=k}^m), \mathrm{Po}(\boldsymbol{\pi})) \qquad (5.2)$$
$$\le 4n\lambda p^{n-m-3} + 6e^{-1}(1 + 2\log^+(\lambda e))(m+1)p^k\sqrt{q(1-p^{m+1-k})},$$

where $\lambda = nqp^k(1 - p^{m+1-k})$ and $\boldsymbol{\pi}\{j\} = nq^2p^j$, $k \le j \le m$. Note that a straightforward application of Corollary 10.J.1 would lead, for $n > 2m+2$, to an estimate of the form

$$d_{TV}(\mathcal{L}(\{W_j\}_{j=k}^m), \mathrm{Po}(\boldsymbol{\pi})) \le \lambda((2k-3)q + 4)p^k,$$

which has advantages if λ is small and m is large, but that we are now primarily concerned with finding ways of exploiting Stein's method when λ is large.

To do this, let $\Xi = \sum_{i=1}^n I_i\delta_{Y_i}$, where $I_i = 1$ whenever $\xi_i = 0$ and

$$k \le Y_i = \min\{j \ge 1 : \xi_{i+j} = 0\} - 1 \le m,$$

so that $\Xi\{j\} = W_j$, $k \le j \le m$, and we wish to estimate $d_{TV}(\mathcal{L}(\Xi), \mathrm{Po}(\boldsymbol{\pi}))$. As before, this is equivalent to bounding $|\mathbb{E}(\mathcal{A}h)(\Xi)|$ when $\mathcal{A}$ is given by (2.15) and h by (1.5) with $\boldsymbol{\lambda} = \boldsymbol{\pi}$ for $f : \mathbb{Z} \to [0,1]$. Arguing as for Theorem 10.I, we can write

$$|\mathbb{E}(\mathcal{A}h)(\Xi)| \qquad (5.3)$$
$$= \left|\sum_{i=1}^n n^{-1}\lambda\sum_{j=k}^m f_j\mathbb{E}\{[h(\Psi_{ij}+\delta_j) - h(\Psi_{ij})] - [h(\Phi_{ij}+\delta_j) - h(\Phi_{ij})]\}\right|,$$

for suitably chosen Ψ_{ij} and Φ_{ij} satisfying $\mathcal{L}(\Psi_{ij}) = \mathcal{L}(\Xi)$ and $\mathcal{L}(\Phi_{ij}+\delta_j) = \mathcal{L}(\Xi \mid I_i = 1, Y_i = j)$, where also

$$f_j = \mathbb{P}[Y_i = j \mid I_i = 1] = qp^{j-k}(1 - p^{m+1-k})^{-1}, \quad k \le j \le m.$$

However, in contrast to the situation in Theorem 10.I, it is no longer the case that $\Phi_{ij} = \Phi_{il}$ if $j \ne l$, and the exploitation of (5.3) becomes more complicated. The idea is to find a further random measure Θ, which is the same for all j and is none the less close to both Φ_{ij} and Ψ_{ij} for each j, and to use a decomposition similar to that used for Theorem 10.I for both

$$[h(\Psi_{ij}+\delta_j) - h(\Psi_{ij})] - [h(\Theta+\delta_j) - h(\Theta)]$$

and

$$[h(\Phi_{ij}+\delta_j) - h(\Phi_{ij})] - [h(\Theta+\delta_j) - h(\Theta)].$$

The details are as follows.

Because of symmetry, we need only consider $i = n$. Let $\Psi_{nj} = \Xi$ be constructed from a realization $\xi_1, \dots, \xi_n$, and $\Phi_{nj} + \delta_j$ from $\xi'_1, \dots, \xi'_n$, where

$$\xi'_n = 0, \quad \xi'_l = 1, 1 \leq l \leq j, \quad \xi'_{j+1} = 0 \quad \text{and} \quad \xi'_l = \xi_l \quad \text{otherwise.}$$

Define also $\mathcal{G}$ to be $\sigma(\xi_i, m+2 \leq i \leq n-1)$, $R = n-1-\max\{i < n : \xi_i = 0\}$, $S = \min\{i > m+1 : \xi_i = 0\} - m - 2$ and Θ to be the configuration generated by $\xi_{m+2+S}, \dots, \xi_{n-1-R}$: thus $\Theta = \sum_{i=m+2}^{n-2-R} I_i \delta_{Y_i}$. Note that $R \geq 0$ counts the consecutive 1's immediately preceding index n and $S \geq 0$ those following index $m+1$. Note also that Θ can be modified to reach Φ_{nj} and Ψ_{nj} by only adding points, whatever the value of $j \leq m$, and that, if $m\pi$ is small, Θ almost always coincides with both Φ_{nj} and Ψ_{nj}.

Introducing the notation

$$\Delta_2 h(\xi, j, l) = h(\xi + \delta_j + \delta_l) - h(\xi + \delta_j) - h(\xi + \delta_l) + h(\xi),$$

we can write

$$\begin{aligned}
&\sum_{j=k}^{m} f_j \mathbb{E}\{[h(\Psi_{nj} + \delta_j) - h(\Psi_{nj})] - [h(\Theta + \delta_j) - h(\Theta)]\} \\
&\quad = \sum_{r,s\geq 0} \mathbb{P}[R = r, S = s] \sum_{j=k}^{m} f_j \mathbb{E}\Big\{ I_{n-r-1} \Delta_2 h(\Theta, j, Y_{n-r-1}) \\
&\qquad + \bar{I}_{m+2+s} \Delta_2 h(\Theta + I_{n-r-1}\delta_{Y_{n-r-1}}, j, \bar{Y}_{m+2+s}) \\
&\qquad + \sum_{i=0}^{m+s-U} I_i \Delta_2 h(\Theta^i, j, Y_i) \,\Big|\, R = r, S = s \Big\},
\end{aligned} \tag{5.4}$$

where Θ^i denotes the configuration generated by $\xi_{m+1+S-U}, \dots, \xi_{n+i-1}$ and 0 in place i,

$$U = \bar{Y}_{m+2+S}, \qquad \bar{Y}_i = \min\{j \geq 1 : \xi_{i-j} = 0\} - 1$$

and $\bar{I}_i = I[\xi_i = 0, k \leq \bar{Y}_i \leq m]$. Evaluating the first term in (5.4) for $0 \leq r \leq k$ yields

$$\begin{aligned}
&\sum_{j=k}^{m} f_j \mathbb{E}\{I_{n-r-1} \Delta_2 h(\Theta, j, Y_{n-r-1}) \mid \mathcal{G}, R = r, S = s\} \\
&\quad = \sum_{j=k}^{m} f_j \sum_{t=k}^{m} p^{t-r} q \Delta_2 h(\Theta, j, t) \\
&\quad = p^{k-r}(1 - p^{m+1-k}) \sum_{j=k}^{m} f_j \sum_{t=k}^{m} f_t \Delta_2 h(\Theta, j, t),
\end{aligned} \tag{5.5}$$

and so, by Lemma 10.2.10,

$$\Big|\sum_{j=k}^{m} f_j \mathbb{E}\{I_{n-r-1}\Delta_2 h(\Theta, j, Y_{n-r-1}) \mid R=r, S=s\}\Big|$$
$$\le \frac{1}{e\lambda}\{1+2\log^+(e\lambda)\}p^{k-r}(1-p^{m+1-k}). \quad (5.6)$$

For $k+1 \le r \le m$, (5.5) is replaced by

$$\sum_{j=k}^{m} f_j \sum_{t=r}^{m} p^{t-r} q \Delta_2 h(\Theta, j, t)$$
$$= (1-p^{m+1-r}) \sum_{j=k}^{m} f_j \sum_{t=r}^{m} \frac{p^{t-r}q}{1-p^{m+1-r}} \Delta_2 h(\Theta, j, t),$$

and Lemma 10.2.10 yields

$$\Big|\sum_{j=k}^{m} f_j \mathbb{E}\{I_{n-r-1}\Delta_2 h(\Theta, j, Y_{n-r-1}) \mid R=r, S=s\}\Big|$$
$$\le \frac{1}{e\lambda}\{1+2\log^+(e\lambda)\}(1-p^{m+1-r})\sqrt{\frac{1-p^{m+1-k}}{p^{r-k}(1-p^{m+1-r})}}. \quad (5.7)$$

For $r > m$, $I_{n-r-1} = 0$.

Turning to the second term, for $0 \le s \le k$ and $t-r \le s$, we get

$$\Big|\sum_{j=k}^{m} f_j \mathbb{E}\{\bar{I}_{m+2+s}\Delta_2 h(\Theta+\delta_t, j, \bar{Y}_{m+2+s}) \mid \mathcal{G}, R=r, S=s, Y_{n-r-1}=t\}\Big|$$
$$= \Big|\sum_{j=k}^{m} f_j \sum_{u=k}^{m} p^{u-s} q \Delta_2 h(\Theta+\delta_t, j, u)\Big|$$
$$\le \frac{1}{e\lambda}\{1+2\log^+(e\lambda)\}p^{k-s}(1-p^{m+1-k}) \quad (5.8)$$

as before (δ_t should be ignored when $t \notin [k, m]$); for $0 \le s \le k$ and $t-r > s$, the expression is replaced by zero if $m+s-t+r+1 < k$, and otherwise by

$$\Big|\sum_{j=k}^{m} f_j \Big\{ \sum_{u=k}^{m+s-t+r} p^{u-s} q \Delta_2 h(\Theta+\delta_t, j, u)$$
$$+ p^{m-t+r+1}\Delta_2 h(\Theta+\delta_t, j, m+s-t+r+1)\Big\}\Big|$$
$$\le \frac{1}{e\lambda}\{1+2\log^+(e\lambda)\}p^{k-s}\sqrt{\frac{1-p^{m+1-k}}{1-p}}, \quad (5.9)$$

which is larger than the estimate in (5.8), and can therefore be used throughout $0 \leq s \leq k$; for $k+1 \leq s \leq m$ and $t-r \leq s$, we obtain

$$\Big|\sum_{j=k}^{m} f_j \sum_{u=s}^{m} p^{u-s} q \Delta_2 h(\Theta + \delta_t, j, u)\Big|$$
$$\leq \frac{1}{e\lambda}\{1 + 2\log^+(e\lambda)\}(1 - p^{m+1-s})\sqrt{\frac{1-p^{m+1-k}}{p^{s-k}(1-p^{m+1-s})}}, \quad (5.10)$$

and for $k+1 \leq s \leq m$ and $t-r > s$ we have the larger estimate

$$\Big|\sum_{j=k}^{m} f_j \Big\{ \sum_{u=s}^{m+s-t+r} p^{u-s} q \Delta_2 h(\Theta + \delta_t, j, u)$$
$$+ p^{m-t+r+1} \Delta_2 h(\Theta + \delta_t, j, m+s-t+r+1)\Big\}\Big|$$
$$\leq \frac{1}{e\lambda}\{1 + 2\log^+(e\lambda)\}\sqrt{\frac{1-p^{m+1-k}}{qp^{s-k}}}. \quad (5.11)$$

For $s > m$, the second term vanishes. Finally, for the last term, defining

$$\mathcal{G}^i = \sigma(\xi_{m+1+S-U}, \ldots, \xi_{n+i-1}),$$

and with $v = m + s - u - i$,

$$\Big|\sum_{j=k}^{m} f_j \mathbb{E}\Big\{ \sum_{i=0}^{m+s-u} I_i \Delta_2 h(\Theta^i, j, Y_i) \mid \mathcal{G}^i, R = r, S = s, \bar{Y}_{m+2+s} = u \Big\}\Big|$$
$$\leq q \sum_{i=0}^{m+s-u-k} \Big|\sum_{j=k}^{m} f_j \Big\{\sum_{l=k}^{v-1} qp^l \Delta_2 h(\Theta^i, j, l) + p^v \Delta_2 h(\Theta^i, j, v)\Big\}\Big|$$
$$\leq \frac{1}{e\lambda}\{1 + 2\log^+(e\lambda)\}(m + 1 - k)p^k\sqrt{q(1 - p^{m+1-k})}. \quad (5.12)$$

To complete the estimation of (5.4), we need the joint distribution of R and S. It is enough to observe that

$$\mathbb{P}[R = r, S = s] = p^{r+s}q^2, \quad r + s + 2 \leq n - m - 2, \quad (5.13)$$

and otherwise to use the inequality

$$\mathbb{P}[R + S + 2 > n - m - 2] \leq (n - m - 2)p^{n-m-3}$$

in conjunction with Lemma 10.1.3(i) to complete the estimate. Combining (5.13) with (5.4) and estimates (5.6), (5.7), (5.9), (5.11) and (5.12) gives a rough upper bound for

$$\Big|\sum_{j=k}^{m} f_j \mathbb{E}\{[h(\Psi_{nj} + \delta_j) - h(\Psi_{nj})] - [h(\Theta + \delta_j) - h(\Theta)]\}\Big|$$

of at most

$$2(n-m-2)p^{n-m-3} + \frac{3}{e\lambda}(1 + 2\log^+(e\lambda))(m+1)p^k\sqrt{q(1-p^{m+1-k})}.$$

A similar argument shows that the same estimate also holds for

$$\left|\sum_{j=k}^{m} f_j \mathbb{E}\{[h(\Phi_{nj}+\delta_j) - h(\Phi_{nj})] - [h(\Theta+\delta_j) - h(\Theta)]\}\right|,$$

and hence 2λ times the estimate bounds $|\mathbb{E}(\mathcal{A}h)(\Xi)|$, giving the required result.

Example 10.5.2 *Short cycles in random permutations.* Let σ be uniformly distributed over the permutations of the elements $1 \le i \le n$. In each cycle of length j in σ, $1 \le j \le f$, choose an element α uniformly at random, and assign to it the value $Y_\alpha = j$: for all other elements β, set $Y_\beta = 0$. Then $W_j = \sum_{i=1}^n I[Y_i = j]$ counts the cycles of length j in σ, $1 \le j \le f$. Direct calculation shows that $\mathbb{P}[Y_i = j] = 1/nj$, and so we define $\pi\{j\} = \mathbb{E}W_j = 1/j$ and $\lambda = \psi(f)$, where $\psi(f) = \sum_{r=1}^f 1/r$, and try approximating $\mathcal{L}(\{W_j\}_{j=1}^f)$ by $\mathrm{Po}(\pi)$, using Lemma 10.2.10. Here, it surprisingly turns out that Lemma 10.2.9 produces as good a result as Lemma 10.2.10, and that neither gives the magic factor λ^{-1}, even though the true order of approximation in fact contains such a factor. Note that a straightforward application of Theorem 2.C gives a bound of order n^{-1} for any fixed W_j, as, for instance, in the case $j = 1$, when the ménage problem is recovered, and that the true order of approximation is very much better: see Arratia and Tavaré (1990).

Here, we use Stein's method to obtain the estimate

$$d_{TV}\left(\mathcal{L}(\{W_j\}_{j=1}^f), \mathrm{Po}(\pi)\right) \le 4\sqrt{\frac{2}{e}}\frac{f\sqrt{\psi(f)}}{n} + 2\left(\frac{f\sqrt{\psi(f)}}{n-f}\right)^2. \quad (5.14)$$

The essential step is to bound

$$\sum_{j=1}^{f} \frac{1}{nj}\mathbb{E}\{[h(\Psi_{1j}+\delta_j) - h(\Psi_{1j})] - [h(\Phi_{1j}+\delta_j) - h(\Phi_{1j})]\},$$

where $\mathcal{L}(\Psi_{1j}) = \mathcal{L}(\Xi)$, $\mathcal{L}(\Phi_{1j}+\delta_j) = \mathcal{L}(\Xi \mid Y_1 = j)$ and the process Ξ is given by $\sum_{i=1}^n I[Y_i \ge 1]\delta_{Y_i}$. To construct Ψ_{1j} and Φ_{1j} close to each other, first realize independent random variables $\xi_1, \ldots, \xi_n$ such that $\xi_i \sim \mathrm{Be}(1/i)$. Let $T_0 = n+1$, and let $T_1, \ldots, T_J$ be the indices i such that $\xi_i = 1$, arranged in decreasing order, and set

$$\Psi_{1j} = \sum_{m=1}^{J} \delta_{T_{m-1}-T_m}.$$

To check that $\mathcal{L}(\Psi_{1j}) = \mathcal{L}(\Xi)$, note that, by direct calculation, $T_0 - T_1$ is uniformly distributed on $\{1, 2, \ldots, n\}$, as is the length of the cycle containing 1, and that, given this value, the subsequent values in both cases are determined as if starting afresh with n replaced by $n - (T_0 - T_1)$. In a similar way, construct Φ_{1j} starting from the variables

$$\xi'_i = \xi_i,\ 1 \le i \le n-j; \qquad \xi'_{n-j+1} = 1; \qquad \xi'_i = 0,\ n-j+2 \le i \le n,$$

and an auxiliary random measure Θ starting from the variables

$$\xi''_i = \xi_i,\ 1 \le i \le n-f; \qquad \xi''_i = 0,\ n-f+1 \le i \le n,$$

in both cases omitting the term $\delta_{T_0-T_1}$.

As in Example 10.5.1, start by estimating

$$\sum_{j=1}^{f} \frac{1}{nj} \mathbb{E}\{[h(\Psi_{1j} + \delta_j) - h(\Psi_{1j})] - [h(\Theta + \delta_j) - h(\Theta)]\}, \tag{5.15}$$

which can be rewritten as

$$\sum_{j=1}^{f} \frac{1}{nj} \mathbb{E}\Big\{\Delta_2 h(\Theta, j, T_0 - T_1) I[T_0 - T_1 \le f] \\ + \sum_{r=1}^{f} \xi_{n-r+1} \Delta_2 h(\Theta_1^r, j, N(r)) I[N(r) \le f]\Big\},$$

where $N(r) = \min\{l \ge 1 : \xi_{n-r+1-l} = 1\}$ and $\Theta_1^r = \Theta + \tilde{\Theta}_1^r$, with $\tilde{\Theta}_1^r$ being the random measure constructed just from $\xi_n, \xi_{n-1}, \ldots, \xi_{n-r+1}$. Now Lemma 10.2.10 would give

$$\Big|\sum_{j=1}^{f} \frac{1}{nj} \mathbb{E}\{\Delta_2 h(\Theta, j, T_0 - T_1) I[T_0 - T_1 \le f]\}\Big| \\ = \Big|\frac{f}{n^2} \psi(f) \sum_{j=1}^{f} \frac{1}{j\psi(f)} \sum_{l=1}^{f} \frac{1}{f} \Delta_2 h(\Theta, j, l)\Big| \\ \le \frac{f\psi(f)\sqrt{\psi(f)}}{n^2} \cdot \frac{1}{e\psi(f)} \{1 + 2\log^+(e\psi(f))\}.$$

Here, the fact that the conditional distribution of $T_0 - T_1$, given that $T_0 - T_1 \le f$, is not the same as $\hat{\pi}$ introduces an unwelcome factor $\sqrt{\psi(f)}$ into the numerator, and, to achieve the same (indeed, slightly better) accuracy, it is simpler to use Lemma 10.2.9 twice to give

$$\Big|\sum_{j=1}^{f} \frac{1}{nj} \mathbb{E}\{\Delta_2 h(\Theta, j, T_0 - T_1) I[T_0 - T_1 \le f]\}\Big| \le \frac{2}{n^2} \sqrt{\frac{2}{e}} f \sqrt{\psi(f)}.$$

For the remainder, from Lemma 10.1.3(ii),

$$\Big|\sum_{r=1}^{f}\sum_{j=1}^{f}\frac{1}{nj}\mathbb{E}\Big\{\xi_{n-r+1}\Delta_2 h(\Theta_1^r, j, N(r))I[N(r)\le f]\Big\}\Big|$$

$$\le \sum_{r=1}^{f}\sum_{j=1}^{f}\frac{1}{nj}\mathbb{E}(\xi_{n-r+1}I[N(r)\le f])$$

$$= \frac{\psi(f)}{n}\sum_{r=1}^{f}\frac{f}{(n-r+1)(n-r)} \le \frac{1}{n}\left(\frac{f\sqrt{\psi(f)}}{n-f}\right)^2.$$

Combining the estimates thus gives

$$n\Big|\sum_{j=1}^{f}\frac{1}{nj}\mathbb{E}\{[h(\Psi_{1j}+\delta_j)-h(\Psi_{1j})]-[h(\Theta+\delta_j)-h(\Theta)]\}\Big|$$

$$\le 2\sqrt{\frac{2}{e}}\left\{\frac{f\sqrt{\psi(f)}}{n}\right\}+\left(\frac{f\sqrt{\psi(f)}}{n-f}\right)^2.$$

The estimation of

$$\sum_{j=1}^{f}\frac{1}{nj}\mathbb{E}\{[h(\Phi_{1j}+\delta_j)-h(\Phi_{1j})]-[h(\Theta+\delta_j)-h(\Theta)]\}$$

$$= \sum_{j=1}^{f}\frac{1}{nj}\mathbb{E}\Big\{\Delta_2 h(\Theta, j, N(j))I[N(j)\le f]$$

$$+\sum_{r=j+1}^{f}\xi_{n-r+1}\Delta_2 h(\Theta_2^r, j, N(r))I[N(r)\le f]\Big\},$$

where $\Theta_2^r = \Theta + \tilde{\Theta}_2^r$ and $\tilde{\Theta}_2^r$ is constructed from $\xi_{n-j}, \xi_{n-j-1}, \ldots, \xi_{n-r+1}$, is accomplished in similar fashion, and yields the same contribution to the error estimate, so that (5.14) follows.

Note that Estimate (5.14) is of order $\lambda^{\frac{1}{2}} f/n$, whereas, if the magic factor λ^{-1} of Lemma 10.2.10 could have been more cleverly exploited, a bound of order f/n (up to a factor $\log\psi(f)$) might have been attained, as claimed in Barbour's contribution to the discussion of Arratia, Goldstein and Gordon (1990). As it happens, there is a simple argument, outlined there, which combines (5.14) and a coupling based on the construction of Ψ_{1j} to establish an error estimate of $2f/n$ for this multivariate Poisson approximation. The same argument also applies to the non-uniform model with density $p(\sigma) = \theta^{c(\sigma)}\{\theta(1+\theta)\ldots(n-1+\theta)\}^{-1}$, where $c(\sigma)$ is the number of cycles in σ and $\theta > 0$; now

$$\xi_i \sim \mathrm{Be}\left(\frac{\theta}{i-1+\theta}\right);\quad \mathbb{E}W_j = \frac{\theta n(n-1)\ldots(n-j+1)}{j(n-1+\theta)(n-2+\theta)\ldots(n-j+\theta)},$$

and the corresponding bound is given by

$$\frac{\theta(1+\theta)f}{n-1+\theta}, \quad \theta \geq 1; \qquad \theta f\left\{\frac{1}{n-f-1}+\frac{\theta}{n-1+\theta}\right\}, \quad \theta < 1,$$

see Arratia, Barbour and Tavaré (1991). The example serves particularly well to illustrate that Poisson process approximation by Stein's method has yet to reach its final form, and that this chapter provides no more than a guide to what may in future be possible.

Appendix

A.1 Probability metrics

There are many ways to define a distance between probability measures. In this book, we are mainly concerned with the total variation distance, but we occasionally also consider the Wasserstein and Hellinger distances: we refer to Zolotarev (1983) and to Reiss (1989) for further possible choices of distance. Our definitions are framed in terms of a general state space $\mathcal{X}$, although we are mostly concerned with distributions over $\mathbb{Z}^+$.

Total variation distance

Let $(\mathcal{X}, \mathcal{A})$ be any measurable space. The *total variation distance* d_{TV} between two probability measures μ_1 and μ_2 on $\mathcal{X}$ is defined to be

$$d_{TV}(\mu_1, \mu_2) = \sup_{A \in \mathcal{A}} |\mu_1(A) - \mu_2(A)|. \tag{1.1}$$

Some authors choose a different normalization, and define the total variation distance to be twice that defined in (1.1). The definition may be rewritten in several equivalent forms

$$\begin{aligned} d_{TV}(\mu_1, \mu_2) &= \sup_{0 \le f \le 1} \left| \int f \, d\mu_1 - \int f \, d\mu_2 \right| \\ &= \tfrac{1}{2} \sup_{|f| \le 1} \left| \int f \, d\mu_1 - \int f \, d\mu_2 \right| \\ &= \sup_{f \in \mathcal{F}_{TV}} \left| \int f \, d\mu_1 - \int f \, d\mu_2 \right| \\ &= \tfrac{1}{2} \|\mu_1 - \mu_2\|, \end{aligned} \tag{1.2}$$

where $\mathcal{F}_{TV} = \{I_A, \, A \in \mathcal{A}\}$ and where $\|\cdot\|$ is the standard norm on signed measures. Furthermore, if μ_1 and μ_2 are both absolutely continuous with respect to some measure λ, and have densities f_1 and f_2 respectively, then

$$\begin{aligned} d_{TV}(\mu_1, \mu_2) &= \tfrac{1}{2} \int |f_1 - f_2| \, d\lambda \\ &= 1 - \int \min(f_1, f_2) \, d\lambda; \end{aligned} \tag{1.3}$$

note that such a λ always exists, since λ may for example be taken to be $\mu_1 + \mu_2$. It is easily seen that d_{TV} is a complete metric on the space of all probability measures on $\mathcal{X}$, that $0 \leq d_{TV} \leq 1$, and that $d_{TV}(\mu_1, \mu_2) = 1$ if and only if μ_1 and μ_2 are mutually singular.

If the state space $\mathcal{X}$ is a discrete space, for example $\mathbb{Z}^+$, a sequence (X_n) of random variables taking values in $\mathcal{X}$ converges in distribution to μ if and only if $d_{TV}(\mathcal{L}(X_n), \mu) \to 0$. On the other hand, if $\mathcal{X}$ is not discrete, $\mathbb{R}$ for example, convergence in total variation is stronger than convergence in distribution, and is then at times too strong to be useful.

If $\mathcal{X}$ is a separable metric space, and X_1 and X_2 are any $\mathcal{X}$-valued random variables defined on the same probability space with distributions given by μ_1 and μ_2, then it is clear that $\mathbb{P}[X_1 \neq X_2] \geq d_{TV}(\mu_1, \mu_2)$. On the other hand, given μ_1 and μ_2, it is possible to construct X_1 and X_2 so that equality holds, a *maximal coupling* of μ_1 and μ_2. To see this, suppose that $0 < \delta = d_{TV}(\mu_1, \mu_2) < 1$, and let X, Y_1 and Y_2 be $\mathcal{X}$-valued random variables on a common probability space, with distributions $(1-\delta)^{-1}(\mu_1 \wedge \mu_2)$, $\delta^{-1}(\mu_1 - \mu_2)^+$ and $\delta^{-1}(\mu_2 - \mu_1)^+$ respectively: suppose also that $I \sim \mathrm{Be}(1-\delta)$ is independent of (X, Y_1, Y_2). Then the random variables X_1, X_2 defined by

$$(X_1(\omega), X_2(\omega)) = \begin{cases} (X(\omega), X(\omega)) & \text{if } I(\omega) = 1; \\ \\ (Y_1(\omega), Y_2(\omega)) & \text{if } I(\omega) = 0, \end{cases}$$

have the required properties. Consequently,

$$d_{TV}(\mu_1, \mu_2) = \min \mathbb{P}[X_1 \neq X_2], \tag{1.4}$$

where the minimum ranges over all couplings (X_1, X_2) of μ_1 and μ_2.

Wasserstein distance

Suppose that the state space is a separable metric space $(\mathcal{X}, d)$ equipped with its Borel σ-field, and consider only probability measures μ on $\mathcal{X}$ such that $\int d(x, x_0)\, d\mu(x) < \infty$ for some, and then for every, $x_0 \in \mathcal{X}$. If, for example, $\mathcal{X} = \mathbb{Z}^+$ or $\mathbb{R}$ and $d = |\cdot|$ as usual, the latter condition says that the expectation of μ is finite. The *Wasserstein distance* d_W (also known as the Dudley, Fortet–Mourier or Kantorovich distance) is then defined by

$$d_W(\mu_1, \mu_2) = \sup_{f \in \mathcal{F}_W} \left| \int f\, d\mu_1 - \int f\, d\mu_2 \right|, \tag{1.5}$$

where

$$\mathcal{F}_W = \{f : \mathcal{X} \to \mathbb{R},\ |f(x) - f(y)| \leq d(x, y)\}, \tag{1.6}$$

the set of Lipschitz functions. Note that when $\mathcal{X} = \mathbb{Z}^+$, $\mathcal{F}_W$ can be expressed as $\{f : \Delta f \leq 1\}$, the form used in the definition of Chapter 1.

It can be shown that

$$d_W(\mu_1, \mu_2) = \inf \mathbb{E}d(X_1, X_2), \tag{1.7}$$

where the infimum is taken over all couplings (X_1, X_2) of μ_1 and μ_2, and that the infimum is attained if $\mathcal{X}$ is complete, for example if $\mathcal{X}$ is $\mathbb{Z}^+$ or $\mathbb{R}$: see Rachev (1984) Section 2.2. In particular, if $\mathcal{X}$ is $\mathbb{Z}^+$ or $\mathbb{R}$,

$$d_W(\mu_1, \mu_2) = \min \mathbb{E}|X_1 - X_2|, \tag{1.8}$$

the minimum being taken over all couplings of μ_1 and μ_2.

For random variables on $\mathbb{Z}^+$ or $\mathbb{R}$, we have the formulae

$$d_W(\mathcal{L}(X), \mathcal{L}(Y)) = \sum_{j \geq 0} |\mathbb{P}[X \leq j] - \mathbb{P}[Y \leq j]| \tag{1.9}$$

and

$$d_W(\mathcal{L}(X), \mathcal{L}(Y)) = \int_{-\infty}^{\infty} |\mathbb{P}[X \leq x] - \mathbb{P}[Y \leq x]| \, dx \tag{1.10}$$

respectively. In these cases, $d_W(\mathcal{L}(X_n), \mu) \to 0$ if and only if the random variables X_n converge in distribution to μ and are uniformly integrable. On $\mathbb{Z}^+$, $d_W \geq d_{TV}$, but on $\mathbb{R}$ the two distances are not comparable.

Note that the Wasserstein distance, unlike the total variation and Hellinger distances, is not preserved under changes of scale. In fact, if X and Y are real valued random variables and $c > 0$,

$$d_W(\mathcal{L}(cX), \mathcal{L}(cY)) = c\, d_W(\mathcal{L}(X), \mathcal{L}(Y)). \tag{1.11}$$

Hellinger distance

The *Hellinger distance* d_H, like the total variation distance, is defined for any pair of probability measures on any measurable space. If μ_1 and μ_2 are both absolutely continuous with respect to a third measure λ and have densities f_1 and f_2 respectively, define

$$d_H(\mu_1, \mu_2) = \left(\tfrac{1}{2} \int (\sqrt{f_1} - \sqrt{f_2})^2 \, d\lambda \right)^{1/2}. \tag{1.12}$$

As observed earlier, such a λ always exists, and the integral in (1.12) does not in fact depend on the particular choice of λ. Some authors prefer to take the Hellinger distance to be $\sqrt{2}\, d_H$. The definition (1.12) can be rewritten as

$$d_H^2(\mu_1, \mu_2) = 1 - \int \sqrt{f_1 f_2} \, d\lambda. \tag{1.13}$$

The metric d_H on the set of all probability measures on $\mathcal{X}$ is complete, $0 \le d_H \le 1$ and $d_H(\mu_1, \mu_2) = 1$ if and only if μ_1 and μ_2 are mutually singular. Furthermore, it is easily seen that

$$d_H^2(\mu_1, \mu_2) \le d_{TV}(\mu_1, \mu_2) \le \sqrt{2} d_H(\mu_1, \mu_2). \tag{1.14}$$

Consequently, the Hellinger distance goes to zero if and only if the total variation distance does, although the rates may differ by a power. In particular, convergence in the Hellinger distance is equivalent to convergence in distribution on a discrete space such as $\mathbb{Z}^+$, but not, for example, on $\mathbb{R}$.

One advantage of the Hellinger distance is that, when considering product measures, it gives better bounds in terms of the marginals than the other two distances.

Proposition A.1.1 *For $i = 1, \ldots, n$, let μ_i and ν_i be probability measures on $\mathcal{X}_i$, and consider the product measures $\prod_1^n \mu_i$ and $\prod_1^n \nu_i$ on $\prod_1^n \mathcal{X}_i$. In the Wasserstein case, take the metric $d(\{x_i\}, \{y_i\}) = \sum d_i(x_i, y_i)$ on $\prod_1^n \mathcal{X}_i$, and assume that the $d_W(\mu_i, \nu_i)$ are defined. Then*

$$d_{TV}\left(\prod_1^n \mu_i, \prod_1^n \nu_i\right) \le \sum_{i=1}^n d_{TV}(\mu_i, \nu_i); \tag{1.15}$$

$$d_W\left(\prod_1^n \mu_i, \prod_1^n \nu_i\right) = \sum_{i=1}^n d_W(\mu_i, \nu_i); \tag{1.16}$$

$$d_H\left(\prod_1^n \mu_i, \prod_1^n \nu_i\right) \le \left\{\sum_{i=1}^n d_H^2(\mu_i, \nu_i)\right\}^{1/2}, \tag{1.17}$$

in fact,

$$d_H^2\left(\prod_1^n \mu_i, \prod_1^n \nu_i\right) = 1 - \prod_1^n \{1 - d_H^2(\mu_i, \nu_i)\}. \tag{1.18}$$

□

In particular, if $\{X_i\}$ and $\{Y_i\}$ are two sequences of independent and identically distributed random variables, then $d_{TV}(\mathcal{L}(\{X_i\}_{i=1}^n), \mathcal{L}(\{Y_i\}_{i=1}^n))$ and $d_W(\mathcal{L}(\{X_i\}_{i=1}^n), \mathcal{L}(\{Y_i\}_{i=1}^n))$ can in general only be estimated by n times the respective distances between the marginals, whereas the factor n can be replaced by $\sqrt{n}$ when estimating $d_H(\mathcal{L}(\{X_i\}_{i=1}^n), \mathcal{L}(\{Y_i\}_{i=1}^n))$: these estimates cannot in general be improved.

Recall that if $\boldsymbol{\lambda}$ is any finite measure on a space $\mathcal{X}$, a Poisson process with intensity $\boldsymbol{\lambda}$ is a random element of the set $\tilde{\mathcal{X}}$ of integer valued discrete measures on $\mathcal{X}$, and that the Poisson process may be constructed as $\sum_1^N \delta_{X_i}$, where $N \sim \mathrm{Po}(\lambda)$, $X_1, X_2, \ldots$ are independent and each X_i has the distribution $\lambda^{-1}\boldsymbol{\lambda}$ on $\mathcal{X}$, where λ denotes $\boldsymbol{\lambda}(\mathcal{X})$. Define the Hellinger distance by (1.12) for any two finite measures, even if their total masses are not one.

Proposition A.1.2 *Let $\boldsymbol{\lambda}$ and $\boldsymbol{\mu}$ be two finite measures on a measurable space $\mathcal{X}$. Then*

$$d_H^2(\mathrm{Po}(\boldsymbol{\lambda}), \mathrm{Po}(\boldsymbol{\mu})) = 1 - \exp\{-d_H^2(\boldsymbol{\lambda}, \boldsymbol{\mu})\},$$

and thus

$$d_H(\mathrm{Po}(\boldsymbol{\lambda}), \mathrm{Po}(\boldsymbol{\mu})) \le d_H(\boldsymbol{\lambda}, \boldsymbol{\mu}).$$

Proof Let $\boldsymbol{\nu} = \boldsymbol{\lambda} + \boldsymbol{\mu}$, and let $\boldsymbol{\lambda}' = \lambda^{-1}\boldsymbol{\lambda}$, $\boldsymbol{\mu}' = \mu^{-1}\boldsymbol{\mu}$ and $\boldsymbol{\nu}' = \nu^{-1}\boldsymbol{\nu}$ be the correspondingly normalized measures. The measure $\mathrm{Po}(\boldsymbol{\lambda})$ is absolutely continuous with respect to $\mathrm{Po}(\boldsymbol{\nu})$, and, on the set of measures $\{\sum_1^n \delta_{x_i}\}$ with total mass n, its density is given by

$$\frac{d\mathrm{Po}(\boldsymbol{\lambda})}{d\mathrm{Po}(\boldsymbol{\nu})}\Big(\sum_1^n \delta_{x_i}\Big) = \frac{\mathrm{Po}(\lambda)\{n\}}{\mathrm{Po}(\nu)\{n\}} \prod_1^n \frac{d\boldsymbol{\lambda}'}{d\boldsymbol{\nu}'}(x_i) = e^{\nu-\lambda} \prod_1^n \frac{d\boldsymbol{\lambda}}{d\boldsymbol{\nu}}(x_i).$$

Consequently, using (1.13),

$$\begin{aligned}
1 - {}& d_H^2(\mathrm{Po}(\boldsymbol{\lambda}), \mathrm{Po}(\boldsymbol{\mu})) \\
&= \sum_{n\ge 0} \int \cdots \int \exp\{\nu - (\lambda+\mu)/2\} \\
&\quad \times \prod_1^n \Big(\frac{d\boldsymbol{\lambda}}{d\boldsymbol{\nu}}(x_i)\frac{d\boldsymbol{\mu}}{d\boldsymbol{\nu}}(x_i)\Big)^{1/2} e^{-\nu}\frac{\nu^n}{n!}\, d\boldsymbol{\nu}'(x_1)\ldots d\boldsymbol{\nu}'(x_n) \\
&= \sum_{n\ge 0} e^{-(\lambda+\mu)/2}\frac{1}{n!}\Big(\int \Big(\frac{d\boldsymbol{\lambda}}{d\boldsymbol{\nu}}\frac{d\boldsymbol{\mu}}{d\boldsymbol{\nu}}\Big)^{1/2} d\boldsymbol{\nu}\Big)^n \\
&= \exp\Big\{-\frac{\lambda+\mu}{2} + \int \Big(\frac{d\boldsymbol{\lambda}}{d\boldsymbol{\nu}}\frac{d\boldsymbol{\mu}}{d\boldsymbol{\nu}}\Big)^{1/2} d\boldsymbol{\nu}\Big\}.
\end{aligned}$$

□

Remark A.1.3 Formula (1.18) is valid also for infinite products. It follows that Proposition A.1.2 extends to σ-finite intensities, if the definitions are appropriately modified.

A.2 Poisson and binomial estimates

We collect here some technical facts about the Poisson and binomial distributions, used in the proofs in the preceding chapters.

We begin by observing that $(1-y)^c \le 1 - cy(1-y)^{c-1}$ whenever $c \ge 1$ and $0 < y < 1$, from Taylor's theorem, and hence that

$$(1-y)^c \le \{1 + cy/(1-y)\}^{-1} = 1 - cy/\{1 + (c-1)y\}. \tag{2.1}$$

On the other hand, if $0 \le c < 1$, Taylor's theorem immediately yields the inequality $(1-y)^c \le 1 - cy$. These two inequalities may be combined to give the simple estimate

$$y^{-1}[1 - (1-y)^c] \ge c/(1+cy), \tag{2.2}$$

valid for $c > 0$, $0 < y < 1$. Note that the right hand side of (2.2) is a decreasing function of y and an increasing function of c, so that the inequality remains true if, on the right hand side, a larger value is substituted for y or a smaller for c.

This elementary inequality is used in the following proposition. The aim is to show that the cumulative tail probabilities of the Poisson distribution are dominated by the largest included point probability, provided that this is small enough, and to establish the correction required if not.

Proposition A.2.1 *If* $\Delta = \lambda^{-1/2}(m - \lambda) \geq 1$,

(i) $\quad \mathrm{Po}(\lambda)\{r\} \leq (1 + \lambda^{-1/2}\Delta)^{-(r-m)}\mathrm{Po}(\lambda)\{m\}, \qquad r \geq m;$

(ii) $\quad \mathrm{Po}(\lambda)\{[m,\infty)\} \geq \frac{1}{4}\mathrm{Po}(\lambda)\{m\}(1 + \lambda^{1/2}/\Delta).$

If $\Delta \leq -1$,

(iii) $\quad \mathrm{Po}(\lambda)\{r\} \leq (1 + \lambda^{-1/2}\Delta)^{m-r-1}\mathrm{Po}(\lambda)\{m-1\}, \qquad r \leq m-1;$

(iv) $\quad \mathrm{Po}(\lambda)\{[0,m-1]\} \geq \frac{2}{7}\mathrm{Po}(\lambda)\{m-1\}[(\lambda^{1/2}/|\Delta|) \vee \frac{7}{2}].$

Proof Part (i) is the simple inequality $\mathrm{Po}(\lambda)\{r\} \leq (\lambda/m)^{r-m}\mathrm{Po}(\lambda)\{m\}$, and (iii) is proved similarly. For (ii) there is nothing to prove if $\Delta > \lambda^{1/2}/3$. If $\Delta \leq \lambda^{1/2}/3$, we use the straightforward inequality

$$\mathrm{Po}(\lambda)\{[m,\infty)\} \geq \mathrm{Po}(\lambda)\{m\}\{1 - (M^{-1}\lambda)^k\}/\{1 - M^{-1}\lambda\} \tag{2.3}$$

for any $M \geq m$ and $k \leq M - m + 1$. Taking

$$M = m + \Delta\lambda^{1/2}/2 = \lambda + 3\Delta\lambda^{1/2}/2$$

and $k = [\Delta\lambda^{1/2}/2] + 1$ as large as possible, this implies from (2.2) that

$$\begin{aligned}\mathrm{Po}(\lambda)\{[m,\infty)\} &\geq \mathrm{Po}(\lambda)\{m\}\Delta\lambda^{1/2}\{2 + \Delta\lambda^{1/2}(1 - M^{-1}\lambda)\}^{-1} \\ &= \mathrm{Po}(\lambda)\{m\}(3 + 2\lambda^{1/2}/\Delta)\Delta^2\{4 + 3\Delta^2 + 6\Delta\lambda^{-1/2}\}^{-1}.\end{aligned}$$

Now, using $\Delta \geq 1$, it follows that, with $\theta = \lambda^{1/2}/\Delta \geq 3$,

$$\mathrm{Po}(\lambda)\{[m,\infty)\} \geq \frac{3 + 2\theta}{7 + 6/\theta}\mathrm{Po}(\lambda)\{m\} \geq \frac{1+\theta}{4}\mathrm{Po}(\lambda)\{m\},$$

as required.

The proof of (iv) is similar. Only the case $|\Delta| \leq 2\lambda^{1/2}/7$ needs any argument, and then, by a similar comparison,

$$\mathrm{Po}(\lambda)\{[0,m-1]\} \geq \mathrm{Po}(\lambda)\{m-1\}\{1 - (\lambda^{-1}M)^k\}/\{1 - \lambda^{-1}M\} \tag{2.4}$$

for $M = m + \Delta\lambda^{1/2}/2 > 0$ and $k = [-\Delta\lambda^{1/2}/2] + 1$, leading to the estimate

$$\mathrm{Po}(\lambda)\{[0,m-1]\} \geq \frac{2}{3}\frac{\lambda^{1/2}}{|\Delta|}\mathrm{Po}(\lambda)\{m-1\}\left(1 - [1 - \tfrac{3}{2}|\Delta|\lambda^{-1/2}]^{|\Delta|\lambda^{1/2}/2}\right).$$

Application of (2.2) now proves the result. □

Similar inequalities can be proved if the Poisson distribution is replaced by the binomial distribution Bi (n,p). We consider only the case $p \leq 1/2$: the results for $p > 1/2$ follow in the obvious way.

Proposition A.2.2 *Setting $\lambda = np$ and $\Delta = \lambda^{-1/2}(m-\lambda)$, the following estimates hold when $p \leq 1/2$:*

(i) *for $\Delta \geq 1$,*

$$\mathrm{Bi}\,(n,p)\{[m,\infty)\} \geq \frac{1}{6}\mathrm{Bi}\,(n,p)\{m\}(1+\lambda^{1/2}/\Delta);$$

(ii) *for $\Delta \leq -1$,*

$$\mathrm{Bi}\,(n,p)\{[0,m-1]\} \geq \frac{2}{11}\frac{\lambda^{1/2}}{|\Delta|}\mathrm{Bi}\,(n,p)\{m-1\}.$$

Proof The proof of (i) closely follows the proof of Proposition A.2.1(ii). If $\Delta > \lambda^{1/2}/5$, there is nothing to prove, and this in particular covers any $m \geq 2n/3$. Otherwise, use an estimate analogous to (2.3), in which $M^{-1}\lambda$ is replaced by $(n-M)p/\{M(1-p)\}$, taking $M = m+\Delta\lambda^{1/2}/2$, which is less than n when $m < 2n/3$ and $p \leq 1/2$, and with k as before. The subsequent reduction to the convenient expression in the proposition is accomplished using (2.2).

The proof of (ii) similarly follows the proof of Proposition A.2.1(iv). It is clearly enough to consider $|\Delta| \leq 2\lambda^{1/2}/3$, and also to take $|\Delta|\lambda^{1/2} \geq 2$, since $\Delta \leq -1$. An estimate analogous to (2.4) is constructed, with $\lambda^{-1}M$ replaced by $M(1-p)/\{(n+1-M)p\}$, where M and k are chosen as before, noting that $M \geq 0$ if $|\Delta| \leq 2\lambda^{1/2}/3$. An application of inequality (2.2), with

$$y = 1 - \frac{M(1-p)}{(n+1-M)p} \leq \frac{\{p+3|\Delta|\lambda^{1/2}/2\}}{\{\lambda(1-p)\}} \leq 7\lambda^{-1/2}|\Delta|/2$$

in $|\Delta|\lambda^{1/2} \geq 2 \geq 4p$, completes the proof. □

We now turn to upper bounds for the point and tail probabilities of the Poisson and binomial distributions.

Proposition A.2.3 *The following inequalities hold for $m \in \mathbb{Z}^+$, $\lambda > 0$:*

(i)
$$\begin{aligned}\mathrm{Po}\,(\lambda)\{m\} &\leq \{1 \wedge (2\pi m)^{-1/2}\}\exp\{-(m-\lambda)^2/(\sqrt{m}+\sqrt{\lambda})^2\} \\ &\leq (2\pi m)^{-1/2}\exp\{-(m-\lambda)^2/2(m+\lambda)\};\end{aligned}$$

(ii)
$$\mathrm{Po}\,(\lambda)\{[m,\infty)\} \leq \left(\frac{m+1}{m+1-\lambda}\right)\mathrm{Po}\,(\lambda)\{m\}, \quad m+1 > \lambda;$$

(iii)
$$\mathrm{Po}\,(\lambda)\{[0,m-1]\} \leq \left(\frac{\lambda}{\lambda+1-m}\right)\mathrm{Po}\,(\lambda)\{m-1\}, \quad m < \lambda+1.$$

In addition, there exist constants $c_1, c_2 > 0$ *such that*

(iv) $$\mathrm{Po}(\lambda)\{m\} \ge c_1 m^{-1/2} \exp\{-\lambda^{-1}(m-\lambda)^2\}, \quad m \ge 1;$$
(v) $$\mathrm{Po}(\lambda)\{m\} \ge c_2 \exp\{-m\log(m/\lambda)\}, \quad m \ge (2\lambda \vee 1).$$

Proof Parts (ii) and (iii) are immediate. For (i), only the case $m > 0$ needs argument. Then, by Stirling's formula,

$$\mathrm{Po}(\lambda)\{m\} \le (2\pi m)^{-1/2} \exp\{m - \lambda + m[\log\lambda - \log m]\}.$$

Now, for $y > z > 0$ and for any $0 \le \theta \le 1$,

$$\begin{aligned}\log y - \log z - y^{-1}(y-z) &\ge \int_z^{\theta y + (1-\theta)z} [(\theta y + (1-\theta)z)^{-1} - y^{-1}]\,dt \\ &= \theta(1-\theta)(y-z)^2[y\{\theta y + (1-\theta)z\}]^{-1},\end{aligned}$$

and the choice $\theta = \sqrt{z}/(\sqrt{z} + \sqrt{y})$ with $y = m$ and $z = \lambda$ gives the inequality for $m \ge \lambda$. For $m < \lambda$, use the analogous inequality

$$\begin{aligned}\log y - \log z - z^{-1}(y-z) &\le -\int_{\theta y + (1-\theta)z}^{y} \{z^{-1} - (\theta y + (1-\theta)z)^{-1}\}\,dt \\ &= -\theta(1-\theta)(y-z)^2\{z[\theta y + (1-\theta)z]\}^{-1}\end{aligned}$$

when $0 < z < y$.

For (iv) and (v), if $m \ge 1$, Stirling's formula gives

$$\mathrm{Po}(\lambda)\{m\} \ge c m^{-1/2} \exp\{m - \lambda - m\log(m/\lambda)\}.$$

Writing $m = \lambda + \lambda^{1/2}\Delta$ for $\Delta = \lambda^{-1/2}(m-\lambda)$, and observing that $\log(m/\lambda) \le \lambda^{-1/2}\Delta$ gives (iv). Part (v) follows because, if $m \ge 2\lambda \ge 1$,

$$m^{-1/2}\exp(m-\lambda) \ge (2\lambda)^{-1/2} e^{\lambda} \ge e^{1/2},$$

whereas, for $\lambda < 1/2$ and $m \ge 1$,

$$m^{-1/2}\exp(m-\lambda) \ge \exp(1-\lambda) \ge e^{1/2}. \qquad \square$$

Remark A.2.4 From Propositions A.2.1(ii) and A.2.3(ii), it follows that, for $\lambda^{-1/2}(m-\lambda) \ge 1$, $\mathrm{Po}(\lambda)\{[m,\infty)\} \asymp m(m-\lambda)^{-1}\mathrm{Po}(\lambda)\{m\}$.

Similar inequalities can also be proved for the binomial distribution.

Proposition A.2.5 *If* $0 \le m \le n$ *and* $0 \le p \le 1$, *the following inequalities are true, with* $\lambda = np$.

$$\text{(i)}\quad \mathrm{Bi}\,(n,p)\{m\} \le \{1 \wedge [2\pi m(1-m/n)]^{-1/2}\}\exp\{-(m-np)^2/2(m+np)\};$$

$$\text{(ii)}\quad \mathrm{Bi}\,(n,p)\{[m,\infty)\} \le \frac{(m+1)(1-p)}{m+1-\lambda-p}\mathrm{Bi}\,(n,p)\{m\}, \quad m > \lambda-(1-p);$$

$$\text{(iii)}\quad \mathrm{Bi}\,(n,p)\{[0,m-1]\} \le \frac{\lambda-mp}{\lambda+1-m-p}\mathrm{Bi}\,(n,p)\{m-1\}, \quad m < \lambda+1-p.$$

Proof Again, (ii) and (iii) are straightforward. For (i), Stirling's formula gives the bound

$$\mathrm{Bi}\,(n,p)\{m\} \le [2\pi m(1-m/n)]^{-1/2} \times \exp\bigl\{-n[(p+x)\log(1+x/p)+(1-p-x)\log(1-x/(1-p))]\bigr\},$$

where $x = n^{-1}(m-np) \le 1-p$. The logarithmic inequalities of the proof of Proposition A.2.3 now yield

$$\begin{aligned}(p+x)&\log(1+x/p)+(1-p-x)\log(1-x/(1-p))\\ &\ge \left(\frac{x}{p}\right)^2\frac{(p+x)}{\left(1+\sqrt{1+x/p}\right)^2}\cdot\frac{p}{(p+x)}\\ &\quad+\left(\frac{x}{1-p}\right)^2\frac{(1-p-x)}{\left(1+\sqrt{1-x/(1-p)}\right)^2}\cdot\frac{(1-p)}{(1-p-x)}\\ &\ge \frac{(m-np)^2}{n(\sqrt{np}+\sqrt{m})^2},\end{aligned}$$

completing the proof for $0 < m < n$. The cases $m = 0, n$ are proved similarly. □

The next result is required for Example 3.4.1.

Proposition A.2.6 *Suppose that* $Z \sim \mathrm{Po}\,(\lambda)$. *Then, uniformly in* $\lambda > 0$ *and* $m \ge \lambda + \lambda^{1/2}$,

$$\text{(i)}\quad \mathbb{E}\{(Z-\lambda)I[Z \ge m]\} \asymp (m-\lambda)\mathbb{P}[Z \ge m],$$

$$\text{(ii)}\quad \mathbb{E}\{(Z-\lambda)^2 I[Z \ge m]\} \asymp (m-\lambda)^2\mathbb{P}[Z \ge m];$$

and, uniformly in $\lambda > 0$ *and* $m \le \lambda - \lambda^{1/2}$,

$$\text{(iii)}\quad \mathbb{E}\{(Z-\lambda)I[Z \le m]\} \asymp (\lambda-m)\mathbb{P}[Z \le m],$$

$$\text{(iv)}\quad \mathbb{E}\{(Z-\lambda)^2 I[Z \le m]\} \asymp (\lambda-m)^2\mathbb{P}[Z \le m].$$

Proof The four parts are proved similarly, so we give the details for (ii) only.

It is obvious that $\mathbb{E}\{(Z-\lambda)^2 I[Z \geq m]\} \geq (m-\lambda)^2 \mathbb{P}[Z \geq m]$. Conversely, by Proposition A.2.1(i) and (ii), with $\Delta = \lambda^{-1/2}(m-\lambda)$,

$$
\begin{aligned}
\mathbb{E}\{(Z-\lambda)^2 I[Z \geq m]\} &\leq \mathbb{E}\{(2(m-\lambda)^2 + 2(Z-m)^2) I[Z \geq m]\} \\
&= 2(m-\lambda)^2 \mathbb{P}[Z \geq m] + 2\sum_{r=m}^{\infty} (r-m)^2 \mathbb{P}[Z = r] \\
&\leq 2(m-\lambda)^2 \mathbb{P}[Z \geq m] + 2\sum_{r=m}^{\infty} (r-m)^2 (1+\lambda^{-1/2}\Delta)^{-(r-m)} \mathbb{P}[Z = m] \\
&\leq 2(m-\lambda)^2 \mathbb{P}[Z \geq m] + 4(1-(1+\lambda^{-1/2}\Delta)^{-1})^{-3} \mathbb{P}[Z = m] \\
&= 2(m-\lambda)^2 \mathbb{P}[Z \geq m] + 4(\frac{1+\lambda^{-1/2}\Delta}{\lambda^{-1/2}\Delta})^{-3} \mathbb{P}[Z = m] \\
&\leq 2(m-\lambda)^2 \mathbb{P}[Z \geq m] + 16(1+\lambda^{1/2}/\Delta)^2 \mathbb{P}[Z \geq m].
\end{aligned}
$$

The result follows because $\lambda^{1/2}/\Delta \leq \lambda^{1/2}\Delta = m - \lambda$ and $1 \leq 2(m-\lambda)$. □

The next proposition gives a useful upper bound for Poisson point probabilities.

Proposition A.2.7 *For any* $\lambda > 0$,

$$
\max_{j \geq 0} \mathrm{Po}(\lambda)\{j\} \leq \frac{1}{\sqrt{2e\lambda}}.
$$

Proof For each $j = 0, 1, \ldots$, set

$$
\begin{aligned}
x_j &= \sup_{\lambda > 0} \lambda^{1/2} \mathrm{Po}(\lambda)\{j\} \\
&= \sup_{\lambda > 0} \lambda^{j+1/2} e^{-\lambda}/j! = (j+\tfrac{1}{2})^{j+1/2} e^{-j-1/2}/j!,
\end{aligned}
$$

since the supremum is attained at $\lambda = j + 1/2$. If $j \geq 1$, then

$$
\begin{aligned}
\log(x_j/x_{j-1}) &= (j+\tfrac{1}{2})\log(j+\tfrac{1}{2}) - (j-\tfrac{1}{2})\log(j-\tfrac{1}{2}) - 1 - \log j \\
&= j\left[(1+\tfrac{1}{2j})\log(1+\tfrac{1}{2j}) - (1-\tfrac{1}{2j})\log(1-\tfrac{1}{2j}) - \tfrac{2}{2j}\right] < 0,
\end{aligned}
$$

because the function $(1+x)\log(1+x) - (1-x)\log(1-x) - 2x$ is decreasing on $[0,1)$. Hence the sequence (x_j) is decreasing, and, for any $\lambda, j \geq 0$,

$$
\lambda^{1/2} \mathrm{Po}(\lambda)\{j\} \leq x_j \leq x_0 = \sqrt{\frac{1}{2e}}.
$$

Note also that, by Fourier inversion,

$$\begin{aligned}\mathrm{Po}(\lambda)\{j\} &= \frac{1}{2\pi}\int_{-\pi}^{\pi} \exp\{\lambda(e^{i\theta}-1) - ij\theta\}\, d\theta \\ &\le \frac{1}{2\pi}\int_{-\pi}^{\pi} e^{\lambda(\cos\theta - 1)} d\theta = e^{-\lambda} I_0(\lambda),\end{aligned}$$

for any $j \geq 0$, where I_0 denotes the modified Bessel function of order 0, giving a sharper bound when $\lambda > 4$: see Abramowitz and Stegun (1964) Formulae 9.6.19 and 9.7.1 and Table 9.8. □

Remark A.2.8 The bound is attained at $j = 0$ for $\lambda = 1/2$. For large λ, $\max_j \mathrm{Po}(\lambda)\{j\} \sim (2\pi\lambda)^{-1/2}$, showing that the bound given in Proposition A.2.7 is of the right order, and that the Bessel function bound is sharp for $\lambda \to \infty$.

Proposition A.2.9 *For any* $m \geq 1$,

$$\frac{1}{e\sqrt{m}} \leq \mathrm{Po}(m)\{m\} < \frac{1}{\sqrt{2\pi m}}.$$

Proof Set $x_m = m^{1/2}\mathrm{Po}(m)\{m\} = m^{m+\frac{1}{2}}e^{-m}/m!$. Then direct calculation shows that

$$\begin{aligned}&\log(x_{m+1}/x_m) \\ &\quad = (m + \tfrac{3}{2})\log(m+1) - (m+1) - \log(m+1) - (m + \tfrac{1}{2})\log m + m \\ &\quad = (m + \tfrac{1}{2})\log(1 + \tfrac{1}{m}) - 1 = m\{(1 + \tfrac{1}{2m})\log(1 + \tfrac{1}{m}) - \tfrac{1}{m}\} > 0,\end{aligned}$$

because the function $(1 + x/2)\log(1 + x) - x$ is increasing and thus positive on $(0, \infty)$. Hence the sequence x_m is increasing, and the result follows because $x_1 = e^{-1}$ and $\lim_{m\to\infty} x_m = (2\pi)^{-1/2}$. □

References

ABRAMOWITZ, M. AND STEGUN, I.A. (1964). *Handbook of mathematical functions.* Dover, New York.

ALDOUS, D.J. (1989a). Stein's method in a two-dimensional coverage problem. *Statistics & Probability Letters*, **8**, 307–14.

ALDOUS, D.J. (1989b). *Probability approximations via the Poisson clumping heuristic.* Springer, New York.

ALM, S.E. (1983). On the distribution of the scan statistic of a Poisson process. In *Probability and Mathematical Statistics. Essays in honour of Carl-Gustav Esseen* (ed. A. Gut and L. Holst), pp.1–10. Department of Mathematics, Uppsala University.

ARRATIA, R., BARBOUR, A.D., AND TAVARÉ, S. (1991). Poisson process approximations for the Ewens sampling formula. (to appear)

ARRATIA, R., GOLDSTEIN, L., AND GORDON, L. (1989). Two moments suffice for Poisson approximations: the Chen–Stein method. *Annals of Probability*, **17**, 9–25.

ARRATIA, R., GOLDSTEIN, L., AND GORDON, L. (1990). Poisson approximation and the Chen–Stein method. *Statistical Science*, **5**, 403–4.

ARRATIA, R., GORDON, L., AND WATERMAN, M.S. (1990). The Erdős–Rényi law in distribution, for coin tossing and sequence matching. *Annals of Statistics*, **18**, 539–70.

ARRATIA, R. AND TAVARÉ, S. (1990). The cycle structure of random permutations. *Annals of Probability.* (to appear)

BARBOUR, A.D. (1982). Poisson convergence and random graphs. *Mathematical Proceedings of the Cambridge Philosophical Society*, **92**, 349–59.

BARBOUR, A.D. (1987). Asymptotic expansions in the Poisson limit theorem. *Annals of Probability*, **15**, 748–66.

BARBOUR, A.D. (1988). Stein's method and Poisson process convergence. *Journal of Applied Probability*, Special Volume, **25A**, 175–84.

BARBOUR, A.D. AND BROWN, T.C. (1990). Stein's method and point process approximation. *Stochastic Processes and their Applications.* (to appear)

BARBOUR, A.D. AND BROWN, T.C. (1992). The Stein–Chen method, point processes and compensators. *Annals of Probability.* (to appear)

Barbour, A.D., Chen, L.H.Y., and Loh, W.-L. (1992). Compound Poisson approximation for nonnegative random variables using Stein's method. *Annals of Probability.* (to appear)

Barbour, A.D. and Eagleson, G.K. (1983). Poisson approximation for some statistics based on exchangeable trials. *Advances in Applied Probability*, **15**, 585–600.

Barbour, A.D. and Eagleson, G.K. (1984). Poisson convergence for dissociated statistics. *Journal of the Royal Statistical Society*, **B 46**, 397–402.

Barbour, A.D. and Eagleson, G.K. (1987). An improved Poisson limit theorem for sums of dissociated random variables. *Journal of Applied Probability*, **24**, 586–99.

Barbour, A.D. and Greenwood, P.E. (1991). Rates of Poisson approximation to finite range random fields. *Annals of Applied Probability.* (to appear)

Barbour, A.D. and Hall, P. (1984). On the rate of Poisson convergence. *Mathematical Proceedings of the Cambridge Philosophical Society*, **95**, 473–80.

Barbour, A.D. and Holst, L. (1989). Some applications of the Stein–Chen method for proving Poisson convergence. *Advances in Applied Probability*, **21**, 74–90.

Barbour, A.D., Janson, S., Karoński, M., and Ruciński, A. (1990). Small cliques in random graphs. *Random Structure & Algorithms*, **1**, 403–34.

Barbour, A.D. and Jensen, J.L. (1989). Local and tail approximations near the Poisson limit. *Scandinavian Journal of Statistics*, **16**, 75–87.

Berman, S.M. (1964). Limit theorems for the maximum term in stationary sequences. *Annals of Mathematical Statistics*, **35**, 502–16.

Berman, S.M. (1987). Poisson and extreme value limit theorems for Markov random fields. *Advances in Applied Probability*, **19**, 106–22.

Birkel, T. (1988). On the convergence rate in the central limit theorem for associated processes. *Annals of Probability*, **16**, 1685–98.

Bogart, K.P. and Doyle, P.G. (1986). Non-sexist solution of the ménage problem. *American Mathematical Monthly*, **93**, 514–18.

Bollobás, B. (1985). *Random graphs.* Academic Press, London.

Bortkewitsch, L. von (1898). *Das Gesetz der kleinen Zahlen.* Teubner Verlag, Leipzig.

BROWN, T.C. (1983). Some Poisson approximations using compensators. *Annals of Probability*, **11**, 726–44.

BRUSS, F.T. (1984). Patterns of relative maxima in random permutations. *Annales de la Societé Scientifique de Bruxelles*, **98(I)**, 19–28.

LE CAM, L. (1960). An approximation theorem for the Poisson binomial distribution. *Pacific Journal of Mathematics*, **10**, 1181–97.

CHEN, L.H.Y. (1975a). Poisson approximation for dependent trials. *Annals of Probability*, **3**, 534–45.

CHEN, L.H.Y. (1975b). An approximation theorem for sums of certain randomly selected indicators. *Zeitschrift für Wahrscheinlichkeitstheorie und verwandte Gebiete*, **33**, 69–74.

CHEN, L.H.Y. (1978). Two central limit problems for dependent random variables. *Zeitschrift für Wahrscheinlichkeitstheorie und verwandte Gebiete*, **43**, 223–43.

CHEN, L.H.Y. AND CHOI, K.P. (1990). Some asymptotic and large deviation results in Poisson approximation. National University of Singapore Research Report No. 446.

CHIANG, D.T. AND NIU, S.C. (1981). Reliability of consecutive k-out-of-n:F system. *Institute of Electrical and Electronic Engineers. Transactions on Reliability*, **R-30**, 87–9.

COWAN, R., COLLIS, C.M., AND GRIGG, G.W. (1987). Breakage of double-stranded DNA due to single-stranded nicking. *Journal of Theoretical Biology*, **127**, 229–45.

COWAN, R., CULPIN, D., AND GATES, D. (1990). Asymptotic results for a problem of DNA breakage. *Journal of Applied Probability*, **27**, 433–39.

CHRYSSAPHINOU, O. AND PAPASTAVRIDIS, S. (1988). A limit theorem for the number of non-overlapping occurrences of a pattern in a sequence of independent trials. *Journal of Applied Probability*, **25**, 428–31.

CRESSIE, N. (1977). The minimum of higher order gaps. *Australian Journal of Statistics*, **19**, 132–43.

DEHEUVELS, P. AND PFEIFER. D. (1988). On a relationship between Uspensky's theorem and Poisson approximations. *Annals of the Institute of Statistical Mathematics*, **40**, 671–81.

DEVROYE, L. (1984). A probabilistic analysis of the height of tries and of the complexity of trie sort. *Acta Informatica*, **21**, 229–37.

EHM, W. (1991). Binomial approximation to the Poisson binomial distribution. *Statistics & Probability Letters*, **11**, 7–16.

Esary, J.D., Proschan, F., and Walkup, D.W. (1967). Association of random variables, with applications. *Annals of Mathematical Statistics*, **38**, 1466–74.

Ethier, S.N. and Kurtz, T.G. (1986). *Markov processes: characterization and convergence.* Wiley, New York.

Falk, M. and Reiss, R.-D. (1988). Poisson approximation of empirical processes. Fachbereich Mathematik, Bericht 218, University of Siegen.

Feller, W. (1968). *An introduction to probability theory and its applications,* Vol. 1, (3rd edn). Wiley, New York.

Freedman, D. (1974). The Poisson approximation for dependent events. *Annals of Probability*, **2**, 256–69.

Gastwirth, J.L. and Bhattacharya, P.K. (1984). Two probability models of pyramid or chain letter schemes demonstrating that their promotional claims are unreliable. *Operations Research*, **32**, 527–36.

Godbole, A.P. (1990). Degenerate and Poisson convergence criteria for success runs. *Statistics & Probability Letters*, **10**, 247–55.

Godsil, C.D. and McKay, B.D. (1990). Asymptotic enumeration of Latin rectangles. *Journal of Combinatorial Theory, Series B*, **48**, 19–44.

Griffeath, D. (1975). A maximal coupling for Markov chains. *Zeitschrift für Wahrscheinlichkeitstheorie und verwandte Gebiete*, **31**, 95–106.

Grimmett, G.R. and Hall, R.R. (1990). The asymptotics of random sieves. (In preparation.)

Harris, B. (1989). Poisson limits for generalized random allocation problems. *Statistics & Probability Letters*, **8**, 123–27.

Hodges, J.L. and Le Cam, L. (1960). The Poisson approximation to the Poisson binomial distribution. *Annals of Mathematical Statistics*, **31**, 737–40.

Holst, L. (1979). Asymptotic normality of sum–functions of spacings. *Annals of Probability*, **7**, 1066–72.

Holst, L. (1980a). On multiple covering of a circle with random arcs. *Journal of Applied Probability*, **17**, 284–90.

Holst, L. (1980b). On matrix occupancy, committee, and capture–recapture problems. *Scandinavian Journal of Statistics*, **7**, 139–46.

Holst, L. (1989). On the number of vertices in the complete graph with a given vertex as nearest neighbour. In *Random graphs '89*, Proceedings of Poznań conference. (In press.)

HOLST, L. (1990). A circle covering problem and DNA breakage. *Statistics & Probability Letters*, **9**, 295–98.

HOLST, L. (1991). On the 'Problème des Ménages' from a probabilistic viewpoint. *Statistics & Probability Letters*, **11**, 225–31.

HOLST, L. AND HÜSLER, J. (1984). On the random coverage of the circle. *Journal of Applied Probability*, **21**, 558–66.

HOLST, L., KENNEDY, J., AND QUINE, M. (1988). Rates of Poisson convergence for some coverage and urn problems using coupling. *Journal of Applied Probability*, **25**, 717–24.

JANSON, S. (1984). Bounds on the distributions of extremal values of a scanning process. *Stochastic Processes and their Applications*, **18**, 313–28.

JANSON, S. (1986). Birthday problems, randomly coloured graphs and Poisson limits of sums of dissociated variables. Uppsala University Department of Mathematics Report No. 1986:16.

JANSON, S. (1987). Poisson convergence and Poisson processes with applications to random graphs. *Stochastic Processes and their Applications*, **26**, 1–30.

JANSON, S. (1990). Poisson approximation for large deviations. *Random Structures & Algorithms*, **1**, 221–9.

JOAG-DEV, K., PERLMAN, M.D., AND PITT, L.D. (1983). Association of normal random variables and Slepian's inequality. *Annals of Probability*, **11**, 451–5.

JOAG-DEV, K. AND PROSCHAN, F. (1983). Negative association of random variables, with applications. *Annals of Statistics*, **11**, 286–95.

JOHNSON, N.L. AND KOTZ, S. (1969). *Distributions in statistics: discrete distributions.* Houghton Mifflin, Boston.

JOHNSON, N.L. AND KOTZ, S. (1977). *Urn models and their application.* Wiley, New York.

KAMAE, T., KRENGEL, U., AND O'BRIEN, G.L. (1977). Stochastic inequalities on partially ordered spaces. *Annals of Probability*, **5**, 899–912.

KAROŃSKI, M. (1984). *Balanced subgraphs of large random graphs.* Adam Mickiewicz University Press, Poznań.

KAROŃSKI, M. AND RUCIŃSKI, A. (1987). Poisson convergence and semi-induced properties of random graphs. *Mathematical Proceedings of the Cambridge Philosophical Society*, **101**, 291–300.

KEILSON, J. (1979). *Markov chain models — rarity and exponentiality.* Springer, New York.

KEMP, R. (1984). *Fundamentals of the average case analysis of particular algorithms.* Wiley–Teubner, New York.

KERSTAN, J. (1964). Verallgemeinerung eines Satzes von Prochorow und Le Cam. *Zeitschrift für Wahrscheinlichkeitstheorie und Verwandte Gebiete*, **2**, 173–9.

KNOX, G. (1964). Epidemiology of childhood leukaemia in Northumberland and Durham. *British Journal of Preventive Social Medicine*, **18**, 17–24.

KOLCHIN, V.F., SEVAST'YANOV, B.A., AND CHISTYAKOV, V.P. (1978). *Random allocations.* Winston, Washington DC.

LANKE, J. (1973). Asymptotic results on matching distributions. *Journal of the Royal Statistical Society*, **B 35**, 117–22.

LEADBETTER, M.R., LINDGREN, G., AND ROOTZÉN, H. (1983). *Extremes and related properties of random sequences and processes.* Springer, New York.

LEONOV, V.P. AND SHIRYAEV, A.N. (1959). On a method of calculation of semi-invariants. *Teoriya Veroyatnostei i ee Primeneniya*, **4**, 342–55. (Russian.) Translated as *Theory of Probability and its Applications*, **4**, 319–28.

LIGGETT, T.M. (1985). *Interacting particle systems.* Springer, New York.

LIGGETT, T.M. AND PORT, S.C. (1988). Systems of independent Markov chains. *Stochastic Processes and their Applications*, **28**, 1–22.

MCGINLEY, W.G. AND SIBSON, R. (1975). Dissociated random variables. *Mathematical Proceedings of the Cambridge Philosophical Society*, **77**, 185–8.

MCKAY, B.D. AND WORMALD, N. (1990). Asymptotic enumeration by degree sequence of graphs with degree $o(n^{1/2})$. (to appear)

MICHEL, R. (1988). An improved error bound for the compound Poisson approximation of a nearly homogeneous portfolio. *ASTIN Bulletin*, **17**, 165–9.

MISES, R. VON (1921). Das Problem der Iterationen. *Zeitschrift für Angewandte Mathematik und Mechanik*, **1**, 298–307.

MISES, R. VON (1939). Über Aufteilungs- und Besetzungswahrscheinlichkeiten. *Revue de la Faculté des Sciences de l'Université d'Istanbul*, **4**, 145–63.

Mittal, Y. and Ylvisaker, D. (1975). Limit distributions for the maxima of stationary Gaussian processes. *Stochastic Processes and their Applications*, **3**, 1–18.

de Montmort, P.R. (1713). *Essay d'analyse sur les jeux de hazard,* (2nd edn). Jacques Quillau, Paris.

Nevzorov, V.B. (1986). Two characterizations using records. In *Stability problems for stochastic models*, Lecture Notes in Mathematics, **1233**, 79–85.

Newman, C.M., Rinott, Y., and Tversky, A. (1983). Nearest neighbors and Voronoi regions in certain point processes. *Advances in Applied Probability*, **15**, 726–51.

Nowicki, K. (1988). Asymptotic Poisson distributions with applications to statistical analysis of graphs. *Advances in Applied Probability*, **20**, 315–30.

Papastavridis, S. (1987). A limit theorem for the reliability of a consecutive k-out-of-n system. *Advances in Applied Probability*, **19**, 746–8.

Papastavridis, S. (1988). A Weibull limit for the reliability of a consecutive k-within-m-out-of-n system. *Advances in Applied Probability*, **20**, 690–2.

Papastavridis, S. (1990). m-consecutive-k-out-of-n:F systems. *Institute of Electrical and Electronic Engineers. Transactions on Reliability*, **39**, 386–8.

Pfeifer, D. (1987). On the distance between mixed Poisson and Poisson distributions. *Statistics & Decisions*, **5**, 367–79.

Pfeifer, D. (1989). Extremal processes, secretary problems and the $1/e$ law. *Journal of Applied Probability*, **26**, 722–33.

Philippou, A.N., Georghiou, C., and Philippou, G.N. (1983). A generalized geometric distribution and some of its properties. *Statistics & Probability Letters*, **1**, 171–5.

Poisson, S.D. (1837). *Recherches sur la probabilité des jugements en matière criminelle et en matière civile, précedées des règles générales du calcul des probabilités.* Bachelier, Paris.

Preston, C.J. (1974). A generalization of the FKG inequalities. *Communications in Mathematical Physics*, **36**, 233–41.

Prohorov, Ju.V. (1953). Asymptotic behaviour of the binomial distribution. *Uspekhi Matematicheskikh Nauk*, **8**, no. 3(55), 135–42. (Russian.)

RACHEV, S.T. (1984). The Monge–Kantorovich mass transference problem and its stochastic applications. *Teoriya Veroyatnostei i ee Primeneniya*, **29**, 625–53. (Russian.) Translated as *Theory of Probability and its Applications*, **29**, 647–76.

REISS, R.-D. (1989). *Approximate distributions of order statistics.* Springer, New York.

ROOTZÉN, H. (1987). A ratio limit theorem for the tails of weighted sums. *Annals of Probability*, **15**, 728–47.

ROSS, S.M. (1982). A simple heuristic approach to simplex efficiency. *European Journal of Operational Research*, **9**, 344–6.

RUCIŃSKI, A. AND VINCE, A. (1985). Balanced graphs and the problem of subgraphs of random graphs. *Congressus Numerantum*, **49**, 181–90.

SERFLING, R.J. (1975). A general Poisson approximation theorem. *Annals of Probability*, **3**, 726–31.

SHORGIN, S.YA. (1977). Approximation of a generalized binomial distribution. *Teoriya Veroyatnostei i ee Primeneniya*, **22**, 867–71. (Russian.) Translated as *Theory of Probability and its Applications*, **22**, 846–50.

SILVERMAN, B.W. AND BROWN, T.C. (1978). Short distances, flat triangles and Poisson limits. *Journal of Applied Probability*, **15**, 815–25.

SMITH, R.L. (1988). Extreme value theory for dependent sequences via the Stein–Chen method of Poisson approximation. *Stochastic Processes and their Applications*, **30**, 317–27.

STEIN, C. (1978). Asymptotic evaluation of the number of Latin rectangles. *Journal of Combinatorial Theory, Series A*, **25**, 38–49.

STEIN, C. (1986). *Approximate computation of expectations.* Institute of Mathematical Statistics Lecture Notes – Monograph Series, Vol. **7**. Hayward, California.

STRASSEN, V. (1965). The existence of probability measures with given marginals. *Annals of Mathematical Statistics*, **36**, 423–39.

SUEN, W.C.S. (1990). A correlation inequality and a Poisson limit theorem for nonoverlapping balanced subgraphs of a random graph. *Random Structure & Algorithms*, **1**, 231–42.

TAKÁCS, L. (1981). On the 'Problème des Ménages'. *Discrete Mathematics*, **36**, 289–97.

USPENSKY, J.V. (1931). On Ch. Jordan's series for probability. *Annals of Mathematics*, **32**, 306–12.

VATUTIN, V.A. AND MIKHAILOV, V.G. (1982). Limit theorems for the number of empty cells in an equiprobable scheme for group allocation of particles. *Teoriya Veroyatnostei i ee Primeneniya*, **27**, 684–92. (Russian.) Translated as *Theory of Probability and its Applications*, **27**, 734–43.

WITTE, H.-J. (1990). A unification of some approaches to Poisson approximation. *Journal of Applied Probability*, **27**, 611–21.

ZOLOTAREV, V.M. (1983). Probability metrics. *Teoriya Veroyatnostei i ee Primeneniya*, **28**, 264–87. (Russian.) Translated as *Theory of Probability and its Applications*, **28**, 278–302.

Index

E

F

G

H

I

J

K

L

M

N

O

P

Q

R

S

T

U

V

W

Y

Z